名师讲坛——Spring 实战开发

（Redis+SpringDataJPA+SpringMVC+SpringSecurity）

李兴华 编著

清華大学出版社

北 京

内容简介

Spring是当今Java开发行业之中的主流技术开源框架，利用Spring框架中IOC&DI与AOP实现机制可以轻松地实现轻量级的Java企业级项目开发。同时简单的代码开发形式与灵活的配置，可以极大地降低开发人员的代码编写难度。基于Spring还可以轻松整合许多Java的标准服务与第三方开发框架，使得项目的开发有良好的规范性。

本书一共分为15章，核心的内容包括Spring框架结构、IOC&DI、AOP、Bean管理、资源管理、表达式语言和定时调度，同时整合了JMS消息服务、Web服务、Redis数据库、JDBC和JPA等服务组件，最后又讲解了流行的SpringMVC以及Spring提供的授权管理组件SpringSecurity。

本书适用于从事Java开发的技术工程师，也适用于Java技术爱好者，同时也可以作为应用型高等院校及培训机构的学习教材。

图书在版编目（CIP）数据

Spring 实战开发：Redis+SpringDataJPA+SpringMVC+SpringSecurity / 李兴华编著. —北京：清华大学出版社，2020.1
（名师讲坛）
ISBN 978-7-302-52278-2

Ⅰ. ①S… Ⅱ. ①李… Ⅲ. ①JAVA语言—程序设计 Ⅳ. ①TP312.8

中国版本图书馆CIP数据核字(2019)第025512号

责任编辑：贾小红
封面设计：魏润滋
版式设计：文森时代
责任校对：马军令
责任印制：刘海龙

出版发行：清华大学出版社
网　　址：http://www.tup.com.cn，http://www.wqbook.com
地　　址：北京清华大学学研大厦A座　　**邮　　编**：100084
社 总 机：010-62770175　　**邮　　购**：010-62786544
投稿与读者服务：010-62776969，c-service@tup.tsinghua.edu.cn
质量反馈：010-62772015，zhiliang@tup.tsinghua.edu.cn
印 装 者：三河市君旺印务有限公司
经　　销：全国新华书店
开　　本：185mm×260mm　　**印　　张**：25.25　　**字　　数**：680千字
版　　次：2020年1月第1版　　**印　　次**：2020年1月第1次印刷
定　　价：79.80元

产品编号：080275-01

前　言

我们在用心做事，做最好的教育，写最好的原创图书。

很早以前，我就想写一本关于 Spring 开发框架的书。但由于日常教学与课程研发的工作量实在太大，这一想法迟迟未能得以实现。也许是机缘巧合，2018 年我抽出了许多时间，把自己的课堂笔记进行了细致整理，顺便将我 15 年来使用 Spring 的心得与技术感悟写了下来，于是有了本书。全书写完的那一刻，有一种轻松的感觉。写完本书后，我将继续回到培训教学与课程研发的繁忙工作之中。

写书是一件造福后来者的事情，一本书的作者实际上担负着知识传承的作用。我希望跟所有学生分享我的技术心得，所以对于本书坚持了我一贯的做法——全部是原创内容。我认为，中国缺少真正的本土原创好书，但不缺那种靠简单的抄袭或疯狂的复制+粘贴堆砌起来的书。

在我十多年的教学生涯之中，发现很多学生都在不断重复着这样一种尴尬和窘境：辛辛苦苦买来一本书，却发现书的内容质量不过关，缺少合理的知识结构体系，技术描述晦涩难懂，所以只看了几眼就扔到了角落里。这样不仅是对金钱的浪费，更是对纸张与环境的浪费及破坏。对于贫困家庭走出来的我而言，非常清楚地知道“学习成本”有多么高，所以我要写真正的原创图书，而非那些靠复制+粘贴拼凑字数的“假书”。这一原则陪伴我走过了十年的创作生涯。

Spring 是一个庞大的开发框架，其庞大不仅体现在强大的 Spring 技术本身，更体现在它能整合各类服务组件这一重任上。这就需要从业者掌握大量的开发技术，而这也正是本书创作的难点所在。为了方便读者学习，我在讲解 Spring 技术时还讲解了一些第三方技术，如 ActiveMQ 消息组件、Redis 数据库、JPA 开发框架、WebService 技术、缓存技术等，并讲解了这些技术与 Spring 开发框架的整合。这样做的主要目的，是为了给读者一个可以直接使用的开发技术架构（见下图），读者理解起来会更容易，后期上手项目开发也会更轻松。

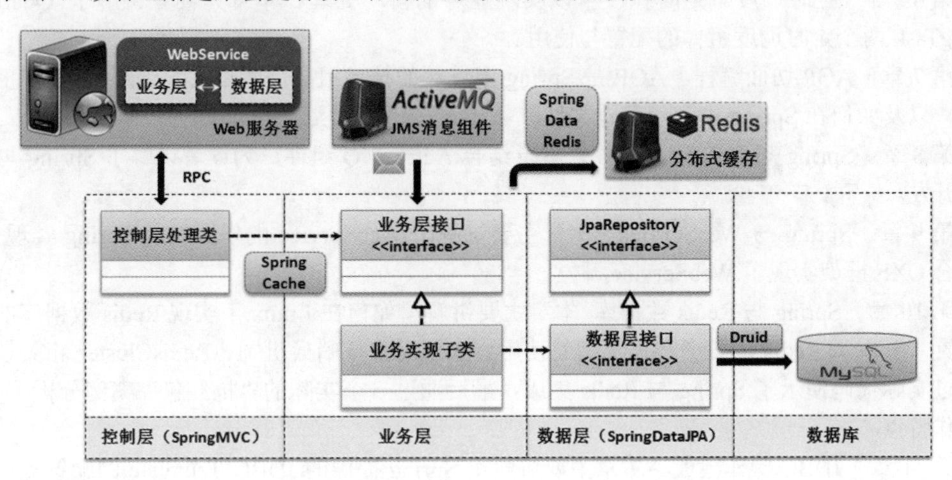

本书特色

☑ 15 年 Spring 开发与教学培训经验总结，准确把握学习脉搏，深入分析重点难点。

☑ 使用 Eclipse + Maven 进行项目开发，与真实项目开发环境完美对接。

☑ 全书 200 个实际案例（附赠源代码），全面分析了 Spring 的各项应用技术，读者可轻松掌握 Spring 开发的核心要领。

☑ 详细讲解了 Redis 数据库的各项使用技术以及 SpringDataRedis 开发技术。

☑ 详细讲解了 JPA 开发框架的各项开发技术，并且深入分析了 SpringDataJPA 技术的使用案例。

☑ Expression Language（EL）模块：表达式语言支持，支持访问和修改属性值，方法调用，支持访问及修改数组、容器和索引器、命名变量，支持算术和逻辑运算，支持从 Spring 容器获取 Bean，也支持列表透明、选择和一般的列表聚合等。利用表达式语言，可以更加灵活地控制配置文件。

☑ 提供了当前流行的高并发抢红包处理案例与 SpringMVC + SpringDataJPA 整合案例。

本书章节安排

第 1 章　Spring 开发框架概述。作为起始章，本章为读者分析了传统 Java EE 开发架构的设计组成与弊端，并介绍了 Spring 开发框架的组成。

第 2 章　控制反转。本章通过几段具体程序，分析了 Spring 的设计理念和控制反转技术的使用，并结合 Junit 讲解了 SpringTest 的使用。

第 3 章　Bean 管理。本章主要讲解了 Spring 开发框架提供的 Bean 配置，分析了各种数据类型的配置以及 Spring 中提供的 Bean 管理机制。

第 4 章　Spring 资源管理。本章主要讲解了 Spring 提供的资源处理标准，利用它可实现各种资源文件的读取；同时介绍了如何结合配置文件并利用表达式实现资源定位。

第 5 章　Spring 表达式语言。Spring 中，最强大的部分就是字符串支持能力。本章详细分析了 Spring 表达式的执行流程以及各种操作语法，并结合 Spring 配置文件，通过实例说明了 SpEL 语言的应用。

第 6 章　定时调度。定时调度可以实现业务的自动处理，本章为读者讲解了 QuartZ 与 SpringTask 两个定时调度组件的配置与使用。

第 7 章　AOP 切面编程。AOP 是 Spring 中重要的技术组成，本章主要讲解了 AOP 的主要概念，以及如何在 Spring 中利用配置文件或 Annotation 注解实现 AOP 编程控制。

第 8 章　Spring 与 JMS 消息组件。本章结合 ActiveMQ 组件，为读者讲解了 Spring 与 JMS 整合开发。

第 9 章　Spring 与 WebService。本章主要讲解了 WebService 的作用以及 Spring 实现处理，并结合 CXF 框架实现了 WebService 开发。

第 10 章　Spring 与 Redis 数据库。本章主要讲解了如何在 Linux 下实现 Redis 数据库的安装与配置，分析了 Redis 中各个数据类型的使用、主从配置、哨兵机制、RedisCluster 相关技术，并通过实际代码演示了 Spring 与 Redis 集成，最后通过一个实际的“抢红包”案例分析了 Redis 的操作特点。

第 11 章　JDBC 操作模板。本章主要讲解了 Spring 提供的 JDBC Template 的配置与使用，

分析了 SpringCache 相关技术的使用，以及如何与 EHCache 或 Redis 结合实现缓存管理。

第 12 章　Spring 事务管理。本章为读者分析了 Spring 提供的事务处理架构，并讲解了基于 AOP 实现的事务管理。

第 13 章　SpringDataJPA。本章主要讲解了 JPA 开发框架的使用、Query 查询、缓存配置、数据关联等相关技术，同时讲解了如何利用 SpringDataJPA 实现数据层开发。

第 14 章　SpringMVC。本章主要讲解了 SpringMVC 的处理架构及各项实现技术，最后通过一个完整案例讲解了 SpringMVC + SpringDataJPA 的联合开发应用。

第 15 章　SpringSecurity。本章主要讲解了 Spring 中提供的认证与授权检测框架的配置与使用，结合 SpringMVC 实现了登录认证、授权检测、CSRF 访问控制、Session 管理和 RememberMe 操作实现。

寄语读者

本书全篇由笔者根据实践项目与教学经验总结而来，虽经过再三斟酌和审校，仍难免存在技术理解上的偏差和解释不到位的地方，欢迎读者批评指正。您的宝贵建议将帮助我们修正此书，大家一起努力，将传道、授业、解惑贯彻到底。

本书用到的程序源代码，读者可扫描图书封底的“文泉云盘”二维码获取其下载方式，也可登录清华大学出版社网站（www.tup.com.cn）进行下载。技术学习部分，读者可登录魔乐科技官网（http://www.mldn.cn）及沐言优拓官网（http://www.yootk.com）进行学习，也可登录笔者的新浪微博进行留言交流。

最后，希望本书成为您的良师益友。祝您读书快乐！

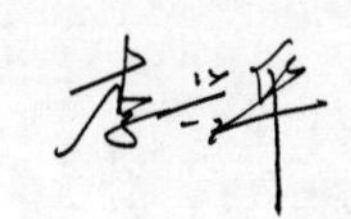

目 录

Contents

第 1 章

Spring 开发框架概述

通过本章学习，可以达到以下目标：

1. 了解传统设计开发中存在的问题与缺陷。
2. 了解 Spring 开发框架的主要特点。
3. 了解 Spring 开发框架的主要组成模块。

Spring 是 Java EE 企业级项目开发中最为重要的开发框架，可以说，Java 项目开发离不开 Spring 的支持。本章将为读者揭开 Spring 的神秘面纱。

1.1 Spring 的产生背景

Java 是大型分布式企业项目开发的首选语言，我们所熟悉的大部分系统平台都是以 Java 为核心进行搭建的。经过二十多年的发展，Java 语言的性能得到了不断完善，更有大量的技术公司围绕 Java 推出了种类繁多的开源项目，以帮助解决 Java 开发中存在的各类困难。

Java 从产生到成为主流技术，其主要依靠的就是 Java EE 企业级平台开发，并且围绕着企业开发这一领域不断地制定技术标准。正因为有了这些技术标准，Java 开发与设计才变得更加灵活，也才有更多的人参与到这些标准的实现技术当中。如图 1-1 所示为 Java EE 开发的标准架构。

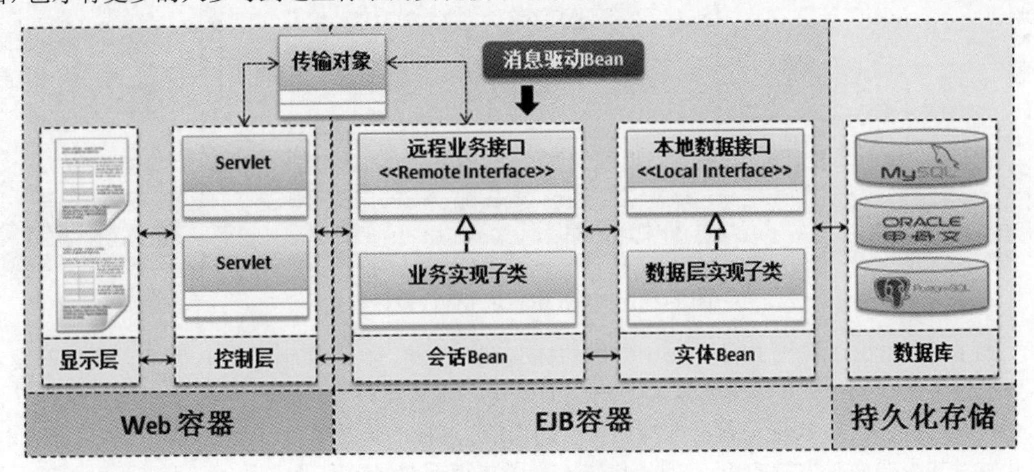

图 1-1　Java EE 标准架构

通过图 1-1 可以发现，Java EE 中采用的是 MVC 设计模式，要求通过 EJB（Enterprise Java

Bean，企业 JavaBean）实现业务中心的搭建，由会话 Bean（Remote Interface，远程接口）定义业务方法，由实体 Bean（Local Interface，本地接口）负责数据层的实现并与 SQL 数据库进行 JDBC 操作。EJB 容器的存在，可以极大地简化 JDBC 的重复性处理，同时可以更好地管理实例化对象，使开发者只关注于业务实现。这样的技术架构虽然设计理念领先，但一个非常现实的问题是 EJB 的使用成本过高。EJB 最终未能得到广泛的应用。

提示：EJB 未能得到广泛应用的原因。

虽然 EJB 技术提供了超前的设计理念，但在进行技术选型的时候，企业却不得不面对成本限制问题。

- 稳定的 EJB 容器（如 WebSphere、WebLogic 等）价格昂贵。当然，用户也可以选用免费的 JBoss。
- EJB 服务器硬件价格不菲，由于需要维护对象状态，所以需要极大的性能开销。

最为重要的是，随着技术的发展，EJB 的不足之处也逐渐被人们认识到。虽然 EJB 3.x 提供的 JPA 标准为后续的实体层开发技术带来了一个新的方向，但遗憾的是并没有得到广泛的应用。

不管 Java 技术未来向何处发展，EJB 这一技术理念的出现，为整个行业带来了新的领航方向。

Java EE 开发标准强调的是 MVC 设计模式的应用以及业务层的定义。由于 EJB 没有普及开来，所以开发者需要在 Web 端手动模拟 EJB 技术实现方案。长久以来，Java 开发的设计结构如图 1-2 所示。

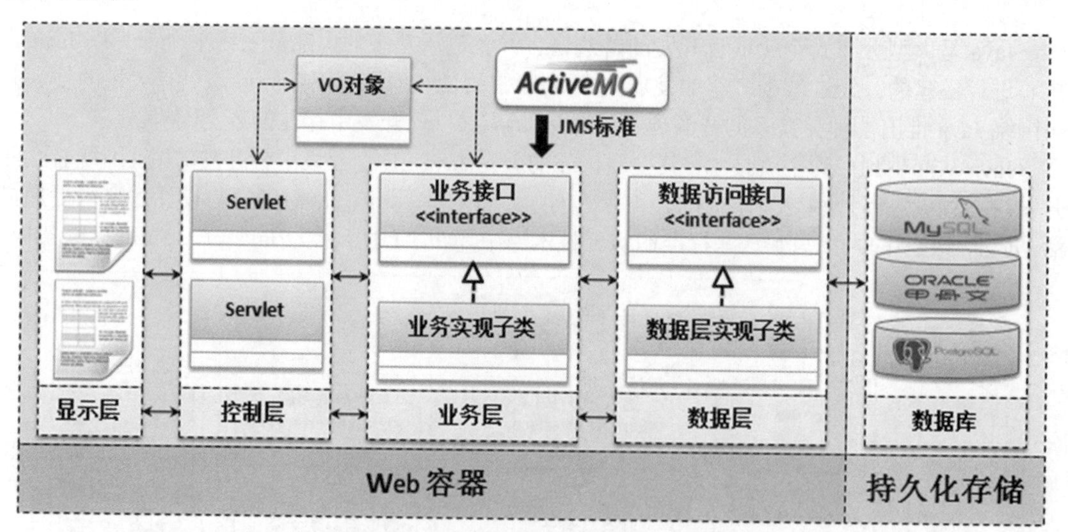

图 1-2　在 Web 中实现 MVC 设计模式

图 1-2 中给出了现代 Java 开发中常用的标准设计结构。由于取消了 EJB 容器，所以开发者需要通过手动代码方式实现业务层与数据层的结构，还需要利用开源的 JMS 消息服务组件（以 ActiveMQ 为代表作）实现消息组件与业务层的通信。这样的设计可以有效降低项目的运行成本，但开发者却需要面对大量的 Java 原生代码。为了防止不同层之间形成耦合，还需要引入工厂设计模式，进行对象获取。为了方便业务层实现事务控制，还需要引入代理设计模式，如图 1-3 所示。这样的开发，同样会造成成本的大幅攀高。

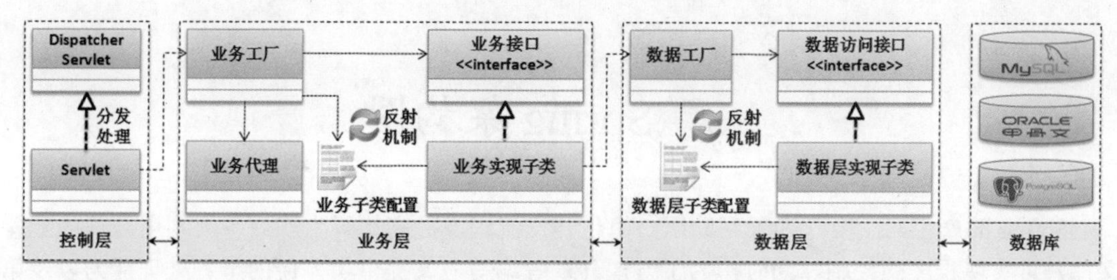

图 1-3　MVC 设计模式应用

通过图 1-3 可以发现，缺少了 EJB 容器支持之后，开发者不仅需要手动进行对象管理，还需要依据反射机制进行一系列的先期架构设计，同时还需要考虑解耦合与多线程等种种问题。如果可以有一个容器，能像 EJB 那样进行对象管理，帮助开发者免去软件架构设计的困难，将极大地提高开发效率。正是在这样的背景下，Spring 开发框架应运而生。

1.2　Spring 简介

Spring 是一个面向对象设计层面的开发框架，其本身提供了一个完善的设计容器，利用此容器可以帮助开发者实现对象管理、线程同步处理、依赖关系配置等。该框架由 Pivotal 公司提供，由 Rod Johnson（见图 1-4）主持设计开发。读者如果想了解更多有关 Spring 开发框架的信息，可以登录 Spring 官方网站（https://spring.io）进行查看，如图 1-5 所示（截图为笔者写作时的官网截图）。

图 1-4　Rod Johnson

图 1-5　Spring 首页

Spring 开发框架的核心设计理念为“使用最本质的技术进行开发”。也就是说，开发者不应该关注代码底层的细节处理（如对象管理、线程分配等），而只应该完成代码的核心功能。为了实现这个目标，在 Spring 开发框架中提供了 IoC 和 AOP 两项核心技术。

☑ **IoC（Inversion of Control，控制反转）**：实例化对象控制，可利用依赖注入（Dependency Injection）与依赖查找（Dependency Lookup）实现类对象之间的引用配置。

☑ **AOP（Aspect Oriented Programming，面向切面编程）**：利用切面表达式可进行代码的织入处理，实现代理设计。

提示：关于 IoC 与 AOP 的简单理解。

Spring 的主要功能可以总结如下：Spring 核心=工厂设计模式+代理设计模式，所以 IoC 可以简单理解为工厂设计模式，AOP 可以简单理解为代理设计模式。只不过比起原生代码，利用 Spring 处理会更加方便，功能也更加强大。

1.3 Spring 架构图

Spring 的产生，主要是为了帮助用户简化开发流程，提高代码生产效率。利用合理的配置文件，可实现程序的控制。同时，为了方便开发者编写，又提供了方便的事务处理能力以及第三方框架整合能力。Spring 的整体架构如图 1-6 所示。

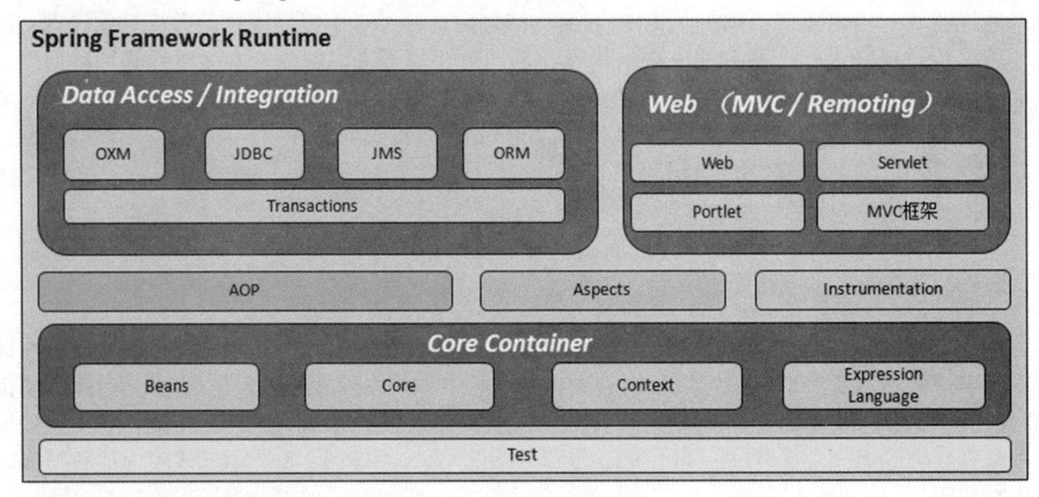

图 1-6　Spring 整体架构

1．核心容器

核心容器（Core Container）包括 Beans 模块、Core 模块、Context 模块和 Expression Language 模块。

- ☑ **Beans 模块：**提供框架的基础部分，主要用于实现控制反转（依赖注入）功能。其中，Bean Factory 是容器的核心部分，其本质是工厂设计模式实现，提倡面向接口编程，对象间的关系由框架通过配置关系进行管理，所有的依赖都由 Bean Factory 来维护。
- ☑ **Core 模块：**封装了框架依赖的最底层部分，包括资源访问、类型转换和其他的常用工具类。
- ☑ **Context 模块：**以 Core 和 Bean 模块为基础，集成 Beans 模块功能并添加资源绑定、数据验证、国际化、Java EE 支持、容器生命周期等，核心接口是 ApplicationContext。
- ☑ **Expression Language（EL）模块：**表达式语言支持，支持访问和修改属性值，方法调用，支持访问及修改数组、容器和索引器、命名变量，支持算术和逻辑运算，支持从 Spring 容器获取 Bean，也支持列表透明、选择和一般的列表聚合等。利用表达式语言，可以更加灵活地控制配置文件。

2．切面编程模块

切面编程模块包含 AOP 模块、Aspects 模块和 Instrumentation 模块。

- ☑ **AOP（Aspect Oriented Programming）模块：**符合 AOP Alliance 规范的面向切面编程实现，提供了如日志记录、权限控制、性能统计等通用功能和业务逻辑分离技术，能动态地把这些功能添加到需要的代码中，从而降低业务逻辑和通用模块的耦合。

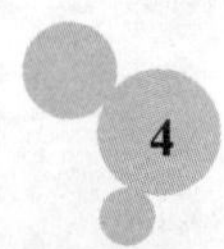

- ☑ **Aspects 模块：** 提供了 AspectJ 的集成，利用 AspectJ 表达式可以方便地实现切面管理。
- ☑ **Instrumentation 模块：** 是 Java 5 之后提供的特性。使用 Instrumentation，开发者可以构建一个代理，用来监测运行在 JVM 上的程序。监测一般是通过在执行某个类文件之前，对该类文件的字节码进行适当修改进行的。

3．数据访问/集成模块

数据访问/集成（Data Access/Integration）模块包含事务管理模块、JDBC 模块、ORM 模块、OXM 模块和 JMS 模块。

- ☑ **事务管理模块：** 用于 Spring 事务管理操作，只要是 Spring 管理的对象，都可以利用此事务模块进行控制。支持编程和声明式两类方式的事务管理。
- ☑ **JDBC 模块：** 提供了 JDBC 的操作模板，利用这些模板可以消除传统冗长的 JDBC 编码和必需的事务控制，同时可以使用 Spring 管理事务，无须额外控制事务。
- ☑ **ORM 模块：** 提供了实体层框架的无缝集成，包括 Hibernate、JPA、MyBatis 等，同时可以使用 Spring 实现事务管理，无须额外控制事务。
- ☑ **OXM 模块：** 提供了 Object/XML 映射，可以将 Java 对象映射成 XML 数据，或者将 XML 数据映射成 Java 对象。Object/XML 映射实现包括 JAXB、Castor、XMLBeans 和 XStream。
- ☑ **JMS 模块：** 用于 JMS（Java Messaging Service）组件整合，提供了一套消息“生产者-消费者”处理模型。JMS 可以用于在两个应用程序之间或分布式系统中，实现消息处理与异步通信。

4．Web（MVC / Remoting）模块

Web（MVC / Remoting）模块包含 Web 模块、Servlet 模块、MVC 框架模块和 Porlet 模块。

- ☑ **Web 模块：** 提供了基础 Web 功能，如多文件上传、集成 IoC 容器、远程过程访问（RMI、Hessian、Burlap）以及 Web Service 支持，并提供了 RestTemplate 类来进行 Restful Services 访问。
- ☑ **Servlet 模块：** 提供了 Spring MVC Web 框架实现。Spring MVC 框架提供了基于注解的请求资源注入，可以更简单地进行数据绑定、数据验证和一套非常易用的 JSP 标签，完全无缝地与其他 Spring 技术进行协作。
- ☑ **MVC 框架模块：** 提供了与常用 MVC 开发框架的整合，如 Struts、JSF 等。
- ☑ **Porlet 模块：** Portal 是一个基于 Web 的应用，它能提供个性化、单点登录、不同源的内容聚合和信息系统的表示层集中。聚合是整合不同 Web 页面源数据的过程。

5．Test 模块

支持 JUnit 和 Test 测试框架，而且额外提供了一些基于 Spring 的测试功能。例如，可在测试 Web 框架时模拟 HTTP 请求功能。或者启动容器，实现依赖注入管理。

Spring 本身提供了对象管理容器，由于 Java EE 项目都是构建在 Web 容器之上的，所以在实际开发过程中经常可看到如图 1-7 所示的运行场景。

通过图 1-7 可以发现，在 Spring 框架上进行项目构建过程之中，是以 Spring 的核心容器实现的基本结构，而后利用 Spring 中提供的系列整合技术根据自身项目的需求简化 ORMapping 组件的开发，再利用 AOP 与 AspectJ 实现业务层事务控制，对于 Web 端可以与 Struts、JSF 等常用开发框架整合，也可以直接使用 SpringMVC 进行整合。

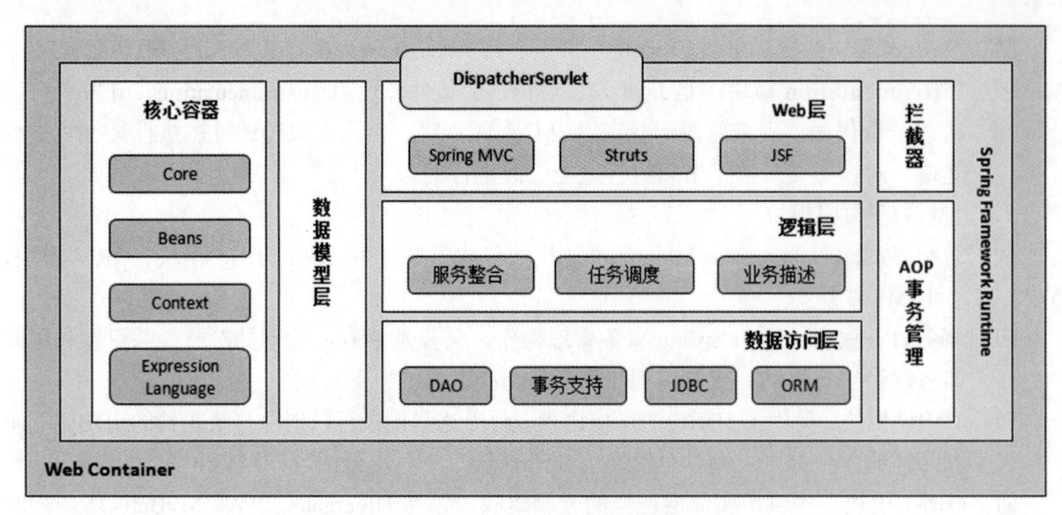

图 1-7 Spring 实际运行场景

1.4 本章小结

1．Spring 产生的主要目的是为了简化 Java EE 开发模型，提供容器机制，实现对象的统一管理。

2．Spring 的核心组成技术是 IoC（工厂设计模式应用）与 AOP（代理设计模式应用）。

3．Spring 提供了各种常用服务的整合能力，可以将所有的服务融合到一个 Spring 容器中进行管理，从而简化配置处理。

第 2 章

控制反转

通过本章学习，可以达到以下目标：

1. 理解控制反转的主要设计目的。
2. 能基于 Maven 搭建 Spring 开发环境。
3. 掌握 IoC 开发与 Spring 基本运行流程。

Spring 的出现，是为了实现“开发者只用关心核心代码”这一目标。也就是说，项目开发不应该围绕着各种底层处理机制来进行，而应该将重点放在业务处理中。为了实现这一目的，Spring 中提供了 IoC（Inversion of Control，控制反转）技术。本章将为读者介绍 IoC 的产生背景以及 Spring 项目的基本结构。

2.1 IoC 产生背景

在实际项目设计过程中，接口是一个重要组成元素，不同层之间的操作需要通过接口来调用。利用接口，可以实现子类的隐藏，也可以更好地描述不同层之间的操作标准。Java 中要想获得接口对象，需要通过关键字 new 来实现。

范例：通过关键字 new 实例化接口对象。

```
package cn.mldn.demo;
interface IMessage {                                             // 定义业务接口
    /**
     * 信息回显处理
     * @param msg 原始信息
     * @return 追加回显后的信息
     */
    public String echo(String msg);
}
class MessageImpl implements IMessage {                          // 定义接口子类
    @Override
    public String echo(String msg) {
        return "【ECHO】" + msg;                                  // 回显处理
    }
```

```
}
public class TestMessageDemo {
    public static void main(String[] args) {
        IMessage message = new MessageImpl();                   // 通过关键字new实例化接口对象
        System.out.println(message.echo("www.mldn.cn"));
    }
}
```

程序执行结果	【ECHO】www.mldn.cn

本程序是一个典型的 Java 原生代码处理操作。new 是 Java 进行对象实例化最基础的关键字，但使用 new 会暴露 IMessage 接口子类，同时会让子类与接口之间产生耦合，如图 2-1 所示。

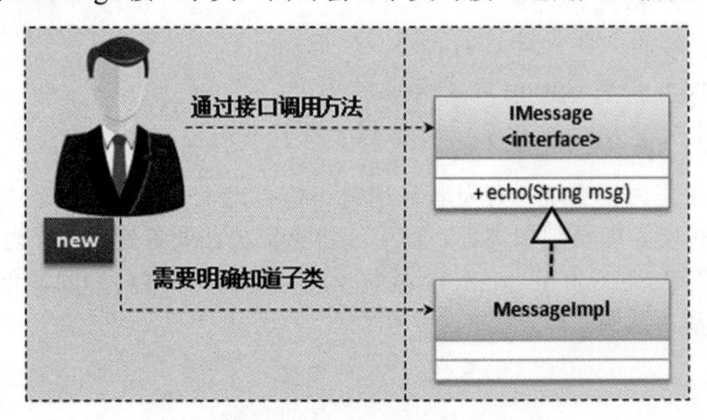

图 2-1　通过关键字 new 实例化接口对象

很明显，不同层之间出现接口子类暴露是一种不合理的设计模式，因为调用者并不需要知道具体子类是哪个，只需要取得接口对象即可。采用工厂设计模式，就可以起到解耦合的目的。

范例：通过工厂设计模式获取接口对象。

```
package cn.mldn.demo;
interface IMessage {                                            // 定义业务接口
    public String echo(String msg);
}
class Factory {
    private Factory() {}                                        // 构造方法私有化
    public static IMessage getInstance(String className) {
        if ("echo".equals(className)) {
            return new MessageImpl() ;                          // 返回接口实例
        }
        return null ;
    }
}
class MessageImpl implements IMessage {                         // 定义接口子类
    @Override
    public String echo(String msg) {
        return "【ECHO】" + msg;                                 // 回显处理
```

```
    }
}
public class TestMessageDemo {
    public static void main(String[] args) {
        IMessage message = Factory.getInstance("echo") ;        // 通过工厂类获取接口对象
        System.out.println(message.echo("www.mldn.cn"));
    }
}
```

程序执行结果	【ECHO】www.mldn.cn

本程序采用工厂设计模式解决了调用者与具体子类之间的耦合关联。调用者只需通过 Factory 类，就可以获取接口对象，不用再关注具体的子类是哪一个。本程序的代码结构如图 2-2 所示。

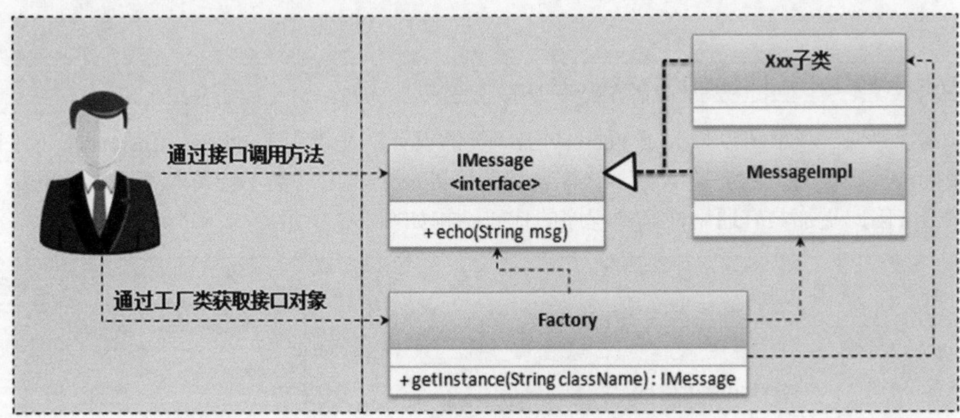

图 2-2　通过工厂类获取接口对象

虽然通过工厂设计模式可以改善接口对象的获取处理，但如果一个接口有无数个子类，且这些子类可随时动态增加，这样的静态工厂设计就会产生问题，会导致大量的修改操作。实际开发中，还需要结合反射机制来改善工厂设计。

范例：利用反射机制改善工厂设计。

```
package cn.mldn.demo;
interface IMessage {                                        // 定义业务接口
    public String echo(String msg);
}
class Factory {
    private Factory() {}                                    // 构造方法私有化
    @SuppressWarnings("unchecked")
    public static <T> T getInstance(String className) {     // 通过反射实例化
        try {
            return (T) Class.forName(className).newInstance() ;
        } catch (Exception e) {
            return null ;
        }
```

```
    }
}
class MessageImpl implements IMessage {                          // 定义接口子类
    @Override
    public String echo(String msg) {
        return "【ECHO】" + msg;                                  // 回显处理
    }
}
public class TestMessageDemo {
    public static void main(String[] args) {                    // 通过工厂类获取接口对象
        IMessage message = Factory.getInstance("cn.mldn.demo.MessageImpl") ;
        System.out.println(message.echo("www.mldn.cn"));
    }
}
```

程序执行结果	【ECHO】www.mldn.cn

本程序通过反射机制修改了 Factory 工厂类的设计，在调用 Factory.getInstance()方法时，必须明确传入子类对象的完整名称。这样设计的好处在于，工厂类不会再与某个具体的接口或子类耦合，因此更加具有通用性。本程序的代码结构如图 2-3 所示。

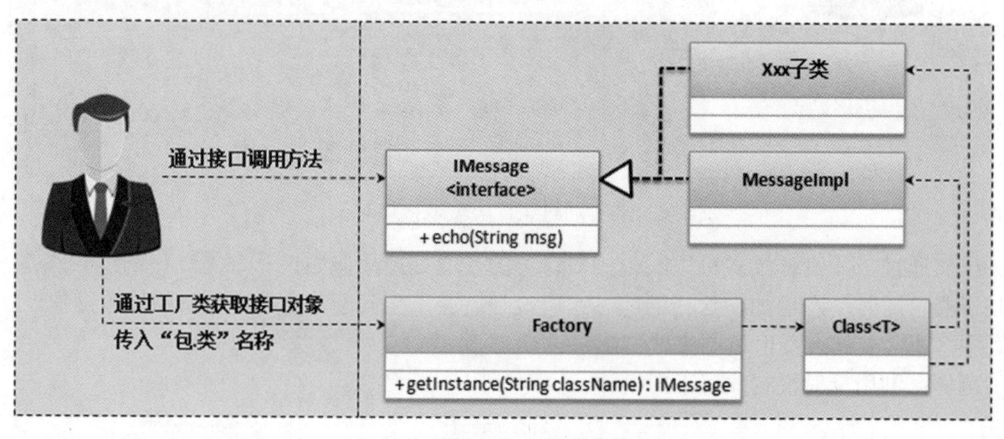

图 2-3　反射与工厂设计相结合

本例利用反射机制成功改良了工厂类的设计，使得程序结构更加清晰，同时避免了程序代码可能产生的耦合问题，但这样的配置依然存在着以下问题：

- ☑ 获取对象时需要传递完整的“包.类”名称。这样的客户端调用显然存在缺陷，最好的解决方案是追加一个配置文件，而后根据某个名称来获取对应的“包.类”名称信息，再通过反射进行加载。
- ☑ 配置文件不应该只简单描述“名称=包.类”关系，还应包含依赖关系配置。
- ☑ 应该更合理地实现多线程管理，避免过多的重复对象产生。

也就是说，要想实现一整套合理的对象管理容器，直接采用原生 Java 代码并不只靠一个简单的反射机制可以解决问题，还需要考虑各种对象的状态与对象管理。Spring 的 IoC 技术可以帮助开发者减少这些设计上的思考，使其将更多精力放在核心代码的开发处理上。

2.2　搭建 Spring 开发环境

下面将使用 Eclipse + Maven 实现 Spring 项目开发。为了配置文件编写方便，首先需要安装 STS（Spring Tool Suite）插件。

提示：关于 STS 插件。

Eclipse 是开源的，需要配置合理的插件使用，开发才会更方便、快捷。不安装插件其实也能够进行开发，但配置过程将烦琐得多。本书建议读者安装 STS 插件，该插件在微服务开发中也有着很好的应用。

需要注意的是，STS 插件需要与 Eclipse 版本相匹配。没有 STS 插件的读者可通过 www.mldn.cn 进行下载。

1.【Eclipse】安装 STS 开发插件。选择【Help】→【Install New Software】菜单命令，如图 2-4 所示。

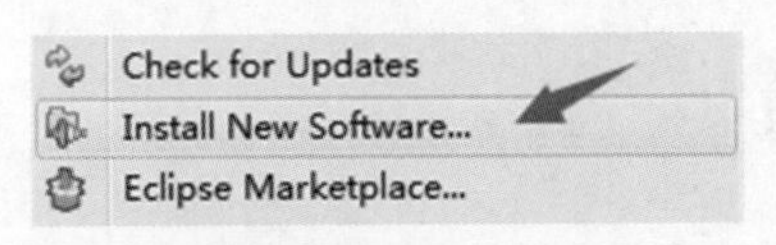

图 2-4　安装新的软件

2.【Eclipse】选择 STS 插件安装包所在的位置，如图 2-5 所示。

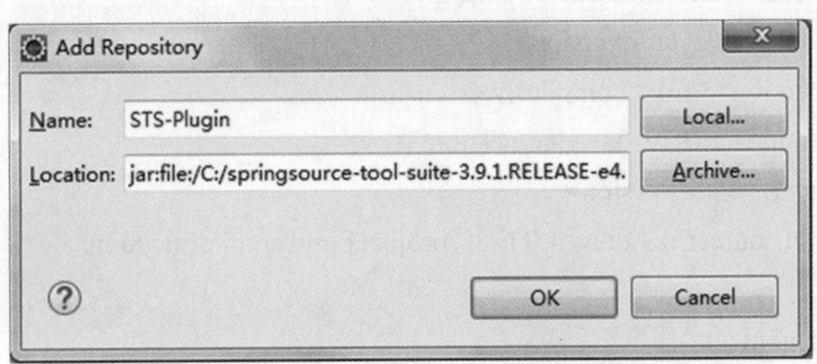

图 2-5　选择 STS 插件安装包位置

3.【Eclipse】选择要安装的组件。由于 STS 开发应用非常广，所以这里选择安装全部组件，如图 2-6 所示。

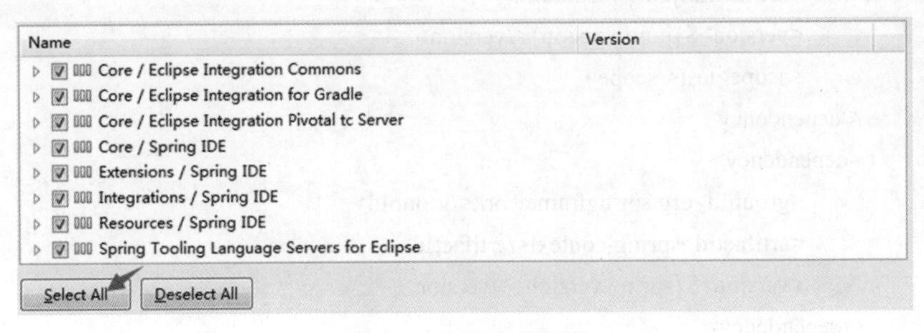

图 2-6　选中全部组件

4.【Eclipse】STS 插件安装完成后，创建一个 Maven 项目 mldnspring，如图 2-7 所示。

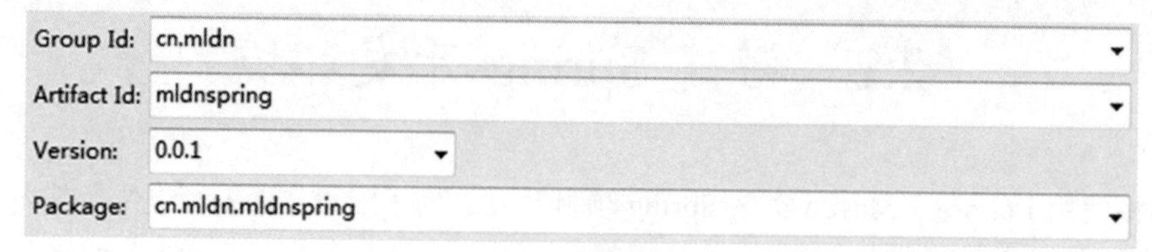

图 2-7　建立新的 Maven 项目

5.【mldnspring 项目】由于 mldnspring 实现的是统一的父 POM 管理，所以修改 pom.xml 配置文件如下。

```
<project xmlns="http://maven.apache.org/POM/4.0.0"
    xmlns:xsi="http://www.w3.org/2001/XMLSchema-instance"
    xsi:schemaLocation="http://maven.apache.org/POM/4.0.0
    http://maven.apache.org/xsd/maven-4.0.0.xsd">
    <modelVersion>4.0.0</modelVersion>
    <groupId>cn.mldn</groupId>
    <artifactId>mldnspring</artifactId>
    <version>0.0.1</version>
    <packaging>pom</packaging>
    <name>mldnspring</name>
    <url>http://maven.apache.org</url>
    <properties>
        <spring.version>5.0.3.RELEASE</spring.version>
        <junti.version>4.12</junti.version>
        <compiler.version>3.6.1</compiler.version>
        <javadoc.version>2.10.4</javadoc.version>
        <jdk.version>1.8</jdk.version>
        <project.build.sourceEncoding>UTF-8</project.build.sourceEncoding>
    </properties>
    <dependencyManagement>
        <dependencies>
            <dependency>
                <groupId>junit</groupId>
                <artifactId>junit</artifactId>
                <version>${junti.version}</version>
                <scope>test</scope>
            </dependency>
            <dependency>
                <groupId>org.springframework</groupId>
                <artifactId>spring-context</artifactId>
                <version>${spring.version}</version>
            </dependency>
```

```
            <dependency>
                <groupId>org.springframework</groupId>
                <artifactId>spring-core</artifactId>
                <version>${spring.version}</version>
            </dependency>
            <dependency>
                <groupId>org.springframework</groupId>
                <artifactId>spring-beans</artifactId>
                <version>${spring.version}</version>
            </dependency>
            <dependency>
                <groupId>org.springframework</groupId>
                <artifactId>spring-context-support</artifactId>
                <version>${spring.version}</version>
            </dependency>
        </dependencies>
    </dependencyManagement>
    <build>
        <finalName>mldnspring</finalName>
        <plugins>
            <plugin>
                <groupId>org.apache.maven.plugins</groupId>
                <artifactId>maven-compiler-plugin</artifactId>
                <version>${compiler.version}</version>
                <configuration>
                    <source>${jdk.version}</source>
                    <target>${jdk.version}</target>
                    <encode>${project.build.sourceEncoding}</encode>
                </configuration>
            </plugin>
        </plugins>
    </build>
</project>
```

本程序使用的 Spring 版本为 5.0.3.RELEASE，其中引入了 Spring 的核心依赖配置。随着开发的深入，还需要引入更多的 Spring 依赖包。

2.3　IoC 开发实现

严格来讲，IoC 并不能称为是一种技术，而是一种设计思想。IoC 产生的主要原因是为了限制使用关键字 new，因此采用统一的容器来进行对象管理。IoC 设计中，重点关注的是组件的依赖性、配置以及生命周期。通过使用 IoC，能够降低组件之间的耦合度，提高类的重用性，更有利于测试，整个产品或系统也更便于集成和配置。

1.【mldnspring 项目】创建一个新的子模块 mldnspring-base，如图 2-8 所示。

Group Id:	cn.mldn
Artifact Id:	mldnspring-base
Version:	0.0.1
Package:	cn.mldn.mldnspring

图 2-8　创建新的子模块

2.【mldnspring-base 项目】修改 pom.xml 配置文件，引入 Spring 依赖库。

```
<dependencies>
    <dependency>
        <groupId>org.springframework</groupId>
        <artifactId>spring-context</artifactId>
    </dependency>
    <dependency>
        <groupId>org.springframework</groupId>
        <artifactId>spring-core</artifactId>
    </dependency>
    <dependency>
        <groupId>org.springframework</groupId>
        <artifactId>spring-beans</artifactId>
    </dependency>
    <dependency>
        <groupId>org.springframework</groupId>
        <artifactId>spring-context-support</artifactId>
    </dependency>
</dependencies>
```

3.【mldnspring-base 项目】建立 IMessage 接口。

```
package cn.mldn.mldnspring.service;
public interface IMessage {
    public String echo(String msg) ;
}
```

4.【mldnspring-base 项目】建立 IMessage 接口实现子类 MessageImpl。

```
package cn.mldn.mldnspring.service.impl;
import cn.mldn.mldnspring.service.IMessage;
public class MessageImpl implements IMessage { // 建立接口实现子类
    @Override
    public String echo(String msg) {            // 覆写接口方法
```

```
        return "【ECHO】" + msg ;
    }
}
```

5.【mldnspring-base 项目】Spring 主要依赖配置文件进行 Bean 管理，所以需要创建一个资源目录 src/main/resources，并将其提升为源代码目录，如图 2-9 所示。

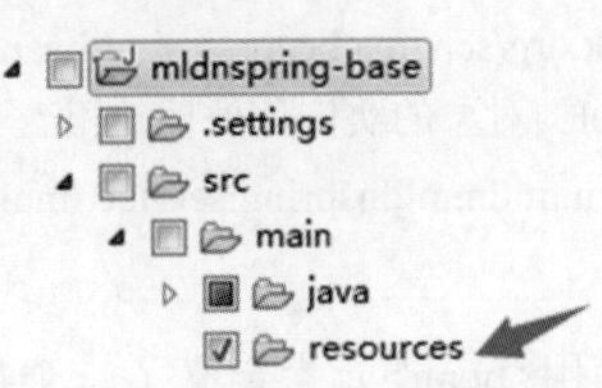

图 2-9　建立 src/main/resources 源代码目录

6.【mldnspring-base 项目】在 src/main/resources 目录中创建 spring/spring-base.xml 配置文件。可通过 STS 提供的组件创建，如图 2-10 所示，而后选择好要保存的路径，如图 2-11 所示。

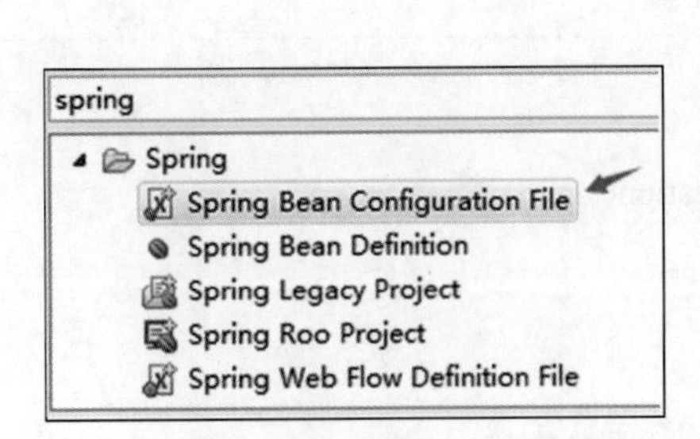

图 2-10　创建 Spring 配置文件

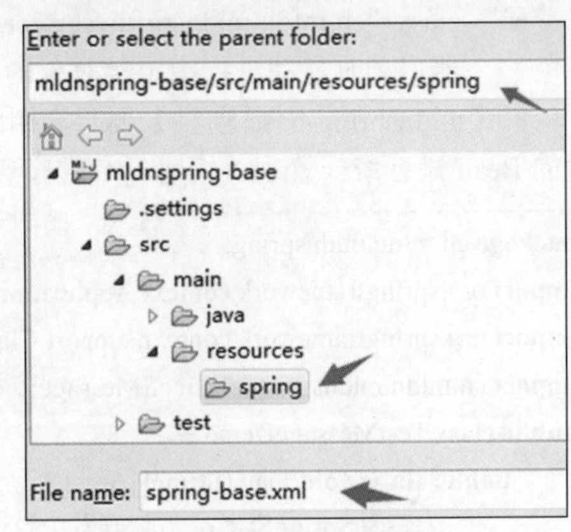

图 2-11　选择配置文件保存目录

创建 Spring 配置文件时，最重要的一步是要选择配置的命名空间。不同的命名空间代表着不同的配置项，由于本例只需要实现 Spring 的核心功能，所以这里导入 beans 命名空间，如图 2-12 所示。

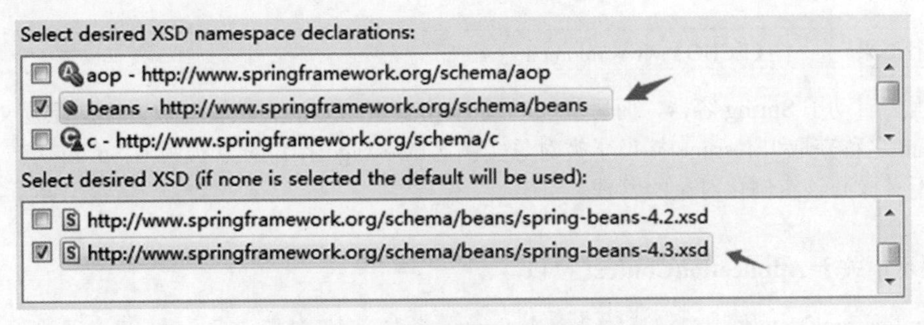

图 2-12　选择配置文件命名空间

7.【mldnspring-base 项目】修改 src/main/resources/spring/spring-base.xml 配置文件，追加 Bean 配置项。

```
<?xml version="1.0" encoding="UTF-8"?>
<beans xmlns="http://www.springframework.org/schema/beans"
    xmlns:xsi="http://www.w3.org/2001/XMLSchema-instance"
    xsi:schemaLocation="
        http://www.springframework.org/schema/beans
        http://www.springframework.org/schema/beans/spring-beans-4.3.xsd">
    <!-- 在Spring中负责MessageImpl类对象的实例化处理，相当于使用反射机制来实例化对象 -->
    <bean id="messageImpl" class="cn.mldn.mldnspring.service.impl.MessageImpl" />
</beans>
```

本程序在 Spring 配置文件中使用<bean>元素定义了一个被 Spring 管理的 Java 对象，该元素属性如下。

- ☑ id="messageImpl"：Bean 的名称，Spring 容器启动后可以根据此名称获取 Spring 管理对象。此名称不允许重复。
- ☑ class="cn.mldn.mldnspring.service.impl.MessageImpl"：Bean 对应的完整类型，Spring 容器启动时自动进行指定类对象的反射实例化处理。

8.【mldnspring-base 项目】要想启用配置文件，需要先启动 Spring 容器，然后才可以对配置的 Bean 进行统一管理。编写一个程序启动类。

```
package cn.mldn.mldnspring;
import org.springframework.context.ApplicationContext;
import org.springframework.context.support.ClassPathXmlApplicationContext;
import cn.mldn.mldnspring.service.IMessage;
public class TestMessageDemo {
    public static void main(String[] args) {
        // 启动Spring容器，实际开发中这一启动过程由Web容器负责
        ApplicationContext ctx = new ClassPathXmlApplicationContext("spring/spring-base.xml");
        IMessage message = ctx.getBean("messageImpl", IMessage.class);          // 获取实例化对象
        System.out.println(message.echo("www.mldn.cn"));
    }
}
```

程序执行结果	【ECHO】www.mldn.cn

本程序启动了 Spring 容器，而后可以通过 ApplicationContext 接口对象，根据配置文件中定义的 Bean 名称获取 IMessage 接口子类对象实例。IMessage 接口对象实例由 Spring 统一管理，开发者不必操心该实例化对象的管理。

提示：关于 ApplicationContext 接口。

ApplicationContext 接口描述的是整个 Spring 容器。该接口有许多子类，读者可通过 Spring 官方网站的 API 文档获取相应信息。如图 2-13 所示列举了几个常用的 ApplicantionContext 子类，通过类名称可以发现，可以使用 ClassPath 加载，也可以使用文件加载，或者通过 Web 加载。

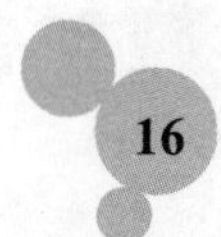

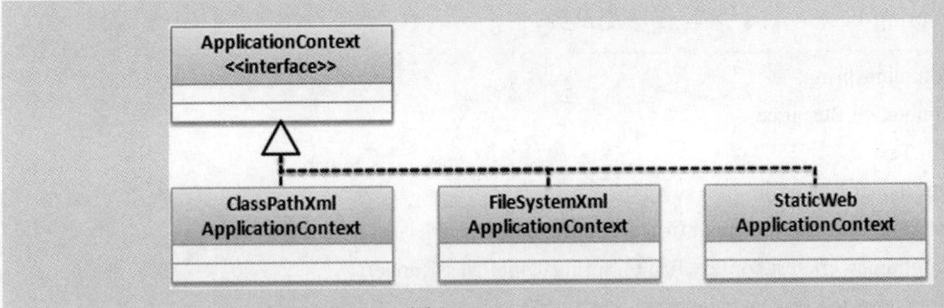

图 2-13　ApplicationContext

实际开发中，用户并不需要关注 ApplicationContext 接口，因为 Spring 容器会随着 Web 容器自动启动。这里列出，是为了便于读者理解和学习。

2.4　SpringTest 测试

前面通过 ApplicationContext 启动了 Spring 容器，并实现了配置文件的加载，但这样处理并不能体现出 Spring 的运行特征。为了更好地还原现实的开发场景，可利用 SpringTest 依赖库和 JUnit 实现测试环境下的 Spring 容器启动，且可以使用@Resource 代替 getBean 方法实现自动注入。

1.【mldnspring 项目】修改 pom.xml 配置文件，追加 spring-test 依赖库。

```
<dependency>
    <groupId>org.springframework</groupId>
    <artifactId>spring-test</artifactId>
    <version>${spring.version}</version>
    <scope>test</scope>
</dependency>
```

2.【mldnspring-base 项目】修改 pom.xml 配置文件，在子模块中引入 Spring 相关测试依赖库。由于这里要基于 JUnit 工具实现测试，所以还需要引入 junit 测试依赖库。

```
<dependency>
    <groupId>JUnit</groupId>
    <artifactId>JUnit</artifactId>
    <scope>test</scope>
</dependency>
<dependency>
    <groupId>org.springframework</groupId>
    <artifactId>spring-test</artifactId>
    <scope>test</scope>
</dependency>
```

3.【mldnspring-base 项目】编写程序测试类。

```
package cn.mldn.mldnspring;
import javax.annotation.Resource;
import org.JUnit.Test;
import org.JUnit.runner.RunWith;
import org.springframework.test.context.ContextConfiguration;
import org.springframework.test.context.JUnit4.SpringJUnit4ClassRunner;
import cn.mldn.mldnspring.service.IMessage;
@ContextConfiguration(locations = { "classpath:spring/spring-base.xml" })      // 进行资源文件定位
@RunWith(SpringJUnit4ClassRunner.class)                                       // 设置使用的测试工具
public class TestMessageDemo {
    @Resource
    private IMessage message ;                                                 // 自动根据类型进行注入
    @Test
    public void testEcho() {
        System.out.println(this.message.echo("www.mldn.cn"));
    }
}
```

程序执行结果	【ECHO】www.mldn.cn

本程序尽可能还原了 Spring 容器的实际运行环境。对它做如下几点说明。

- ☑ @ContextConfiguration：表示 Spring 配置文件所在的目录。本程序通过 classpath 进行加载，由于 src/main/resources 属于源目录，所以目录中保存的所有资源将自动设置在 CLASSPATH 之中。
- ☑ @RunWith(SpringJUnit4ClassRunner.class)：表示要使用的测试工具类型。
- ☑ @Resource：表示资源注入配置。首先会根据类型进行匹配，由于在 spring-base.xml 文件中配置的是 MessageImpl 子类，所以会自动与 IMessage 接口对应实现对象注入。如果有需要，也可以利用具体名称进行注入。

范例：设置具体的 Bean 名称。

```
@Resource(name="messageImpl")                          // 根据配置文件名称注入
private IMessage message ;                             // 自动根据类型进行注入
```

@Resource 注解中，代码 name="messageImpl"中的名称就是 spring-base.xml 配置文件中 <bean>元素 id 属性设置的名称。利用这种模式，可以避免一个接口、多个实例存在时无法匹配的问题。

2.5 本章小结

1．一个良好的项目结构设计不应该产生耦合。要想解耦合，可以采用工厂设计与反射相结合的模式。

2．Spring 开发框架可以帮助用户简化对象管理，避免使用关键字 new 产生大量的对象。

3．Spring 容器启动后，可以自动加载配置文件中定义的 Bean，并可以采用 getBean 方法通过容器获取实例化对象。

4. SpringTest 可以结合各种测试工具，实现 Spring 容器的启动，并且实现 Bean 的自动注入。

5．javax.annotation.Resource 注解可以实现配置文件中 Bean 对象的注入，其首先会根据类型进行自动注入。如果类型相同，也可以使用 name 属性设置具体的注入 Bean 名称。

第 3 章

Bean 管理

通过本章学习，可以达到以下目标：

1. 掌握 Bean 的基本配置与引用关系处理。
2. 掌握 p 命名空间的使用与关联注入。
3. 掌握集合对象的注入。
4. 掌握构造方法的注入与有参构造调用。
5. 掌握 Spring 中 Bean 的实例化处理模式。
6. 掌握 Bean 的初始化与销毁处理。

控制反转是 Spring 的核心理念，利用这一技术可以基于 Spring 实现完整的 Bean 容器管理与引用关联配置，同时也可以更好地实现 Bean 的生命周期控制。本章将为读者讲解 Spring 中 Bean 的创建与各类对象的注入。

3.1 Bean 基本管理

Spring 容器之中，类对象需要交由 Spring 统一管理。在 Spring 配置中，除了可以定义类的基本处理逻辑外，还可以实现属性的注入处理。

1.【mldnspring-base 项目】创建 Emp.java 的程序类，实现常用属性定义。

```
package cn.mldn.mldnspring.vo;
import java.io.Serializable;
import java.util.Date;
@SuppressWarnings("serial")
public class Emp implements Serializable {
	private Long empno ;
	private String ename ;
	private Integer age ;
	private Double salary ;
	private Date hiredate ;
	// setter、getter、toString 略
}
```

2.【mldnspring-base 项目】修改 spring-base.xml 配置文件，追加 Emp 程序类的 Bean 配置。

```
<bean id="hiredateObject" class="java.util.Date"/>                          <!-- 配置系统类 -->
<bean id="emp" class="cn.mldn.mldnspring.vo.Emp">                           <!-- 配置 Bean 对象 -->
    <property name="empno" value="7369"/>
    <property name="ename" value="李兴华"/>
    <property name="salary" value="960.00"/>
    <property name="age" value="18"/>
    <property name="hiredate" ref="hiredateObject"/>                        <!-- 引入其他配置 Bean -->
</bean>
```

本程序实现了 Emp 类的 Bean 配置，并且采用<property>元素设置了类对象中要保存的属性，使用 value 设置了属性的具体数据（只能是常用类型，如 String、int、Integer 等）。由于 hiredate 属于 java.util.Date 类型，所以本例将系统的 Date 类也定义为 Bean 对象，随后通过 ref 进行引用配置。

提示：采用内部 Bean 进行配置。

以上程序使用了两个配置 Bean，并且利用语句<property name="hiredate" ref="hiredateObject"/>实现了 Bean 的依赖关系。如果不想配置为两个，可以采用内部 Bean 定义形式处理。

范例：采用内部 Bean 配置。

```
<property name="hiredate">
    <bean class="java.util.Date"/>                <!-- 内部 Bean -->
</property>
```

内部 Bean 只提供给一个属性使用，所以不用再配置 id 属性。

3.【mldnspring-base 项目】编写测试程序类，实现 Emp 对象注入。

```
@ContextConfiguration(locations = { "classpath:spring/spring-base.xml" })// 进行资源文件定位
@RunWith(SpringJUnit4ClassRunner.class)                                 // 设置要使用的测试工具
public class TestEmp {
    @Resource                                       // Emp 对象只有一个，所以不需要设置 Bean 名称
    private Emp emp ;                               // 注入对象
    @Test
    public void testBean() {
        System.out.println(this.emp);
    }
}
```

程序执行结果	Emp [empno=7369, ename=李兴华, age=18, salary=960.0, hiredate=Mon Jan 29 16:30:33 CST 2019]

本程序通过 Spring 创建了 Emp 的实例化对象，由于所有的属性都已经在配置文件中定义了，所以注入后可以直接获取 Bean 对象的完整信息。

除了可以进行单独的 Bean 对象定义之外，也可以利用 Spring 配置文件实现 Bean 的引用配置。例如，假设某个雇员属于某个部门，可以得到如图 3-1 所示的类关联关系。

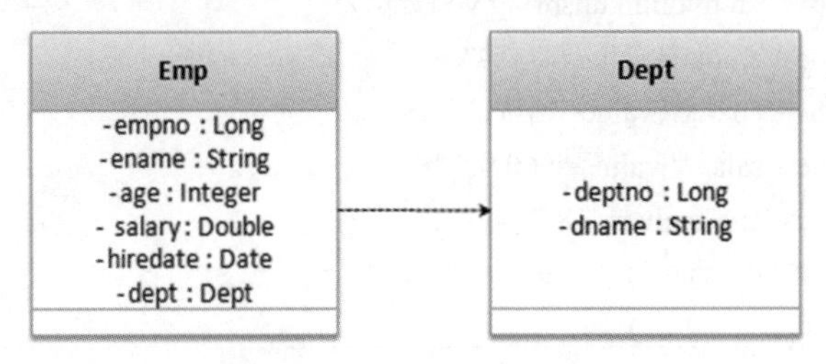

图 3-1　类关联关系

提示：关于数据转型。

进行 Spring 配置时，所有配置内容均是以字符串形式出现的。常用的基本数据类型可以自动实现类型转换，例如，10 会自动转为 int 或 Integer。如果要设置的属性类型是布尔型，则可匹配的内容有多个，如 0（false）和 1（true）、off（false）和 on（true）、true 和 false 等。

4.【mldnspring-base 项目】创建 Dept.java 程序类，同时修改 Emp.java 类，追加关联配置。

Dept.java 类

```java
package cn.mldn.mldnspring.vo;
import java.io.Serializable;
@SuppressWarnings("serial")
public class Dept implements Serializable {
	private Long deptno ;
	private String dname ;
	// setter、getter、toString 略
}
```

Emp.java 类

```java
package cn.mldn.mldnspring.vo;
import java.io.Serializable;
import java.util.Date;
@SuppressWarnings("serial")
public class Emp implements Serializable {
	private Long empno ;
	private String ename ;
	private Integer age ;
	private Double salary ;
	private Date hiredate ;
	private Dept dept ;
}
```

5.【mldnspring-base 项目】修改 spring-base.xml 配置，追加 Bean 关联配置。

```xml
	<bean id="dept" class="cn.mldn.mldnspring.vo.Dept">
		<property name="deptno" value="10"/>
		<property name="dname" value="MLDN 教学研发部"/>
	</bean>
	<bean id="emp" class="cn.mldn.mldnspring.vo.Emp">		<!-- 配置 Bean 对象 -->
		<property name="empno" value="7369"/>
		<property name="ename" value="李兴华"/>
		<property name="salary" value="960.00"/>
		<property name="age" value="18"/>
```

```
        <property name="hiredate">
            <bean class="java.util.Date"/>                    <!--  内部 Bean -->
        </property>
        <property name="dept" ref="dept"/>                    <!--  引入其他配置 Bean -->
    </bean>
```

本程序定义了 Dept 类对象 id="dept"，而后在定义 Emp 类对象时通过 ref 可以实例化 Bean 对象引用，这样就实现了关联配置。

提示：Spring 拥有完善的反射处理机制。

上述配置，如果读者对于反射机制非常熟悉的话，也可以通过 Class、Method 类对象，利用 setter 方法进行反射处理。但要想将其做到通用性，则还需要花费一些功夫。正是因为 Spring 拥有完善的反射处理机制，所以能有效提高开发效率。

6.【mldnspring-base 项目】在编写测试类时，可以通过 Emp 类对象找到 Dept 类对象，也可以直接将 Dept 类对象注入到程序中进行获取。本测试程序将只注入 Emp 类对象。

```
@ContextConfiguration(locations = { "classpath:spring/spring-base.xml" })   // 进行资源文件定位
@RunWith(SpringJUnit4ClassRunner.class)                   // 设置要使用的测试工具
public class TestEmp {
    @Resource                                              // Emp 对象只有一个，不需要设置 Bean 名称
    private Emp emp ;                                      // 注入对象
    @Test
    public void testBean() {
        System.out.println(this.emp);
        System.out.println(this.emp.getDept());            // 获取部门信息
    }
}
```

程序执行结果	Emp [empno=7369, ename=李兴华, age=18, salary=960.0, hiredate=Mon Jan 29 16:46:01 CST 2019] Dept [deptno=10, dname=MLDN 教学研发部]

此时程序实现了关联对象的匹配，而对象之间的依赖关联也可以通过 Spring 配置文件清晰表达。这样，当进行关联关系修改时也会非常方便。

3.2　使用 p 命名空间定义 Bean

Spring 中虽然提倡通过配置文件进行统一的 Bean 管理，但由于有些配置过于重复，所以为了方便 Bean 定义，也可以采用更为简洁的模式来完成，这就是 p 命名空间。

1.【mldnspring-base 项目】Eclipse 中安装了 STS 开发插件之后，对于 spring-base.xml 这样的配置文件，可以直接通过 Namespaces 选项为其追加 p 命名空间，如图 3-2 所示。

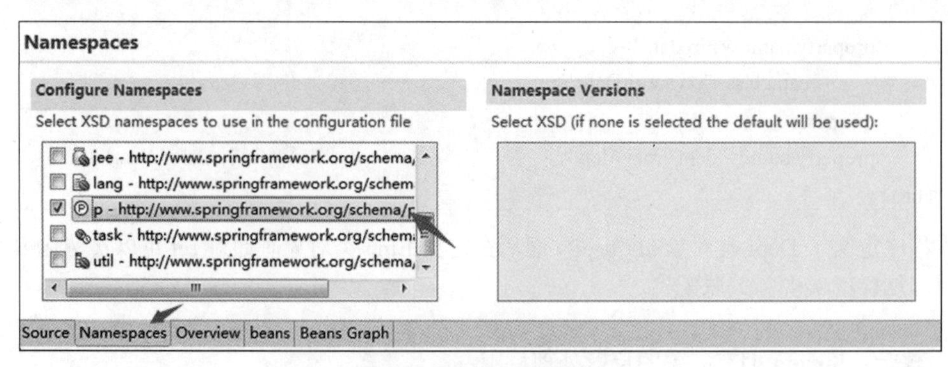

图 3-2　使用 p 命名空间

命名空间配置完成后，可以在 spring-base.xml 配置文件的头部发现如下的命名空间信息：

```
<beans xmlns="http://www.springframework.org/schema/beans"
    xmlns:xsi="http://www.w3.org/2001/XMLSchema-instance"
    xmlns:p="http://www.springframework.org/schema/p"
    xsi:schemaLocation="
            http://www.springframework.org/schema/beans
            http://www.springframework.org/schema/beans/spring-beans-4.3.xsd">
```

2.【mldnspring-base 项目】使用 p 命名空间进行配置。

```
<bean id="hiredate" class="java.util.Date" />              <!-- 定义日期对象 -->
<bean id="dept" class="cn.mldn.mldnspring.vo.Dept"
    p:deptno="10" p:dname="MLDN 教学研发部" />
<bean id="emp" class="cn.mldn.mldnspring.vo.Emp"
    p:empno="7369" p:ename="李兴华" p:salary="960.00" p:age="18"
    p:hiredate-ref="hiredate" p:dept-ref="dept" />         <!-- 配置 Bean 对象 -->
```

本程序进行 Bean 定义时，使用 p 命名空间简化了属性内容与属性引用的设置。

☑　设置属性内容，语法为：p:属性名称=内容。

☑　设置属性引用，语法为：p:属性名称-ref=引用 Bean 名称。

提示：p 命名空间的存在，是为了完善开源项目。

读者可以发现，使用 p 命名空间时，虽然配置项看起来少了许多，但却不如直接使用传统 Bean 进行配置清晰。不得不说，开源项目存在一个弊端，那就是要考虑所有开发者的感受，要尽可能满足一切需求。事实上，Spring 里还有许多功能类似但语法形式不同的配置处理。使用哪个，就看开发者自身的需要了。

3.3　注入集合对象

集合是 Java 项目开发中的重要组成部分，Spring 中也支持常用集合类型的注入，包括数组

（等价于 List 集合）、Set 集合、Map 集合和 Properties 集合。

3.3.1　注入数组对象

数组是一种固定的线性存储集合，Spring 中将数组与 List 集合进行了统一，即配置数组时使用<list>或<array>元素描述，最终的效果是完全相同的。

1.【mldnspring-base 项目】修改 Dept.java 程序类，追加数组属性。

```
@SuppressWarnings("serial")
public class Dept implements Serializable {
    private Long deptno ;
    private String dname ;
    private String[] infos ;                                    // 描述部门信息
    private List<Emp> emps ;                                    // 部门中的员工信息
    // setter、getter、toString 略
}
```

本类中一共可以注入两个数组：部门信息（infos）和部门员工（List）。

2.【mldnspring-base 项目】修改 spring-base.xml 配置文件，追加数组配置。

```
<bean id="dept" class="cn.mldn.mldnspring.vo.Dept">
    <property name="deptno" value="10"/>
    <property name="dname" value="MLDN 教学研发部"/>
    <property name="infos">          <!-- 注入数组对象 -->
        <array>
            <value>魔乐科技在线学习（www.mldn.cn）</value>
            <value>魔乐科技训练营（www.mldnjava.cn）</value>
            <value>极限 IT 训练营（www.jixianit.com）</value>
        </array>
    </property>
    <property name="emps">                        <!-- 注入该部门中的所有雇员信息 -->
        <list>
            <ref bean="empA"/>                    <!-- 引用雇员 Bean 对象 -->
            <ref bean="empB"/>                    <!-- 引用雇员 Bean 对象 -->
        </list>
    </property>
</bean>
<bean id="empA" class="cn.mldn.mldnspring.vo.Emp">       <!-- 配置 Bean 对象 -->
    <property name="empno" value="7369"/>
    <property name="ename" value="李兴华-A"/>
    <property name="salary" value="960.00"/>
    <property name="age" value="18"/>
    <property name="hiredate">
        <bean class="java.util.Date"/>                    <!-- 内部 Bean -->
```

```
        </property>
        <property name="dept" ref="dept"/>                    <!-- 引入其他配置 Bean -->
    </bean>
    <bean id="empB" class="cn.mldn.mldnspring.vo.Emp">        <!-- 配置 Bean 对象 -->
        <property name="empno" value="7839"/>
        <property name="ename" value="李兴华-B"/>
        <property name="salary" value="960.00"/>
        <property name="age" value="19"/>
        <property name="hiredate">
            <bean class="java.util.Date"/>                    <!-- 内部 Bean -->
        </property>
        <property name="dept" ref="dept"/>                    <!-- 引入其他配置 Bean -->
    </bean>
```

本程序定义了两个 Emp 对象，而后将这两个对象直接注入到了 Dept 对象之中（List 集合）。

提示：<list>与<array>元素可以互换。

通过以上配置可以发现，infos 为数组，所以使用<array>元素描述；emps 为 List 集合，所以使用<list>元素描述。Spring 中，这两者之间是可以互相替换的。

范例：修改配置。

```
<bean id="dept" class="cn.mldn.mldnspring.vo.Dept">
    <property name="deptno" value="10"/>
    <property name="dname" value="MLDN 教学研发部"/>
    <property name="infos">          <!-- 注入数组对象 -->
        <list>
            <value>魔乐科技在线学习（www.mldn.cn）</value>
            <value>魔乐科技训练营（www.mldnjava.cn）</value>
            <value>极限 IT 训练营（www.jixianit.com）</value>
        </list>
    </property>
    <property name="emps">                         <!-- 注入该部门所有雇员信息 -->
        <array>
            <ref bean="empA"/>                     <!-- 引用雇员 Bean 对象 -->
            <ref bean="empB"/>                     <!-- 引用雇员 Bean 对象 -->
        </array>
    </property>
</bean>
```

本程序中，infos 使用了<list>元素，emps 使用了<array>元素，虽然标记元素不同，但 Spring 会自动帮助用户进行处理。

3．【mldnspring-base 项目】编写 TestDept 程序类，测试数组对象注入。

```
@ContextConfiguration(locations = { "classpath:spring/spring-base.xml" })   // 进行资源文件定位
@RunWith(SpringJUnit4ClassRunner.class)                                       // 设置要使用的测试工具
public class TestDept {
    @Resource
    private Dept dept ;                                                      // 注入对象
    @Test
    public void testbean() {
        System.out.println("部门编号：" + this.dept.getDeptno() + "、部门名称：" +
                this.dept.getDname());
        System.out.println("部门信息：" + Arrays.toString(this.dept.getInfos()));
        this.dept.getEmps().forEach((emp)->{                                 // 循环输出部门信息
            System.out.println("雇员信息：" + emp);
        });
    }
}
```

程序执行结果	部门编号：10、部门名称：MLDN 教学研发部 部门信息：[魔乐科技在线学习（www.mldn.cn），魔乐科技训练营（www.mldnjava.cn），极限 IT 训练营（www.jixianit.com）] 雇员信息：Emp [empno=7369, ename=李兴华-A, age=18, salary=960.0, hiredate=Mon Jan 29 17:20:51 CST 2019] 雇员信息：Emp [empno=7839, ename=李兴华-B, age=19, salary=960.0, hiredate=Mon Jan 29 17:20:51 CST 2019]

本程序在测试类中启动了 Spring 容器，并且注入了 Dept 对象，由于该对象之中 infos 属于数组，所以可以直接使用 Arrays.toString()进行输出。对于 List 集合，则可以直接利用 Lambda 表达式结合 forEach 方法输出。

提示：List 实例化子类。

通过如上配置可以发现，在 Bean 配置中如果使用<list>或<array>元素进行数组或集合注入，会自动进行 List 集合实例化。在测试类中，可以通过如下代码测试 Spring 使用的 List 子类：

```
this.dept.getEmps().getClass()
```

程序执行后返回的是 class java.util.ArrayList 子类。之所以使用 ArrayList，是因为进行配置文件定义时，配置项的元素个数是固定的，使用 ArrayList 作为默认 List 实例化子类更加合理。

3.3.2　注入 Set 集合

Set 集合的最大特征是不允许存在重复内容。另外，Spring 中为了保证数据的存储顺序，默认使用 LinkedHashSet 子类实例化 Set 集合。

1．【mldnspring-base 项目】修改 Dept 程序类，追加 Set 集合属性。

```
@SuppressWarnings("serial")
public class Dept implements Serializable {
    private Long deptno ;
    private String dname ;
    private Set<String> infos ;                                    // 描述部门信息
}
```

2.【mldnspring-base 项目】编辑 spring-base.xml 配置文件，追加部门配置。

```
<bean id="dept" class="cn.mldn.mldnspring.vo.Dept">
    <property name="deptno" value="10"/>
    <property name="dname" value="MLDN 教学研发部"/>
    <property name="infos">            <!-- 注入 Set 对象 -->
        <set>
            <value>魔乐科技在线学习（www.mldn.cn）</value>
            <value>魔乐科技在线学习（www.mldn.cn）</value>
            <value>魔乐科技训练营（www.mldnjava.cn）</value>
            <value>极限 IT 训练营（www.jixianit.com）</value>
        </set>
    </property>
</bean>
```

由于需要注入的是 Set 集合，所以使用了<set>元素。另外，为了体现 Set 集合不保存重复数据这一特点，这里特意设置了两项相同的内容。

3.【mldnspring-base 项目】编写测试类。

```
@ContextConfiguration(locations = { "classpath:spring/spring-base.xml" })     // 进行资源文件定位
@RunWith(SpringJUnit4ClassRunner.class)                                        // 设置要使用的测试工具
public class TestDept {
    @Resource
    private Dept dept ;                                                        // 注入对象
    @Test
    public void testbean() {
        System.out.println("部门编号：" + this.dept.getDeptno() + "、部门名称：" +
                this.dept.getDname());
        this.dept.getInfos().forEach((info)->{                                 // 循环输出部门信息
            System.out.println("\t|- 信息：" + info);
        });
    }
}
```

程序执行结果	部门编号：10、部门名称：MLDN 教学研发部
	    |- 信息：魔乐科技在线学习（www.mldn.cn）
	    |- 信息：魔乐科技训练营（www.mldnjava.cn）
	    |- 信息：极限 IT 训练营（www.jixianit.com）

执行程序可以发现， Set 集合并没有保存重复的内容；另外，Set 集合中各内容的顺序就是其配置文件定义的顺序。

3.3.3　注入 Map 集合

Map 集合注入的时候需要通过<map>元素进行配置，配置时需要设置元素的 key 与 value。同时，为了保证集合保存顺序，默认会使用 LinkedHashMap 子类。

1.【mldnspring-base 项目】修改 Dept.java 类，追加 Map 集合属性。

```
@SuppressWarnings("serial")
public class Dept implements Serializable {
    private Long deptno ;
    private String dname ;
    private Map<String,Emp> emps ;                    // 描述雇员信息
    // setter、getter、toString 略
}
```

2.【mldnspring-base 项目】编辑 spring-base.xml 配置文件，注入 Map 集合。

```
<bean id="dept" class="cn.mldn.mldnspring.vo.Dept">
    <property name="deptno" value="10"/>
    <property name="dname" value="MLDN 教学研发部"/>
    <property name="emps">                         <!-- 注入 Map 集合对象 -->
        <map>
            <entry key="mldnuser-a" value-ref="empA"/>
            <entry key="mldnuser-b" value-ref="empB"/>
        </map>
    </property>
</bean>
```

本程序为 Map 集合注入了两个雇员的信息。由于 Map 中的 key 类型为 String，所以设置的 key 为普通字符串数据，而 Map 中的 value 则引入了配置 Bean 类。

3.【mldnspring-base 项目】编写测试程序，实现 Map 数据输出。

```
@ContextConfiguration(locations = { "classpath:spring/spring-base.xml" })     // 进行资源文件定位
@RunWith(SpringJUnit4ClassRunner.class)                          // 设置要使用的测试工具
public class TestDept {
    @Resource
    private Dept dept ;                                          // 注入对象
    @Test
    public void testBean() {
        System.out.println("部门编号：" + this.dept.getDeptno() + "、部门名称：" +
                this.dept.getDname());
        this.dept.getEmps().forEach((key,emp)->{                 // 循环输出部门信息
            System.out.println("\t|- key = " + key + "、value = " + emp.getEname());
```

```
        });
    }
}
```

程序执行结果	部门编号：10、部门名称：MLDN 教学研发部 　　|- key = mldnuser-a、value = 李兴华-A 　　|- key = mldnuser-b、value = 李兴华-B

由于 Map 集合中的 key 不允许重复， 所以在本程序中如果设置了重复的 key，则会产生替换处理。通过配置结果也可以发现，Map 集合中的数据保存顺序为配置顺序。

3.3.4　注入 Properties 集合

Properties 是 Hashtable 的子类，在实际开发中用于配置信息的保存处理，其所保存的 key 与 value 全部为 String 型数据。

1．【mldnspring-base 项目】修改 Dept.java 程序类，追加 Properties 属性。

```
@SuppressWarnings("serial")
public class Dept implements Serializable {
    private Long deptno ;
    private String dname ;
    private Properties infos ;          // 描述信息
    // setter、getter、toString 略
}
```

2．【mldnspring-base 项目】修改 spring-base.xml 配置文件，配置 Properties 属性内容。

```
    <bean id="dept" class="cn.mldn.mldnspring.vo.Dept">
        <property name="deptno" value="10"/>
        <property name="dname" value="MLDN 教学研发部"/>
        <property name="infos">          <!-- 注入 Properties 集合对象 -->
            <props>
                <prop key="mldn">魔乐科技在线学习（www.mldn.cn）</prop>
                <prop key="mldnjava">魔乐科技训练营（www.mldnjava.cn）</prop>
                <prop key="jixianit">极限 IT 训练营（www.jixianit.com）</prop>
            </props>
        </property>
    </bean>
```

3．【mldnspring-base 项目】编写测试程序，输出全部的 Properties 信息。

```
@ContextConfiguration(locations = { "classpath:spring/spring-base.xml" })      // 进行资源文件定位
@RunWith(SpringJUnit4ClassRunner.class)                    // 设置要使用的测试工具
public class TestDept {
    @Resource
    private Dept dept ;                                    // 注入对象
    @Test
```

```
    public void testBean() {
        System.out.println("部门编号：" + this.dept.getDeptno() + "、部门名称：" +
            this.dept.getDname());
        this.dept.getInfos().forEach((key,value)->{                    // 循环输出部门信息
            System.out.println("\t|- key = " + key + "、value = " + value);
        });
    }
}
```

程序执行结果	部门编号：10、部门名称：MLDN 教学研发部 |- key = mldnjava、value = 魔乐科技训练营（www.mldnjava.cn） |- key = mldn、value = 魔乐科技在线学习（www.mldn.cn） |- key = jixianit、value = 极限 IT 训练营（www.jixianit.com）

由于 Properties 是 Hashtable 子类，所以可以直接使用 forEach 输出。通过输出结果可以发现，Properties 中保存的顺序是无序的。

提示：Properties 与框架整合。

在 Bean 管理中，有关集合注入最为重要的就是 Properties 属性，如果需要进行 Spring 框架整合处理，例如整合 Hibernate、MyBatis 等，可以通过 Properties 进行相关属性的定义。

3.4　注入构造方法

进行 Bean 定义的过程中，默认调用类中的无参构造方法实例化 Bean 对象。考虑到用户使用方便，Spring 中也提供了有参构造的 Bean 配置，此时需要通过配置传递相应配置参数。

1.【mldnspring-base 项目】定义 Dept 程序类，取消无参构造，并追加有参构造。

```
@SuppressWarnings("serial")
public class Dept implements Serializable {
    private Long deptno ;
    private String dname ;
    public Dept(Long deptno,String dname) {
        this.deptno = deptno ;
        this.dname = dname ;
    }
    // setter、getter、toString 略
}
```

2.【mldnspring-base 项目】修改 spring-base.xml 配置文件，追加构造方法调用。

```
<bean id="dept" class="cn.mldn.mldnspring.vo.Dept">
    <constructor-arg index="0"   value="10"/>                    <!-- 第一个参数 -->
    <constructor-arg index="1"   value="MLDN 教学研发部"/>       <!-- 第二个参数 -->
</bean>
```

本程序在定义构造方法时，利用 index 配置了要设置的构造方法的参数顺序，这样会自动将设置的内容按照顺序传递到构造方法之中，以实现有参构造对象实例化。

提示：可以采用配置顺序替代 index。

在使用<constructor-arg>元素定义构造方法参数顺序时，也可以直接依靠配置顺序实现构造方法参数的配置。

范例：省略 index 属性。

```
<bean id="dept" class="cn.mldn.mldnspring.vo.Dept">
        <constructor-arg value="10"/>                    <!-- 第一个参数 -->
        <constructor-arg value="MLDN 教学研发部"/>      <!-- 第二个参数 -->
</bean>
```

本程序取消了 index 属性，因此会按照配置顺序进行构造方法参数传递。

3.【mldnspring-base 项目】Spring 可根据顺序进行构造方法的参数匹配，但很多开发者认为直接设置具体的参数名称会比较方便。要想实现这一目的，需要修改 Dept 类的构造方法。

```java
package cn.mldn.mldnspring.vo;
import java.beans.ConstructorProperties;
import java.io.Serializable;
@SuppressWarnings("serial")
public class Dept implements Serializable {
        private Long deptno ;
        private String dname ;
        @ConstructorProperties(value= {"paramDeptno","paramDname"})
        public Dept(Long deptno,String dname) {
                this.deptno = deptno ;
                this.dname = dname ;
        }
        // setter、getter、toString 略
}
```

本程序使用了@ConstructorProperties 注解并按照顺序定义了构造方法中的参数名称。这样，在配置文件中就可以通过注解中的参数名称进行内容设置。

4.【mldnspring-base 项目】修改 spring-base.xml 配置文件，利用参数名称传递构造方法参数。

```
<bean id="dept" class="cn.mldn.mldnspring.vo.Dept">
        <constructor-arg name="paramDeptno" value="10"/>                    <!-- 第一个参数 -->
        <constructor-arg name="paramDname" value="MLDN 教学研发部"/>      <!-- 第二个参数 -->
</bean>
```

此时，可以通过配置的参数名称实现构造方法的参数注入。

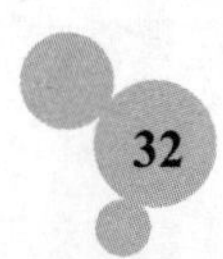

3.5　自动匹配

在 Spring 中除了可以通过名称实现 Bean 的关联之外，也可以利用自动匹配形式找到指定的类型并进行关联。之前使用过的@Resource 就属于这种自动匹配的处理形式。进行自动匹配时可以使用两种模式：根据名称（byName）匹配和根据类型（byType）匹配。

1.【mldnspring-base 项目】定义要使用的程序类。

Dept.java 类

```
@SuppressWarnings("serial")
public class Dept implements Serializable {
    private Long deptno ;
    private String dname ;
    // setter、getter、toString 略
}
```

Emp.java 类

```
@SuppressWarnings("serial")
public class Emp implements Serializable {
    private Long empno ;
    private String ename ;
    private Dept dept ;
    // setter、getter、toString 略
}
```

此时，Emp 程序中定义的 Dept 类型的属性名称为 dept，这样就可以根据名称自动进行匹配，利用反射获取相应的对象实例后实现内容设置。

2.【mldnspring-base 项目】定义 spring-base.xml 配置文件，根据类型进行 Bean 的自动匹配。

```
<bean id="mydept" class="cn.mldn.mldnspring.vo.Dept">                <!-- 名称与属性不同 -->
    <property name="deptno" value="10"/>
    <property name="dname" value="MLDN 教学研发部"/>
</bean>
<bean id="emp" class="cn.mldn.mldnspring.vo.Emp" autowire="byType"><!-- 配置 Bean 对象 -->
    <property name="empno" value="7369"/>
    <property name="ename" value="李兴华"/>
</bean>
```

本程序在定义 Emp 对象时使用 autowire="byType"进行了自动匹配，因此会自动根据类型找到合适的 Bean 并进行注入。

3.【mldnspring-base 项目】编写测试类。

```
@ContextConfiguration(locations = { "classpath:spring/spring-base.xml" })    // 进行资源文件定位
@RunWith(SpringJUnit4ClassRunner.class)                                        // 设置要使用的测试工具
public class TestEmp {
    @Resource
    private Emp emp ;                                                          // 注入对象
    @Test
    public void testBean() {
        System.out.println(this.emp);
        System.out.println(this.emp.getDept());
```

```
    }
}
```

程序执行结果	Emp [empno=7369, ename=李兴华] Dept [deptno=10, dname=MLDN 教学研发部]

本程序在进行配置的时候并没有为 Emp 类设置具体的属性，但由于采用了自动类型匹配，所以可实现部门对象的注入。

4.【mldnspring-base 项目】除了根据类型匹配外，也可以根据名称实现匹配，这样即便有多个同类型 Bean 存在，也可以根据名称匹配注入。

```
<bean id="mydept" class="cn.mldn.mldnspring.vo.Dept">                    <!-- 名称与属性不同 -->
    <property name="deptno" value="10"/>
    <property name="dname" value="MLDN 教学研发部"/>
</bean>
<bean id="dept" class="cn.mldn.mldnspring.vo.Dept">                      <!-- 名称与属性相同 -->
    <property name="deptno" value="20"/>
    <property name="dname" value="极限 IT 学习部"/>
</bean>
<bean id="emp" class="cn.mldn.mldnspring.vo.Emp" autowire="byName"><!-- 配置 Bean 对象 -->
    <property name="empno" value="7369"/>
    <property name="ename" value="李兴华"/>
</bean>
```

本程序定义了两个 Dept 对象（名称分别为 mydept 与 dept），由于在定义 Emp 对象时使用的是根据名称自动注入（autowire="byName"），因此会自动将 20 号部门的对象信息注入到 Emp 对象之中。

5.【mldnspring-base 项目】在实际的开发过程之中，有可能存在多个相同类型的 Bean 配置，如果已经配置了根据类型自动注入，则这里可以使用 primary="true"来配置优先选择项。

```
<bean id="mydept" class="cn.mldn.mldnspring.vo.Dept" primary="true"><!-- 名称与属性不同 -->
    <property name="deptno" value="10"/>
    <property name="dname" value="MLDN 教学研发部"/>
</bean>
<bean id="dept" class="cn.mldn.mldnspring.vo.Dept">                      <!-- 名称与属性相同 -->
    <property name="deptno" value="20"/>
    <property name="dname" value="极限 IT 学习部"/>
</bean>
<bean id="emp" class="cn.mldn.mldnspring.vo.Emp" autowire="byType"><!-- 配置 Bean 对象 -->
    <property name="empno" value="7369"/>
    <property name="ename" value="李兴华"/>
</bean>
```

本程序在定义 Emp 对象时采用了根据类型自动匹配，由于此时设置了两个 Dept 对象，所以默认情况下应出现无法匹配的错误信息。由于在 mydept 对象定义时使用了 primary="true"属性，因此这里并不会产生冲突，而会直接将此对象注入到 Emp 对象之中。

 **提示：除了可设置优先选择项，也可以设置取消候选项。**

对于以上的配置处理，如果不想使用 primary="true"属性，也可以直接将某一个配置的 Bean 取消候选注入资格。使用 autowire-candidate="false"可取消某个类的注入资格。

范例：取消类的注入资格。

```
<bean id="mydept" class="cn.mldn.mldnspring.vo.Dept">
    <property name="deptno" value="10"/>
    <property name="dname" value="MLDN 教学研发部"/>
</bean>
<bean id="dept" class="cn.mldn.mldnspring.vo.Dept"
        autowire-candidate="false">
    <property name="deptno" value="20"/>
    <property name="dname" value="极限 IT 学习部"/>
</bean>
<bean id="emp" class="cn.mldn.mldnspring.vo.Emp" autowire="byType">
    <property name="empno" value="7369"/>
    <property name="ename" value="李兴华"/>
</bean>
```

在本程序中定义了两个 Dept 对象，由于 dept 取消了候选配置资格，所以即便根据类型自动匹配，也无法使用此对象。

自动匹配在实际开发中非常重要，并且应用广泛。虽然在 Spring 中提供了一系列的主选与候选控制，但是从实际开发来讲，应尽量避免重名类的定义。

3.6 Bean 的实例化管理

默认情况下，只要在 Spring 配置文件中定义了对象，容器就会自动对其进行实例化处理，而后就可以通过容器直接获取实例化对象，这样开发者就不用再关注对象的创建与回收过程。但是对象是什么时候被创建的？又被创建了多少次呢？本节将针对 Spring 中的 Bean 管理进行演示说明。

1.【mldnspring-base 项目】为了方便观察，首先定义 Dept 程序类，并且提供构造方法输出。

```
@SuppressWarnings("serial")
public class Dept implements Serializable {
    private Long deptno ;
    private String dname ;
    public Dept() {
        System.out.println("****** 【Dept 对象实例化】 " + super.toString() + " ******");
    }
    // setter、getter 略
```

```
    @Override
    public String toString() {
        return "部门编号：" + this.deptno + "、部门名称：" + this.dname + "、对象信息：" +
                super.toString() ;
    }
}
```

本程序在类中定义了构造方法，并且调用了父类的 toString 方法，以获取对象的编号信息，观察实例化对象的个数。

2.【mldnspring-base 项目】在 spring-base.xml 配置文件中定义 Bean。

```
<bean id="dept" class="cn.mldn.mldnspring.vo.Dept">
    <property name="deptno" value="10"/>
    <property name="dname" value="MLDN 教学研发部"/>
</bean>
```

3.【mldnspring-base 项目】编写一个主类，实现容器启动，但不通过容器获取任何的 Bean 对象。

```
package cn.mldn.mldnspring;
import org.springframework.context.ApplicationContext;
import org.springframework.context.support.ClassPathXmlApplicationContext;
public class StartApplication {
    public static void main(String[] args) {
        ApplicationContext ctx = new ClassPathXmlApplicationContext("spring/spring-base.xml");
    }
}
```

程序执行结果	****** 【Dept 对象实例化】 cn.mldn.mldnspring.vo.Dept@5a61f5df ******

此时，程序只启动了 Spring 容器，并没有任何调用。控制台中可以发现此时已成功实现了 Bean 对象的实例化，所以可以证明，当 Spring 容器启动时会自动根据配置文件的定义实现 Bean 的反射实例化。

4.【mldnspring-base 项目】创建 3 个线程对象，并且实现 Dept 实例化对象获取，观察对象产生个数。

```
package cn.mldn.mldnspring;
import org.springframework.context.ApplicationContext;
import org.springframework.context.support.ClassPathXmlApplicationContext;
import cn.mldn.mldnspring.vo.Dept;
public class StartApplication {
    public static void main(String[] args) {
        ApplicationContext ctx = new ClassPathXmlApplicationContext("spring/spring-base.xml");
        for (int x = 0 ; x < 3 ; x ++) {
            int temp = x ;
            new Thread(()->{
```

```
                Dept dept = ctx.getbean("dept",Dept.class) ;                        // 获得程序类
                dept.setDeptno(dept.getDeptno() + temp);                            // 修改部门信息
                dept.setDname(dept.getDname() + " - " + temp);                      // 修改部门信息
                System.out.println(Thread.currentThread().getName() + "、dept = " + dept);
            }) .start() ;
        }
    }
}
```

程序执行结果	****** 【Dept 对象实例化】 cn.mldn.mldnspring.vo.Dept@5a61f5df ****** Thread-1、dept = 部门编号：11、部门名称：MLDN 教学研发部 - 0、对象信息：cn.mldn.mldnspring.vo.Dept@5a61f5df Thread-3、dept = 部门编号：13、部门名称：MLDN 教学研发部 - 0 - 2、对象信息：cn.mldn.mldnspring.vo.Dept@5a61f5df Thread-2、dept = 部门编号：11、部门名称：MLDN 教学研发部 - 0、对象信息：cn.mldn.mldnspring.vo.Dept@5a61f5df

可以发现，默认情况下 Spring 对配置文件中定义的 Bean 只会实例化一次。而后不管如何获取 Bean 对象，都只会获得相同的 Bean，即 Spring 中默认采用的是单例设计模式。

5.【mldnspring-base 项目】如果不希望 Spring 中的对象采用单例设计模式，可以在 Bean 配置中进行修改。

```
<bean id="dept" class="cn.mldn.mldnspring.vo.Dept" scope="prototype">
    <property name="deptno" value="10"/>
    <property name="dname" value="MLDN 教学研发部"/>
</bean>
```

此时程序中配置了 scope="prototype"，表示在 Spring 容器启动时不会自动进行 Bean 的实例化，只在需要的时候实例化新的 Bean 对象。

提示：不要改变默认的单例配置。

Spring 的核心功能就是对 Bean 进行合理管理，所以在实际开发中很少会见到取消单例配置的处理操作。如果要在 Bean 中配置单例，可以采用如下默认配置。

```
<bean id="dept" class="cn.mldn.mldnspring.vo.Dept" scope="singleton">
    <property name="deptno" value="10"/>
    <property name="dname" value="MLDN 教学研发部"/>
</bean>
```

本配置中使用 scope="singleton"明确表示此 Bean 为单例设计。当然，这也是默认配置。

6.【mldnspring-base 项目】重新启动测试程序类，观察单例取消后的信息输出。

```
****** 【Dept 对象实例化】 cn.mldn.mldnspring.vo.Dept@3ff21858 ******
****** 【Dept 对象实例化】 cn.mldn.mldnspring.vo.Dept@67d60b18 ******
****** 【Dept 对象实例化】 cn.mldn.mldnspring.vo.Dept@2ebb6f9e ******
```

```
Thread-2、dept = 部门编号：11、部门名称：MLDN 教学研发部 - 1、对象信息：cn.mldn.mldnspring. vo.
Dept@67d60b18
Thread-1、dept = 部门编号：10、部门名称：MLDN 教学研发部 - 0、对象信息：cn.mldn.mldnspring. vo.
Dept@3ff21858
Thread-3、dept = 部门编号：12、部门名称：MLDN 教学研发部 - 2、对象信息：cn.mldn.mldnspring. vo.
Dept@2ebb6f9e
```

此时创建了 3 个实例化对象，即每次通过容器获取对象时都需要创建新的实例化对象返回。

7.【mldnspring-base 项目】进行单例设计时，有饿汉式与懒汉式两种类型。Spring 默认采用的是饿汉式单例设计，这样在 Spring 容器启动时可以自动实现 Bean 的实例化处理。如果有需要，也可将其修改为懒汉式加载，即延迟进行初始化。

```
<bean id="dept" class="cn.mldn.mldnspring.vo.Dept" lazy-init="true">
    <property name="deptno" value="10"/>
    <property name="dname" value="MLDN 教学研发部"/>
</bean>
```

本配置中使用了 lazy-init="true"，这样只有在第一次获取 Bean 对象时才会被实例化处理。Spring 的存在使得懒汉设计模式中不需要再考虑多线程的同步处理问题。

3.7 Bean 的初始化与销毁

对于 Java 的程序类，系统提供了默认的初始化与销毁处理方法：初始化采用构造方法实现，销毁处理使用 finliaze 方法完成。这两个方法，一个是对象实例化之后的调用，一个是对象被回收前的调用。Spring 开发框架中，由于所有的对象都归 Spring 管理，所以无法使用 Java 中的初始化与销毁方法，但开发者可以定义自己的初始化和销毁的操作，在这一基础上又追加了自定义初始化与销毁的方法设置。

1.【mldnspring-base 项目】定义 Message 类，并自定义初始化与销毁方法。

```
package cn.mldn.mldnspring.vo;
public class Message {
    public void send(String msg) {
        System.out.println("****** 【消息发送】" + msg + " ******");
    }
    public void initMessage() {
        System.out.println("【Message 初始化-initMessage】建立要进行消息发送的连接通道。");
    }
    public void destroyMessage() {
        System.out.println("【Message 销毁-destroyMessage】消息发送完毕，关闭发送通道处理。");
    }
}
```

2.【mldnspring-base 项目】修改 spring-base.xml 配置文件，追加初始化与销毁处理。

```
<bean id="message" class="cn.mldn.mldnspring.vo.Message"
        init-method="initMessage" destroy-method="destroyMessage" />
```

本程序中使用了 init-method 定义了 Bean 的初始化方法，destroy-method 定义了 Bean 的销毁方法。

3.【mldnspring-base 项目】编写程序启动 Spring 容器。默认情况下，会自动调用配置中 init-method 所定义的方法进行初始化。如果想观察销毁方法调用，则必须明确地进行注册销毁处理。

```
package cn.mldn.mldnspring;
import org.springframework.context.support.ClassPathXmlApplicationContext;
import cn.mldn.mldnspring.vo.Message;
public class StartApplication {
    public static void main(String[] args) {
        ClassPathXmlApplicationContext ctx =
            new ClassPathXmlApplicationContext("spring/spring-base.xml");
        Message msg = ctx.getbean("message", Message.class);          // 获得程序 Bean
        msg.send("www.mldn.cn");
        ctx.registerShutdownHook();                                    // 调用销毁方法
    }
}
```

程序执行结果	【Message 类构造方法】cn.mldn.mldnspring.vo.Message@35fb3008 【Message 初始化-initMessage】建立要进行消息发送的连接通道。 ****** 【消息发送】www.mldn.cn ****** 【Message 销毁-destroyMessage】消息发送完毕，关闭发送通道处理。

可以发现，在 Spring 中定义的初始化方法是在构造方法调用后才会执行，同时只有执行了容器销毁后，ctx.registerShutdownHook()才可以明确发现销毁方法调用。

提示：销毁操作会由容器自动完成。

在进行 Spring 项目整合开发时，可以利用销毁处理来释放资源。例如，在与 ORMapping 开发框架整合时，可以利用销毁操作调用数据库关闭处理。大部分情况下，销毁操作会由 Spring 自行处理，开发者并不需要进行关注。

3.8　基于 Annotation 配置管理

Spring 中，所有的 Bean 都必须通过配置文件进行管理。这样处理的优势在于可以利用配置文件实现程序控制，劣势在于大型项目中可能出现配置文件过多的情况。为了弥补这一设计缺陷，Spring 提供了 Annotation 配置支持。

提示：关于配置文件与 Annotation。

Spring 早期版本比较强调配置文件与程序相分离的设计原则。随着 Spring 项目的不断增多，开发人员面临着大量配置文件的维护工作，所以后期的 Spring 版本提供了 Annotation 注解配置。利用这些注解配置和一些规则，可以避免配置文件过多，但配置文件并不会彻底消失。为了平衡配置文件的数量，Spring 还提供了配置 Bean 类，即可通过 Java 类实现配置。这一点在本章后面的部分可以看见。

为了方便读者理解 Spring 注解配置与实际开发的关联关系，下面将通过一个简单的业务层与数据层（不使用数据库）调用模拟形式，来进行讲解。调用关系如图 3-3 所示。

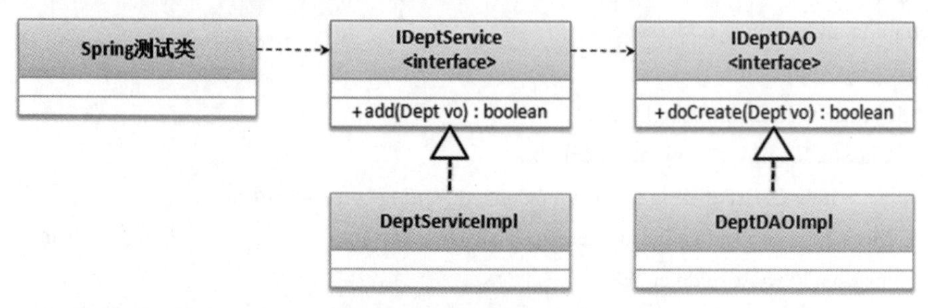

图 3-3　实际调用还原

3.8.1　context 扫描配置

要想使用 Spring 中的注解配置，需要先配置注解类的扫描包。也就是说，配置包下的所有注解都会自动生效，扫描包的配置则需要在项目中引入 context 命名空间，如图 3-4 所示。

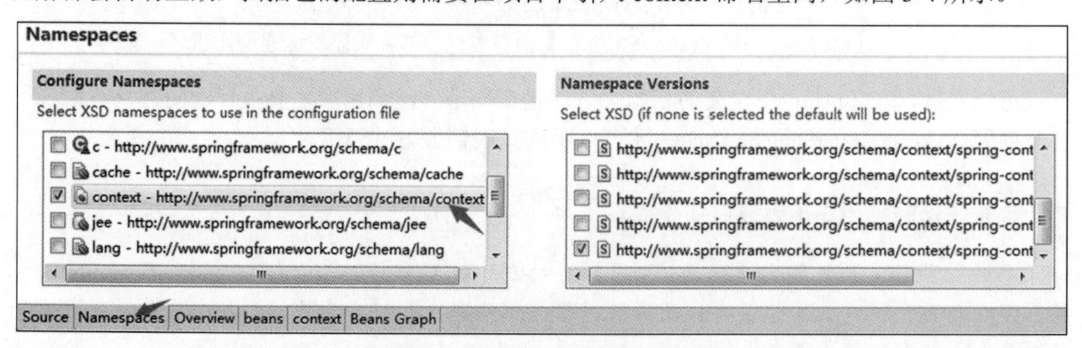

图 3-4　配置中引入 context 注解

引入 context 配置之后，还需要使用<context:component-scan>元素配置程序的扫描包，这样配置包以及其子包下的所有程序类就都可以实现注解配置了。

范例：【mldnspring-base 项目】修改 spring-base.xml 配置文件。

```
<?xml version="1.0" encoding="UTF-8"?>
<beans xmlns="http://www.springframework.org/schema/beans"
    xmlns:xsi="http://www.w3.org/2001/XMLSchema-instance"
```

```
    xmlns:p="http://www.springframework.org/schema/p"
    xmlns:context="http://www.springframework.org/schema/context"
    xsi:schemaLocation="
        http://www.springframework.org/schema/beans
        http://www.springframework.org/schema/beans/spring-beans-4.3.xsd
        http://www.springframework.org/schema/context
        http://www.springframework.org/schema/context/spring-context-4.3.xsd">
    <context:annotation-config />                           <!-- 开启扫描配置，可以不定义 -->
    <context:component-scan base-package="cn.mldn.mldnspring" />   <!-- 此包下的所有注解自动生效 -->
</beans>
```

早期 Spring 版本中，要使用注解配置，需要先定义<context:annotation-config />元素，表示启用注解。随着版本提升，现在已经不需要进行配置了。在定义<context:component-scan>元素的扫描包时可以同时定义多个扫描包，使用“,”进行分隔。

提示：程序开发包。

本程序定义的扫描基础包为 cn.mldn.mldnspring，所以后续的 DAO 实现类必须定义在 cn.mldn.mldnspring.dao.impl 子包中，业务层实现类定义在 cn.mldn.mldnspring.service.impl 子包中，这样就可以自动扫描这些类上的注解，从而实现自动配置。

3.8.2　资源扫描与注入

注解配置环境搭建完成后，如果想采用注解形式实现 Bean 配置，还需要在配置类上使用如下的 4 个注解（全部等价于<bean>功能）。

☑　**定义组件**：org.springframework.stereotype.Component。

☑　**数据层注解**：org.springframework.stereotype.Repository。

☑　**业务层注解**：org.springframework.stereotype.Service。

☑　**控制层注解**：org.springframework.stereotype.Controller。

提示：4 个注解功能相同。

@Repository、@Service、@Controller 3 个注解，如果开发者打开源代码，会发现其中都包含了@Component 注解（等价于配置文件中的<bean>元素定义，自动交由 Spring 管理）。实际上这 4 个注解的功能是完全相同的，Spring 中为了描述的准确性，才设计了不同的名称。

1.【mldnspring-base 项目】定义 IDeptDAO 接口。

```
package cn.mldn.mldnspring.dao;
import cn.mldn.mldnspring.vo.Dept;
public interface IDeptDAO {
    public boolean doCreate(Dept vo) ;
}
```

2.【mldnspring-base 项目】定义 IDeptDAO 接口实现子类，该类将通过注解配置。

```
package cn.mldn.mldnspring.dao.impl;
import org.springframework.stereotype.Repository;
import cn.mldn.mldnspring.dao.IDeptDAO;
import cn.mldn.mldnspring.vo.Dept;
@Repository
public class DeptDAOImpl implements IDeptDAO {
    @Override
    public boolean doCreate(Dept vo) {
        System.out.println("【DeptDAO】增加部门，" + vo);
        return true;
    }
}
```

由于数据层需要进行持久化处理，所以本程序使用了@Repository 注解进行 Bean 的定义。

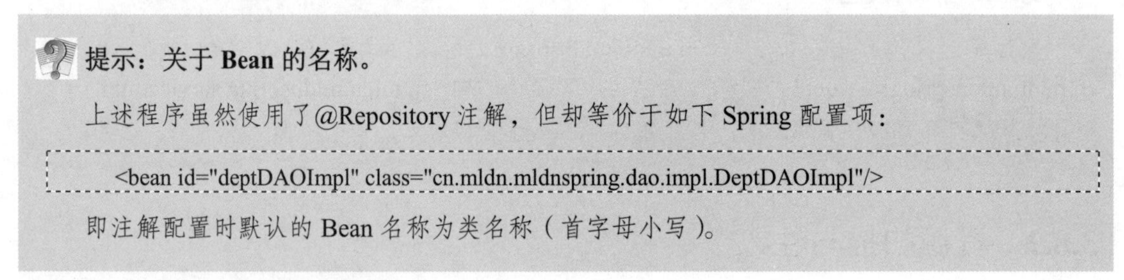

提示：关于 Bean 的名称。

上述程序虽然使用了@Repository 注解，但却等价于如下 Spring 配置项：

```
<bean id="deptDAOImpl" class="cn.mldn.mldnspring.dao.impl.DeptDAOImpl"/>
```

即注解配置时默认的 Bean 名称为类名称（首字母小写）。

3.【mldnspring-base 项目】定义 IDeptService 业务接口。

```
package cn.mldn.mldnspring.service;
import cn.mldn.mldnspring.vo.Dept;
public interface IDeptService {
    public boolean add(Dept vo) ;
}
```

4.【mldnspring-base 项目】定义 IDeptService 接口实现子类。

```
package cn.mldn.mldnspring.service.impl;
import javax.annotation.Resource;
import org.springframework.stereotype.Service;
import cn.mldn.mldnspring.dao.IDeptDAO;
import cn.mldn.mldnspring.service.IDeptService;
import cn.mldn.mldnspring.vo.Dept;
@Service
public class DeptServiceImpl implements IDeptService {
    @Resource                                          // 自动注入
    private IDeptDAO deptDAO ;
    @Override
```

```
	public boolean add(Dept vo) {
		System.out.println("************ 执行业务层方法 ************");
		return this.deptDAO.doCreate(vo);
	}
}
```

由于 DeptDAOImpl 子类上使用了@Repository 注解，所以该类对象将由 Spring 自动管理。在业务层中使用@Resource 注解实现了 IDeptDAO 子类对象的注入处理。DeptServiceImpl 类上也提供有@Service 注解，所以该类对象也将被 Spring 自动管理。

5.【mldnspring-base 项目】编写测试类，注入 IDeptService 接口实例。

```
package cn.mldn.mldnspring;
import javax.annotation.Resource;
import org.junit.Test;
import org.junit.runner.RunWith;
import org.springframework.test.context.ContextConfiguration;
import org.springframework.test.context.junit4.SpringJUnit4ClassRunner;
import cn.mldn.mldnspring.service.IDeptService;
import cn.mldn.mldnspring.vo.Dept;
import junit.framework.TestCase;
@ContextConfiguration(locations = { "classpath:spring/spring-base.xml" })// 进行资源文件定位
@RunWith(SpringJUnit4ClassRunner.class)                                  // 设置要使用的测试工具
public class TestDeptService {
	@Resource
	private IDeptService deptService;                                    // 注入业务对象
	@Test
	public void testAdd() {
		Dept vo = new Dept();                                            // 创建 VO 类对象
		vo.setDeptno(10L);                                               // 设置属性
		vo.setDname("MLDN 教学研发部");                                   // 设置属性
		TestCase.assertTrue(this.deptService.add(vo));                   // 测试业务方法
	}
}
```

程序执行结果	************ 执行业务层方法 ************ 【DeptDAO】增加部门，部门编号：10、部门名称：MLDN 教学研发部

测试程序会自动启动 Spring 容器，加载 spring-base.xml 配置文件，并根据 context 扫描包配置自动获取 IDeptService 接口实例，这样测试类中就可以直接调用业务方法了。这种做法与 Spring 在实际开发之中的处理形式非常类似。

3.8.3　@Autowired 注解

进行资源注入的时候，我们使用的是 javax.annotation.Resource 注解，但此注解并不是 Spring 的官方注解。Spring 开发框架中还有一个 org.springframework.beans.factory.annotation.Autowired 注解，在不出现重名 Bean 的情况下，两者的效果是完全一样的。

范例：【mldnspring-base 项目】修改之前的程序类，使用@Autowired 注解替代@Resource 注解。

【业务实现子类】**DeptServiceImpl.java** 类	【测试类】**TestDeptService.java** 类
@Autowired **private** IDeptDAO deptDAO ;	@Autowired **private** IDeptService deptService;

此时的程序依然可以正常执行，即可以根据类型自动实现 Bean 对象的注入管理。但当 Bean 有两个不同的实例化对象时，将无法准确地进行注入处理。

范例：【mldnspring-base 项目】修改 spring-base.xml 配置文件，采用手动方式增加一个 DeptServiceImpl 子类配置。

```
<bean id="deptServiceNew" class="cn.mldn.mldnspring.service.impl.DeptServiceImpl"/>
```

这里，程序采用两种配置方式，定义了两个 DeptServiceImpl 子类对象。这种情况下，不管使用的是@Autowired 还是@Resource 类型注入，执行时都会出现如下错误提示信息：

```
org.springframework.beans.factory.UnsatisfiedDependencyException: Error creating bean with name
'cn.mldn.mldnspring.TestDeptService': Unsatisfied dependency expressed through field 'deptService'; nested
exception is org.springframework.beans.factory.NoUniquebeanDefinitionException: No qualifying bean of type
'cn.mldn.mldnspring.service.IDeptService' available: expected single matching bean but found 2:
deptServiceImpl,deptServiceNew
```

该错误信息明确表示存在两个同样类型的 IDeptService 对象，Spring 无法区分要注入的是哪个对象。对于该问题，有以下 3 种解决方案。

解决方案 1：使用@Autowired 注解，并采用优先选择配置。

前面在讲 Bean 配置的时候，曾经讲过自动匹配处理，可以在配置文件中使用 primary="true" 进行优先选择配置，此配置可以直接在类中使用**@Primary** 注解完成。

```
@Service
@Primary
public class DeptServiceImpl implements IDeptService {}
```

此时将优先选择注解配置的 Bean 类，这样就不会出现 Bean 冲突问题。

解决方案 2：使用@Resource 注解，定义引入 Bean 名称。

```
@Resource(name="deptServiceNew")              // 注入指定名称的 Bean 对象
private IDeptService deptService;             // 注入业务对象
```

解决方案 3：联合使用@Autowired 注解和@Qualifier 注解。

@Resource 虽然可以解决重名 Bean 的问题，但由于部分开发者认为其并不是 Spring 提供的注解，所以更愿意使用@Autowired。为了解决这个问题，在 Spring 开发框架中还可以使用@Qualifier 注解来标注待注入的对象名称。

```
@Autowired
@Qualifier("deptServiceNew")                  // 注入指定名称的 Bean 对象
private IDeptService deptService;             // 注入业务对象
```

如果程序直接使用@Autowired，将无法确定要导入的是哪个 Bean 对象。使用@Qualifier 可以指明要导入的 Bean 名称，从而避免混淆。

提示：@Resource 注解与@Autowired 注解的区别。

通过分析可以发现，实际上，@Resource=@Autowired+@Qualifier。默认情况下，@Resource 与@Autowired 会根据类型（byType）自动匹配注入对象。类型相同时，可通过名称（byName）进行匹配，@Resource 可直接通过 name 属性设置 Bean 名称，@Autowired 必须结合@Qualifier 注解来设置名称。

3.8.4　使用 Java 类进行配置

项目中，如果觉得配置文件过多易导致配置混乱，可以使用 Java 程序类来实现 Bean 的配置处理。此模式基于 context 扫描包进行配置，同时需要用到@Configuration 注解。

范例：【mldnspring-base 项目】在扫描包中定义配置 Bean。

```
package cn.mldn.mldnspring.config;
import org.springframework.context.annotation.Bean;
import org.springframework.context.annotation.Configuration;
import cn.mldn.mldnspring.dao.IDeptDAO;
import cn.mldn.mldnspring.dao.impl.DeptDAOImpl;
@Configuration                                          // 表示当前类是一个专门用于配置的实现类
public class DAOConfig {
      @Bean(name="deptDAONew")                          // Bean 配置
      public IDeptDAO getDeptDAOInstance() {            // 方法名称可以随便编写
            return new DeptDAOImpl() ;                  // 返回一个实例化对象
      }
}
```

本程序将配置类直接保存到了 context 扫描子包中，由于使用@Configuration 注解，所以会自动将该类中有@Bean 注解的配置项交由 Spring 容器管理。其中，@Bean(name="deptDAONew")注解的作用等同于在配置文件中编写了如下配置项：

```
<bean id="deptDAONew" class="cn.mldn.mldnspring.dao.impl.DeptDAOImpl" />
```

利用 Bean 实现的配置相对简单，在 Spring 微架构开发中有着广泛应用。

3.9　本章小结

1．Spring 中所有被管理的程序类都要通过配置文件或注解进行定义，且在默认情况下会随着容器启动而自动启动。

2．Spring 进行集合注入时，数组与 List 功能等价，使用的是 ArrayList 子类实例化；Map

使用的是 LinkedHashMap 子类实例化；Set 使用的是 LinkedHashSet 子类实例化。

3．Spring 可以直接通过配置文件实现 Bean 的依赖关系管理，也可以利用类型或名称实现自动关联配置。

4．Spring 支持自动扫描配置，可以利用 context 配置的扫描包和适当的注解进行自动配置。

5．资源注入可以使用@Resource 或@Autowired 注解实现。

第 4 章

Spring 资源管理

通过本章学习，可以达到以下目标：

1．理解 Resource 接口的作用与核心方法。
2．理解 ResourceLoader 接口的作用与资源加载。
3．掌握不同资源加载标记的使用。
4．掌握资源路径通配符的使用。

java.io 包是进行文件输入与输出处理的核心支持包，由于 java.io 包在进行资源读取时过于琐碎，为了方便实现资源的读取配置，在 Spring 中提供了 Resource 资源加载支持，利用 Resource 可以方便地实现不同类型资源的加载，使开发更加简便。

4.1 Resource 接口简介

在实际项目开发中，经常需要进行资源数据的加载。所谓资源，就是指定访问路径上的数据信息，这些资源可能来自于配置文件、网络或*.jar 文件。对于资源数据的访问，在 java.io 包里有着完整的类定义，如 InputStream、OutputStream、Reader、Writer 等，但这些类的支持有限。例如，对于如下资源，可能就不好用了。

☑ 读取某个 jar 文件中指定的某一个文件信息。
☑ 读取一批数据信息。例如，可能有无数个 Spring 配置文件都采用了 spring-*.xml 的形式命名。
☑ 在一个目录下可能有一堆的子目录里面都有重名文件。

Spring 中，由于经常需要对资源文件进行读取处理，所以专门为资源的统一访问设计了一个接口 org.springframework.core.io.Resource，该接口的定义如下：

```
public interface Resource extends InputStreamSource
```

Resource 实际上是 InputStreamSource 的子接口，而 InputStreamSource 描述的是输入源，里面只定义了一个 getInputStream 方法，以获取输入流对象。Resource 接口定义的方法如表 4-1 所示。

表 4-1　Resource 接口定义的方法

编号	方法名称	类型	描述
1	public long contentLength() throws IOException	普通	获取资源长度

续表

编号	方法名称	类型	描述
2	public boolean exists()	普通	资源是否存在
3	public File getFile() throws IOException	普通	返回资源对应的 File 对象
4	public String getFilename()	普通	获取文件的名称
5	public URL getURL() throws IOException	普通	获取资源的完整网络路径
6	public URI getURI() throws IOException	普通	获取资源相对的路径
7	public default boolean isOpen()	普通	文件是否已经被打开
8	public default boolean isFile()	普通	给定的路径是否是文件
9	public InputStream getInputStream() throws IOException	普通	获取资源输入的数据流，是通过 InputStreamSource 父接口继承而来

4.2 读取不同资源

Resource 表示所有资源的统一访问标准。在 Resource 接口中有 4 个常用接口子类：ByteArrayResource（内存资源）、ClassPathResource（CLASSPATH 下定位资源）、FileSystemResource（文件资源）和 UrlResource（网络资源），如图 4-1 所示。

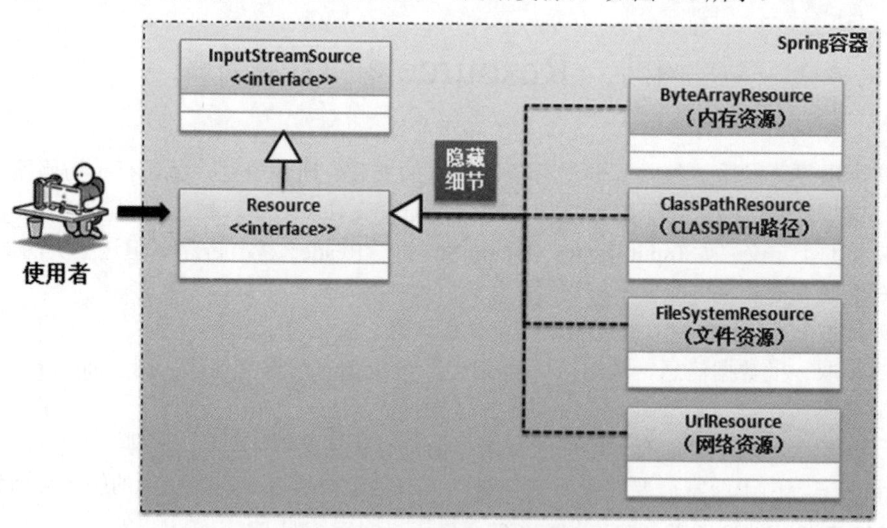

图 4-1 Resource 接口常用子类

1.【mldnspring-base 模块】编写程序，实现内存资源读取。

```
package cn.mldn.mldnspring;
import java.util.Scanner;
import org.springframework.core.io.ByteArrayResource;
import org.springframework.core.io.Resource;
public class ByteArrayResourceDemo {
    public static void main(String[] args) throws Exception {
```

```
        // 创建一个内存，读取Resource资源对象
        Resource resource = new ByteArrayResource("www.mldn.cn".getBytes()) ;
        System.out.println("资源长度：" + resource.contentLength());
        // 读取内容，可使用Resource父接口InputStreamSrouce提供的getInputStream方法完成
        Scanner scan = new Scanner(resource.getInputStream()) ;
        while (scan.hasNext()) {
            System.out.print(scan.next());
        }
        scan.close();
    }
}
```

程序执行结果	资源长度：11 www.mldn.cn

本程序实现了一个内存资源的读取操作，在进行内存资源读取时需要通过 ByteArrayResource 类的构造方法设置要读取的资源内容，而后就可以采用 Resource 接口中的 getInputStream 方法获取数据输入流对象。

2. 【mldnspring-base 模块】进行文件资源读取，使用 FileSystemResource 子类完成。

```
package cn.mldn.mldnspring;
import java.io.File;
import java.util.Scanner;
import org.springframework.core.io.FileSystemResource;
import org.springframework.core.io.Resource;
public class FileSystemResourceDemo {
    public static void main(String[] args) throws Exception {
        // 创建一个文件，读取Resource资源对象
        Resource resource = new FileSystemResource(new File("D:" + File.separator + "mldn.txt")) ;
        System.out.println("资源长度：" + resource.contentLength());
        // 读取内容，可使用Resource父接口InputStreamSrouce提供的getInputStream()方法来完成
        Scanner scan = new Scanner(resource.getInputStream()) ;
        while (scan.hasNext()) {
            System.out.print(scan.next());
        }
        scan.close();
    }
}
```

程序执行结果	资源长度：21 魔乐科技：www.mldn.cn

本程序利用 FileSystemResource 子类实现了本地文件的读取，在构造 FileSystemResource 类对象时传递了要读取的资源路径，而后依然利用 Resource 接口标准实现数据读取。

3.【mldnspring-base 模块】读取 CLASSPATH 资源。

```
package cn.mldn.mldnspring;
import java.util.Scanner;
import org.springframework.core.io.ClassPathResource;
import org.springframework.core.io.Resource;
public class ClassPathResourceDemo {
	public static void main(String[] args) throws Exception {
		// 创建一个文件，读取Resource资源对象
		Resource resource = new ClassPathResource("spring/spring-base.xml") ;
		System.out.println("资源长度：" + resource.contentLength());
		// 读取内容，可使用Resource父接口InputStreamSrouce提供的getInputStream方法来完成
		Scanner scan = new Scanner(resource.getInputStream()) ;
		scan.useDelimiter("\n") ;
		while (scan.hasNext()) {
			System.out.print(scan.next());
		}
		scan.close();
	}
}
```

本程序读取了在源文件夹目录之中的 Spring 配置文件信息，直接使用 ClassPathResource 类简化了 CLASSPATH 路径的访问。

4.【mldnspring-base 模块】读取网络资源，路径为 http://localhost/mldn/mldn-data.txt。

```
package cn.mldn.mldnspring;
import java.util.Scanner;
import org.springframework.core.io.Resource;
import org.springframework.core.io.UrlResource;
public class UrlResourceDemo {
	public static void main(String[] args) throws Exception {
		// 创建一个文件，读取Resource资源对象
		Resource resource = new UrlResource("http://localhost/mldn/mldn-data.txt") ;
		System.out.println("资源长度：" + resource.contentLength());
		// 读取内容，可使用Resource父接口InputStreamSrouce提供的getInputStream()方法完成
		Scanner scan = new Scanner(resource.getInputStream()) ;
		while (scan.hasNext()) {
			System.out.print(scan.next());
		}
		scan.close();
	}
}
```

通过以上 4 个资源读取程序，相信读者已经发现了，Spring 之所以需要设置 Resource 接口标准，是为了对资源访问进行统一管理，即通过 Resource 接口子类弥补 InputStream 类的功能局限。

4.3　ResourceLoader 接口

在 Spring 设计的时候，已经明确地将所有资源统一规划为由 Resource 接口对象负责读取。但在使用过程中却需要面临一个问题，也是 Spring 设计中强调的一个核心思想——解耦合。之前编写的代码都直接采用了子类为父接口实例化的模式来处理，很明显这样的设计是不合理的。为了解决 Resource 读取不同资源的问题，专门又提供了一个 org.springframework.core.io.ResourceLoader 接口。该接口提供两个处理方法，如表 4-2 所示。

表 4-2　ResourceLoader 接口方法

编号	方法名称	类型	描述
1	public ClassLoader getClassLoader()	普通	获取类加载器
2	public Resource getResource(String location)	普通	根据指定的路径返回 Resource 接口实例

通过表 4-2 可以发现，ResourceLoader 接口中提供的 getResource 方法可以直接返回 Resource 接口实例，关系如图 4-2 所示。最关键的是，在调用此方法时可以通过字符串传递访问路径。常用的访问路径标记如表 4-3 所示。

表 4-3　资源定位

编号	资源定位	描述
1	file:路径	本地磁盘资源加载
2	http://路径	网络资源加载
3	classpath:路径	CLASSPATH 资源加载

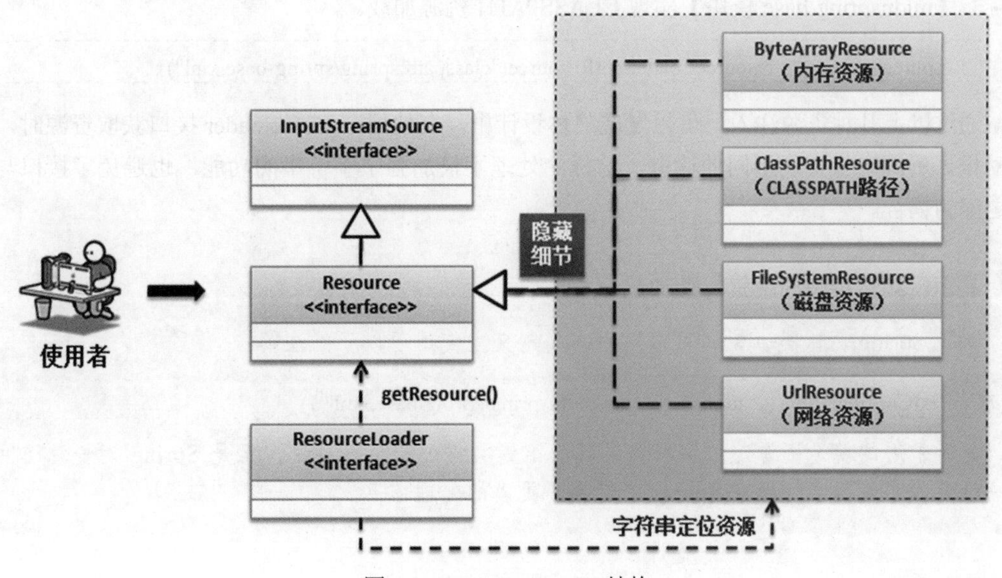

图 4-2　ResourceLoader 结构

1．【mldnspring-base 模块】通过 ResouceLoader 读取文件资源。

```
package cn.mldn.mldnspring;
import java.util.Scanner;
import org.springframework.core.io.DefaultResourceLoader;
import org.springframework.core.io.Resource;
import org.springframework.core.io.ResourceLoader;
public class FileSystemResourceDemo {
    public static void main(String[] args) throws Exception {
        ResourceLoader resourceLoader = new DefaultResourceLoader() ;
        Resource resource = resourceLoader.getResource("file:d:/mldn.txt") ;    // 文件资源
        System.out.println("资源长度：" + resource.contentLength());
        Scanner scan = new Scanner(resource.getInputStream()) ;
        while (scan.hasNext()) {
            System.out.print(scan.next());
        }
        scan.close();
    }
}
```

本程序使用 DefaultResourceLoader 子类为 ResourceLoader 接口进行实例化，随后通过 file:d:/mldn.txt 字符串实现了磁盘资源的加载。

2．【mldnspring-base 模块】实现网络资源加载。

```
Resource resource = resourceLoader.getResource("http://localhost/mldn/mldn-data.txt") ;
```

3．【mldnspring-base 模块】实现 CLASSPATH 资源加载。

```
Resource resource = resourceLoader.getResource("classpath:spring/spring-base.xml") ;
```

通过以上几种资源注入，可以发现整体设计中，在使用 resourceloader 接口获取资源时，将自动根据资源定位实现不同的读取，这样的处理不仅加强了字符串的功能，也避免了接口与子类之间的耦合。

提示：测试类中的配置文件加载。

对于 Spring 测试类，读者可以发现在类定义上使用了以下的注解：

```
@ContextConfiguration(locations = { "classpath:spring/spring-base.xml" })
```

此时采用的就是资源定位字符串实现的配置文件加载，可以说，正是 Spring 对资源访问的统一设计，才使得字符串在 Spring 中包含了更多的处理信息。

4.4　资源注入

明确了字符串与资源定位的联系后，就可以依据 Spring 容器进行 ResourceLoader 自行管理。也就是说，用户在编写的时候只需要定义好相应的字符串，Spring 就会为指定的字符串找到匹配的资源类，并自动进行实例化处理。

1.【mldnspring-base 模块】建立一个资源配置 Bean。

```
package cn.mldn.mldnspring.resource.util;
import org.springframework.core.io.Resource;
public class DefaultResourceBean {
    private Resource resource ;                                    // 直接定义要使用的Bean对象
    public void setResource(Resource resource) {
        this.resource = resource;
    }
    public Resource getResource() {
        return resource;
    }
}
```

类配置中只进行了 Resource 资源对象的接收，资源的类型可以通过字符串进行传递。

2.【mldnspring-base 模块】通过配置文件注入时，Spring 会自行实现 ResourceLoader 接口，所以开发者可根据自身需要通过字符串配置资源路径。可以配置的路径包括 3 种：文件资源、网络资源和 CLASSPATH 资源。

文件资源	<bean id="resourceBean" class="cn.mldn.mldnspring.resource.util.DefaultResourceBean"> <property name="resource" value="file:d:/mldn.txt"/> </bean>
网络资源	<bean id="resourceBean" class="cn.mldn.mldnspring.resource.util.DefaultResourceBean"> <property name="resource" value="http://localhost/mldn/mldn-data.txt"/> </bean>
CLASSPATH资源	<bean id="resourceBean" class="cn.mldn.mldnspring.resource.util.DefaultResourceBean"> <property name="resource" value="classpath:spring/spring-base.xml"/> </bean>

3.【mldnspring-base 模块】编写测试类，获取资源配置。

```
@ContextConfiguration(locations = { "classpath:spring/spring-base.xml" })        // 进行资源文件定位
@RunWith(SpringJUnit4ClassRunner.class)                                        // 设置要使用的测试工具
```

```
public class TestResource {
    @Autowired
    private DefaultResourceBean resourceBean ;                    // 注入资源Bean对象
    @Test
    public void testResource() throws Exception {
        Scanner scan = new Scanner(this.resourceBean.getResource().getInputStream()) ;
        scan.useDelimiter("\n") ;
        while (scan.hasNext()) {
            System.out.print(scan.next());
        }
        scan.close();
    }
}
```

由于所有的资源都要通过 ResourceLoader 获取，因此只需要配置好相应的字符串，就可以在程序中根据 Resource 接口来实现内容加载。

4.5 注入资源数组

要实现多个资源的统一读取，可以采用数组或 List 集合形式来实现一组 Resource 对象的保存。在进行配置时，也可以通过多种资源描述符配置资源访问路径。

1．【mldnspring-base 模块】修改 DefaultResourceBean，保存 Resource 集合。

```
package cn.mldn.mldnspring.resource.util;
import java.util.List;
import org.springframework.core.io.Resource;
public class DefaultResourceBean {
    private List<Resource> resources ;
    public void setResources(List<Resource> resources) {
        this.resources = resources;
    }
    public List<Resource> getResources() {
        return resources;
    }
}
```

2．【mldnspring-base 模块】修改 spring-base.xml 配置文件，配置多个资源路径。

```
<bean id="resourceBean" class="cn.mldn.mldnspring.resource.util.DefaultResourceBean">
    <property name="resources">
        <list>
            <value>http://localhost/mldn/mldn-data.txt</value>
            <value>classpath:spring/spring-base.xml</value>
```

```
                <value>file:d:/mldn.txt</value>
            </list>
        </property>
    </bean>
```

本配置文件中一共定义了 3 个资源读取的路径，分别是网络、CLASSPATH 和文件。

3.【mldnspring-base 模块】编写测试类，实现一组资源的读取。

```
@ContextConfiguration(locations = { "classpath:spring/spring-base.xml" })    // 进行资源文件定位
@RunWith(SpringJUnit4ClassRunner.class)                                      // 设置要使用的测试工具
public class TestResource {
    @Autowired
    private DefaultResourceBean resourceBean ;
    @Test
    public void testResource() throws Exception {
        Iterator<org.springframework.core.io.Resource> iter = this.resourceBean.getResources().iterator() ;
        while (iter.hasNext()) {
            Scanner scan = new Scanner(iter.next().getInputStream()) ;
            scan.useDelimiter("\n") ;
            while (scan.hasNext()) {
                System.out.print(scan.next());
            }
            scan.close();
            System.out.println("*********************************************");
        }
    }
}
```

由于所有的资源都将统一注入到 List 集合中，所以在测试程序类中将直接使用 Iterator 获取所有 Resource 接口对象并实现资源加载。

4.6　路径通配符

为了方便资源读取，Spring 开发框架引用了 Ant 构建工具中所定义的通配符，以实现不同层级或不同名称匹配时的资源加载问题。具体来说，有如下 3 种通配符。

☑ ?：表示可匹配任意的零位或一位字符。例如，spring?.xml 可匹配 spring1.xml、springa.xml、spring.xml 等。

☑ *：表示可匹配零位、一位或多位字符。例如，spring-*.xml 可匹配 spring-service.xml、spring-action.xml 等。

☑ **：表示可匹配任意的目录。

范例：【mldnspring-base 模块】读取指定 spring 目录中所有以 spring-开头的资源信息。

```
<bean id="resourceBean" class="cn.mldn.mldnspring.resource.util.DefaultResourceBean">
    <property name="resources">
        <list>      <!-- 加载spring及其子目录中所有以spring-开头的配置文件 -->
            <value>classpath:spring/**/spring-*.xml</value>
        </list>
    </property>
</bean>
```

除了可以读取当前工作目录中的配置文件资源外，也可以读取所有*.jar 文件中的资源。例如，要想读取 CLASSPATH 下 jar 文件中的资源，需要使用“classpath*:路径”定位格式。

1.【mldnspring-base 模块】修改资源读取类，将 List 集合修改为数组。

```
package cn.mldn.mldnspring.resource.util;
import org.springframework.core.io.Resource;
public class DefaultResourceBean {
    private Resource[] resources ;                              // 直接定义要使用的Bean对象
    public void setResources(Resource[] resources) {
        this.resources = resources;
    }
    public Resource[] getResources() {
        return resources;
    }
}
```

需要注意的是，采用 classpath*的形式读取资源时，返回的一定是一组资源，此时需要采用资源数组进行接收。如果不是资源数组，将出现 java.io.FileNotFoundException 异常。

2.【mldnspring-base 模块】配置 spring-base.xml 文件，读取所有*.jar 文件中的*.MF 文件。

```
<bean id="resourceBean" class="cn.mldn.mldnspring.resource.util.DefaultResourceBean">
    <property name="resources">
        <array>                 <!-- 将会对CLASSPATH和所有*.jar文件进行查找 -->
            <value>classpath*:**/META-INF/*.MF</value>
        </array>
    </property>
</bean>
```

本程序实现了 CLASSPATH 路径下的资源匹配。由于 classpath*会在当前程序的 CLASSPATH 下以及所有的*.jar 文件下进行查询，因此查询效率要比直接使用“classpath:路径”慢许多。

4.7 本章小结

1．为了方便读取资源，Spring 重新设计了资源读取标准——Resource 接口，该接口是 InputStreamSource 的子接口。

2．ResourceLoader 可以结合资源访问字符串，返回指定资源位置的 Resource 对象，以统一资源读取处理。

3．利用配置文件实现的资源访问配置，实际上就是对 ResourceLoader 接口功能的包装。

4．进行资源访问时，可以使用“classpath:路径”“file:路径”“http://路径”形式加载不同的资源。

5．Spring 资源读取支持 Ant 路径匹配模式。可以采用“?”匹配任意一个字符，也可以使用“*”匹配零个、一个或多个字符。如果要进行目录匹配，则可以使用“**”完成。

第 5 章

Spring 表达式语言

通过本章学习，可以达到以下目标：

1. 理解 SpEL 的主要目的。
2. 理解 SpEL 的处理流程。
3. 理解基本表达式、Class 表达式、集合表达式的使用。
4. 理解集合与变量处理操作。
5. 理解配置文件与 SpEL 的处理关系。

在 Sprin 中最大的特征实际上就在于字符串的处理加强，不仅针对于字符串的类型转换与资源配置，实际上可以利用字符串描述更多的内容，而这就是 Spring 表达式的作用，本章将为读者分析 Spring 表达式的基本形式与实际应用。

5.1 Spring 表达式基本定义

在 Spring 中为了方便开发者进行解耦和设计，可以使用字符串来进行一些特殊含义的描述，但是考虑到实际开发的复杂性，Spring 开发框架进一步加强了字符串的功能，推出了 Spring 表达式语言（Spring Expression Language，SpEL），通过 SpEL 使得字符串不仅仅可以实现一些基础的计算功能，还以根据表达式实现对象实例化或者是进行一些复杂的配置文件的编写。

范例：【mldnspring-base 模块】定义 Spring 表达式。

```
package cn.mldn.mldnspring;
import org.springframework.expression.EvaluationContext;
import org.springframework.expression.Expression;
import org.springframework.expression.ExpressionParser;
import org.springframework.expression.spel.standard.SpelExpressionParser;
import org.springframework.expression.spel.support.StandardEvaluationContext;
public class SpELFirstDemo {
	public static void main(String[] args) {
		String str = "(\"Hello \" + \"World !!!\").substring(6,12)" ;		// 表达式定义
		// 1. 定义一个专属的表达式解析工具
		ExpressionParser parser = new SpelExpressionParser() ;		// 定义一个Spring表达式解析器
		// 2. 定义一个表达式的处理类
```

```
        Expression exp = parser.parseExpression(str) ;                  // 从字符串里面解析出内容
        // 3. 进行最终表达式的计算
        EvaluationContext context = new StandardEvaluationContext() ;
        // 4. 通过表达式进行结果的计算
        System.out.println(exp.getValue(context));
    }
}
```

程序执行结果	World

本程序实现了一个最基础的表达式处理，主要功能是字符串截取处理。最为关键的是字符串中所有的程序代码都可以正常执行，而这就是 SpEL 的核心所在。要想实现这一功能，需要如下辅助类的支持。

☑ 表达式解析器：org.springframework.expression.ExpressionParser。
  |- 主要是负责为给定的表达式进行内容的解析操作处理的接口标准。
  |- 利用 SpEL 的标准表达式处理子类 SpelExpressionParser 为 ExpressionParser 接口实例化。

☑ 表达式执行类：org.springframework.expression.Expression。
  |- 通过此类可以得到最终计算结果。

☑ 表达式计算：org.springframework.expression.EvaluationContext。
  |- 计算处理上下文，主要可以实现一些表达式变量的处理，并且实现表达式计算。

范例：【mldnspring-base 模块】在表达式中使用变量。

```
package cn.mldn.mldnspring;
import org.springframework.expression.EvaluationContext;
import org.springframework.expression.Expression;
import org.springframework.expression.ExpressionParser;
import org.springframework.expression.spel.standard.SpelExpressionParser;
import org.springframework.expression.spel.support.StandardEvaluationContext;
public class SpELFirstDemo {
    public static void main(String[] args) {
        String str = "(\"Hello \" + \"World !!!\").substring(#start,#end)" ;    // 表达式定义
        ExpressionParser parser = new SpelExpressionParser() ;              // Spring表达式解析器
        Expression exp = parser.parseExpression(str) ;                      // 解析表达式
        EvaluationContext context = new StandardEvaluationContext() ;       // 上下文计算
        context.setVariable("start", 6);                                    // 定义变量
        context.setVariable("end", 12);                                     // 定义变量
        System.out.println(exp.getValue(context));                          // 表达式计算
    }
}
```

程序执行结果	World

本程序实现了与之前类似的功能，在定义表达式的时候采用了变量形式#start、#end，这样就可以在进行计算前设置变量内容，以实现表达式计算。

5.2 表达式解析原理

SpEL 的核心功能就是提高字符串的处理能力。为了帮助用户更好地理解表达式的工作原理，下面将通过一个具体程序进行分析。

范例：【mldnspring-base 模块】编写一个数学计算表达式。

```
package cn.mldn.mldnspring;
import org.springframework.expression.EvaluationContext;
import org.springframework.expression.Expression;
import org.springframework.expression.ExpressionParser;
import org.springframework.expression.spel.standard.SpelExpressionParser;
import org.springframework.expression.spel.support.StandardEvaluationContext;
public class SpELFirstDemo {
    public static void main(String[] args) {
        String str = "10 + 20" ;                                          // 表达式字符串
        ExpressionParser parser = new SpelExpressionParser() ;            // Spring表达式解析器
        Expression exp = parser.parseExpression(str) ;                    // 解析表达式
        EvaluationContext context = new StandardEvaluationContext() ;     // 上下文计算
        Integer result = (Integer) exp.getValue(context) ;                // 表达式计算
        System.out.println(result);
    }
}
```

程序执行结果	30

本程序定义了一个简单的数学计算表达式，该表达式在 Spring 中的处理流程如图 5-1 所示。

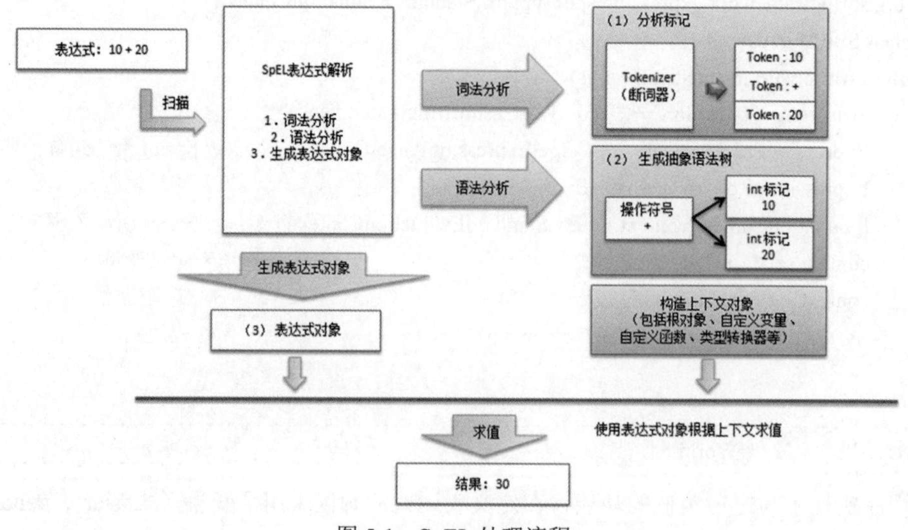

图 5-1　SpEL 处理流程

在整个 SpEL 的处理过程中，要针对给定的标记进行识别，而后根据识别后的结果进行相应内容的转换处理。由于表达式中可能进行各种变量内容的设置，所以还需要有一个上下文的变量环境。最终才可以计算出一个完整的结果。

提示：开发者只关注表达式语句编写。

通过一系列的范例可以发现，如果要想实现 SpEL 表达式的使用，除了要编写表达式，还需要编写一系列的处理类才可以执行完毕，这些都属于表达式的处理过程。在实际开发中如果结合 Spring 容器进行配置，开发者只需要编写表达式即可，对于处理过程中的操作类可以暂时忽略。

5.3　自定义分隔符

为了便于使用表达式语言，需要为表达式设置边界分隔符。定义边界分隔符可以使用 org.springframework.expression.ExpressionParser 接口来实现。该接口的定义如下：

```
public interface ExpressionParser {
    public Expression parseExpression(String expressionString) throws ParseException;
    public Expression parseExpression(String expressionString, ParserContext context)
        throws ParseException;
}
```

在该接口中，定义的 parseExpression 方法可以接收一个 ParserContext 接口对象，此接口可以实现表达式的边界定义。ParserContext 接口定义如下：

```
public interface ParserContext {
    public boolean isTemplate();                                    // 是否使用此模板
    public String getExpressionPrefix();                            // 分隔符前缀
    public String getExpressionSuffix();                            // 分隔符后缀
    ParserContext TEMPLATE_EXPRESSION = new ParserContext() {      // 默认分隔符，#{}
        @Override
        public boolean isTemplate() {
            return true;
        }
        @Override
        public String getExpressionPrefix() {
            return "#{";
        }
        @Override
        public String getExpressionSuffix() {
            return "}";
        }
```

```
    };
}
```

范例：【mldnspring-base 模块】自定义表达式分隔符。

```
package cn.mldn.mldnspring;
import org.springframework.expression.EvaluationContext;
import org.springframework.expression.Expression;
import org.springframework.expression.ExpressionParser;
import org.springframework.expression.ParserContext;
import org.springframework.expression.spel.standard.SpelExpressionParser;
import org.springframework.expression.spel.support.StandardEvaluationContext;
public class SpELFirstDemo {
    public static void main(String[] args) {
        String str = "#[10 + 20]" ;                                    // 表达式字符串
        ExpressionParser parser = new SpelExpressionParser() ;         // Spring表达式解析器
        Expression exp = parser.parseExpression(str,new ParserContext() {
            @Override
            public boolean isTemplate() {
                return true ;
            }
            @Override
            public String getExpressionSuffix() {
                return "]";
            }
            @Override
            public String getExpressionPrefix() {
                return "#[";
            }
        }) ;                                                           // 解析表达式
        EvaluationContext context = new StandardEvaluationContext() ;  // 上下文计算
        Integer result = (Integer) exp.getValue(context) ;             // 表达式计算
        System.out.println(result);
    }
}
```

由于已明确定义了要使用的边界分隔符，所以在实际使用的时候会自动剔除掉指定的边界符号，之后再进行正常的表达式解析处理。

5.4 基本表达式

基本表达式描述的是简单的程序逻辑。例如，定义了一个数字 1，则字面表达式解析后的内

容就是 1。同样，也可以通过表达式实现关系运算、逻辑运算以及三目运算处理。

5.4.1　字面表达式

字面表达式指的是直接在字符串中定义字符串或基本数据类型（数字、布尔、字符），通过字面表达式可以直接看到数据本身的内容。如表 5-1 所示为常见字面表达式的使用。

表 5-1　字面表达式

编号	表达式	操作范例	计算结果
1	字符串	String str = "'Hello ' + 'World'" ;	Hello World
		String str = "\"Hello \" + \"World\"" ;	
2	数值	String str = "1" ;	1
		String str = "1.1" ;	1.1
		String str = "1.1E10" ;	11000000000.00
3	布尔型	String str = "true" ;	true
4	null描述	String str = "null" ;	null

范例：【mldnspring-base 模块】处理科学计数法。

```
package cn.mldn.mldnspring;
import org.springframework.expression.EvaluationContext;
import org.springframework.expression.Expression;
import org.springframework.expression.ExpressionParser;
import org.springframework.expression.spel.standard.SpelExpressionParser;
import org.springframework.expression.spel.support.StandardEvaluationContext;
public class SpELDemo01 {
    public static void main(String[] args) {
        String str = "1.1E10" ;                                          // 表达式字符串
        ExpressionParser parser = new SpelExpressionParser() ;           // Spring表达式解析器
        Expression exp = parser.parseExpression(str) ;                   // 解析表达式
        EvaluationContext context = new StandardEvaluationContext() ;    // 上下文计算
        Double result = exp.getValue(context,Double.class) ;             // 表达式计算
        System.out.printf("%11.2f",result);                              // 格式化输出
    }
}
```

程序执行结果	11000000000.00

本程序采用了科学计数法实现了表达式的定义，随后为了方便显示内容，使用了 printf 格式化显示。

5.4.2　数学表达式

数学表达式描述的是基本的加、减、乘、除等计算处理，如表 5-2 所示。

表 5-2　数学表达式

编号	表达式	操作范例	计算结果
1	四则运算	String str = "1 + 2 - 3 * 4 / 5";	1
2	求模	String str = "10 % 3"; String str = "10 mod 3";或String str = "10 MOD 3";	1
3	幂运算	String str = "10 ^ 3";	1000
4	除法	String str = "10 DIV 3";	3

5.4.3　关系表达式

在定义关系表达式时，除了可以判断大小与相等关系外，还可以实现范围的判断，如表 5-3 所示。

表 5-3　关系表达式

编号	表达式	操作范例	计算结果
1	相等判断	String str = "10 == 10"; String str = "10 EQ 10";	true
2	不等判断	String str = "10 != 10"; String str = "10 NE 10";	false
3	大于	String str = "10 > 10"; String str = "10 GT 10";	false
4	大于等于	String str = "10 >= 10"; String str = "10 GE 10";	true
5	小于	String str = "10 < 10"; String str = "10 LT 10";	false
6	小于等于	String str = "10 <= 10"; String str = "10 LE 10";	true
7	区间判断	String str = "10 BETWEEN {5,20}";	true

5.4.4　逻辑表达式

逻辑判断主要有 3 个处理逻辑：与、或、非。利用逻辑表达式可以连接多个关系表达式，如表 5-4 所示给出了一些常见逻辑表达式的使用。

表 5-4　逻辑表达式

编号	表达式	操作范例	计算结果
1	与操作	String str = "'a' == 'a' && 10 > 5"; String str = "'a' == 'a' AND 10 > 5";	true
2	或操作	String str = "'a' == 'a' || 10 > 5"; String str = "'a' == 'a' OR 10 > 5";	true
3	非操作	String str = "NOT('a' == 'a' && 10 > 5)";	false

在表 5-4 中，由于非操作是在整体逻辑计算之后进行处理的，所以使用“( )”改变了计算的优先级。

5.4.5　三目运算操作

三目运算符（?:）组成的表达式是一种基于判断的赋值表达式，不仅在开发中广泛使用，在 SpEL 中也同样支持。基本的三目运算操作如表 5-5 所示。

表 5-5　三目运算操作

编号	表达式	操作范例	计算结果
1	基础三目	String str = "1 > 2 ? 'Hello' : \"World\"";	World
2	null 处理	String str = "null == null ? 'Hello' : \"World\"";	Hello
3	true 处理	String str = "true ? 'Hello' : \"World\"";	Hello

范例：【mldnspring-base 模块】利用三目运算符构建一个基础的三目运算操作。

```
package cn.mldn.mldnspring;
import org.springframework.expression.ExpressionParser;
import org.springframework.expression.spel.standard.SpelExpressionParser;
public class SpELDemo02 {
    public static void main(String[] args) {
        String exp = "1 > 2 ? 'Hello':'World'" ;
        ExpressionParser parser = new SpelExpressionParser();
        String result = parser.parseExpression(exp).getValue(String.class);
        System.out.println(result);
    }
}
```

程序执行结果	World

本程序利用三目运算符实现了判断与赋值处理，其基本的流程与程序区别不大。SpEL 中除了这种三目运算符之外，还从 Groovy 语言引入了用于简化的三目运算符——Elivis 运算符。其基本结构为“表达式 1?:表达式 2”，当表达式 1 为非 null 时，返回表达式 1；当表达式 1 为 null 时，返回表达式 2。

范例：【mldnspring-base 模块】使用 Elivis 运算符。

```
package cn.mldn.mldnspring;
import org.springframework.expression.ExpressionParser;
import org.springframework.expression.spel.standard.SpelExpressionParser;
public class SpELDemo03 {
    public static void main(String[] args) {
        ExpressionParser parser = new SpelExpressionParser();
        String resultA = parser.parseExpression(
                "null ?: \"www.mldn.cn\"").getValue(String.class);
        String resultB = parser.parseExpression(
```

```
            "'mldn' ?: \"www.mldnjava.cn\"").getValue(String.class);
        System.out.println(resultA);                                    // 为null，返回表达式2
        System.out.println(resultB);                                    // 不为null，返回表达式1
    }
}
```

程序执行结果	www.mldn.cn mldn

5.4.6 字符串处理表达式

在 SpEL 中，最大的特点是可以直接通过表达式的描述实现类中方法的调用，下面将通过字符串的操作进行演示。

范例：【mldnspring-base 模块】进行字符串的指定字符获取（charAt）。

```
package cn.mldn.mldnspring;
import org.springframework.expression.EvaluationContext;
import org.springframework.expression.Expression;
import org.springframework.expression.ExpressionParser;
import org.springframework.expression.spel.standard.SpelExpressionParser;
import org.springframework.expression.spel.support.StandardEvaluationContext;
public class SpELDemo04 {
    public static void main(String[] args) {
        String str = "'helloworld'[1]";
        ExpressionParser parser = new SpelExpressionParser();
        Expression exp = parser.parseExpression(str);
        EvaluationContext context = new StandardEvaluationContext(exp);
        char value = exp.getValue(context,Character.class);
        System.out.println(value);
    }
}
```

程序执行结果	e

在 SpEL 中通过索引访问，可以代替 String 类中的 charAt 方法的使用。

范例：【mldnspring-base 模块】实现字符串的大写转换。

```
package cn.mldn.mldnspring;
import org.springframework.expression.EvaluationContext;
import org.springframework.expression.Expression;
import org.springframework.expression.ExpressionParser;
import org.springframework.expression.spel.standard.SpelExpressionParser;
import org.springframework.expression.spel.support.StandardEvaluationContext;
public class SpELDemo05 {
    public static void main(String[] args) {
        String str = "'www.mldn.cn'.toUpperCase()";
```

```
        ExpressionParser parser = new SpelExpressionParser();
        Expression exp = parser.parseExpression(str);
        EvaluationContext context = new StandardEvaluationContext(exp);
        String result = exp.getValue(context,String.class);
        System.out.println(result);
    }
}
```

程序执行结果	WWW.MLDN.CN

范例：【mldnspring-base 模块】字符串替换。

```
package cn.mldn.mldnspring;
import org.springframework.expression.EvaluationContext;
import org.springframework.expression.Expression;
import org.springframework.expression.ExpressionParser;
import org.springframework.expression.spel.standard.SpelExpressionParser;
import org.springframework.expression.spel.support.StandardEvaluationContext;
public class SpELDemo06 {
    public static void main(String[] args) {
        String str = "'www.mldn.cn'.toUpperCase().replaceAll('mldn','mldnjava')";
        ExpressionParser parser = new SpelExpressionParser();
        Expression exp = parser.parseExpression(str);
        EvaluationContext context = new StandardEvaluationContext(exp);
        String result = exp.getValue(context,String.class);
        System.out.println(result);
    }
}
```

程序执行结果	WWW.mldnjava.CN

通过程序可以发现，在定义字符串方法调用时，可以直接采用代码链的形式实现方法的调用。对于开发者而言，只是在编写一个简单的字符串。

5.4.7　正则匹配运算

正则在程序开发中有着重要的地位，利用正则可以方便地实现数据的匹配与字符串的相应操作。在 SpEL 中同样支持正则运算功能。

范例：【mldnspring-base 模块】使用正则进行验证。

```
package cn.mldn.mldnspring;
import org.springframework.expression.EvaluationContext;
import org.springframework.expression.Expression;
import org.springframework.expression.ExpressionParser;
import org.springframework.expression.spel.standard.SpelExpressionParser;
import org.springframework.expression.spel.support.StandardEvaluationContext;
public class SpELDemo07 {
```

```
    public static void main(String[] args) {
        String str = "'999' matches '\\d{3}'";                    // 正则运算
        ExpressionParser parser = new SpelExpressionParser();
        Expression exp = parser.parseExpression(str);
        EvaluationContext context = new StandardEvaluationContext(exp);
        boolean result = exp.getValue(context,Boolean.class);
        System.out.println(result);
    }
}
```

程序执行结果	true

5.5 Class 表达式

Class 的主要功能是实现反射处理，在 SpEL 中可以直接通过 Class 表达式实现对象处理以及类中成员或方法的调用。Class 表达式的处理形式如表 5-6 所示。

表 5-6 Class 表达式

编号	表达式	操作范例	计算结果
1	获取 Class	String str = "T(java.lang.String)";	Class<String>
		String str = "T(java.util.Date)";	Class<Date>
2	静态属性	String str = "T(Integer).MAX_VALUE";	2147483647
3	静态方法	String str = "T(Integer).parseInt('567')";	567
4	对象实例化	String str = "new java.util.Date()";	Tue Oct 17 10:31:35 CST 2019
5	instanceof	String str = "'mldnjava' instanceof T(String)";	true

下面针对表 5-6 给出的表达式操作，做几个详细的说明。

范例：【mldnspring-base 模块】获取 Class 类型对象。

```
package cn.mldn.mldnspring;
import org.springframework.expression.EvaluationContext;
import org.springframework.expression.Expression;
import org.springframework.expression.ExpressionParser;
import org.springframework.expression.spel.standard.SpelExpressionParser;
import org.springframework.expression.spel.support.StandardEvaluationContext;
public class SpELClassDemo01 {
    public static void main(String[] args) {
        String str = "T(java.lang.String)";                    // 获取String的Class对象
        ExpressionParser parser = new SpelExpressionParser();
        Expression exp = parser.parseExpression(str);
        EvaluationContext context = new StandardEvaluationContext(exp);
        Class<?> cls = exp.getValue(context,Class.class);
        System.out.println(cls.getName());
```

```
	}
}
```

程序执行结果	java.lang.String

本程序获取了 String 类的 Class 对象。这种做法与 Class.forName 功能类似，但显然基于 SpEL 处理更加规范。

范例：【mldnspring-base 模块】调用静态方法。

```
package cn.mldn.mldnspring;
import org.springframework.expression.EvaluationContext;
import org.springframework.expression.Expression;
import org.springframework.expression.ExpressionParser;
import org.springframework.expression.spel.standard.SpelExpressionParser;
import org.springframework.expression.spel.support.StandardEvaluationContext;
public class SpELClassDemo02 {
	public static void main(String[] args) {
		String str = "T(Integer).parseInt('567')";				// 调用静态方法
		ExpressionParser parser = new SpelExpressionParser();
		Expression exp = parser.parseExpression(str);
		EvaluationContext context = new StandardEvaluationContext(exp);
		Integer result = exp.getValue(context,Integer.class);
		System.out.println(result * 2);
	}
}
```

程序执行结果	1134

范例：【mldnspring-base 模块】表达式实例化对象。

```
package cn.mldn.mldnspring;
import java.util.Date;
import org.springframework.expression.EvaluationContext;
import org.springframework.expression.Expression;
import org.springframework.expression.ExpressionParser;
import org.springframework.expression.spel.standard.SpelExpressionParser;
import org.springframework.expression.spel.support.StandardEvaluationContext;
public class SpELClassDemo03 {
	public static void main(String[] args) {
		String str = "new java.util.Date()";				// 实例化Date对象
		ExpressionParser parser = new SpelExpressionParser();
		Expression exp = parser.parseExpression(str);
		EvaluationContext context = new StandardEvaluationContext(exp);
		Date result = exp.getValue(context,Date.class);
		System.out.println(result);
	}
}
```

通过 Class 表达式可以实现反射调用，最为关键的是可以利用字符串的结构实现指定类对象的实例化处理。这一点大大加强了字符串的功能，也加强了 Spring 配置文件的功能。

5.6 表达式变量操作

在表达式中可以直接定义出具体的操作数据信息，也可以通过变量的形式进行定义。但是所有的变量要求在表达式计算前设置好相应的数据。

范例：【mldnspring-base 模块】进行变量定义。

```
package cn.mldn.mldnspring;
import org.springframework.expression.EvaluationContext;
import org.springframework.expression.Expression;
import org.springframework.expression.ExpressionParser;
import org.springframework.expression.spel.standard.SpelExpressionParser;
import org.springframework.expression.spel.support.StandardEvaluationContext;
public class SpELVarDemo01 {
    public static void main(String[] args) {
        String str = "#myvar1 + #myvar2";                                  // 定义两个变量
        ExpressionParser parser = new SpelExpressionParser();
        Expression exp = parser.parseExpression(str);
        EvaluationContext context = new StandardEvaluationContext(exp);
        context.setVariable("myvar1", "Hello ");                           // 设置变量内容
        context.setVariable("myvar2", "World!");                           // 设置变量内容
        String result = exp.getValue(context,String.class);
        System.out.println(result);
    }
}
```

程序执行结果	Hello World!

本程序在定义表达式时，定义了变量 myvar1 和 myvar2，因此在计算前必须明确设置这两个变量的内容。需要注意的是，由于此时并没有明确标记这两个变量的具体类型，所以可以动态决定。例如，可以将变量的内容设置为整数。

```
context.setVariable("myvar1", 100);                                // 设置变量内容
context.setVariable("myvar2", 200);                                // 设置变量内容
Integer result = exp.getValue(context,Integer.class);
```

对比两个程序可以发现，通过变量设计，表达式的计算更加灵活，结构也更加清晰。

在 SpEL 中除了可采用自定义变量之外，还默认支持 root 根变量。当表达式只需要一个变量的时候，可以采用此类做法。利用根变量，可以在 StandardEvaluationContext 子类实例化的时候直接传递操作数据。

范例：【mldnspring-base 模块】使用根变量。

```
package cn.mldn.mldnspring;
import org.springframework.expression.EvaluationContext;
import org.springframework.expression.Expression;
import org.springframework.expression.ExpressionParser;
import org.springframework.expression.spel.standard.SpelExpressionParser;
import org.springframework.expression.spel.support.StandardEvaluationContext;
public class SpELVarDemo03 {
    public static void main(String[] args) {
        String str = "#root=='mldn' ? 'Hello MLDN' : '大家好'";                    // 定义两个变量
        ExpressionParser parser = new SpelExpressionParser();
        Expression exp = parser.parseExpression(str);
        EvaluationContext context = new StandardEvaluationContext("mldn");    // 根变量赋值
        String result = exp.getValue(context,String.class) ;
        System.out.println(result);
    }
}
```

程序执行结果	Hello MLDN

这里使用的是 root 根变量，在创建 StandardEvaluationContext 对象时可以通过传递的参数设置根变量的内容。

在 SpEL 中除了可以在表达式中定义变量名称外，也可以定义方法名称。但是这个方法名称要想使用，需要通过方法引用的形式来完成处理。

提示：关于 Java 8 与 SpEL 的方法引用。

在 Java 8 中提供方法引用的处理支持，其属于 Java 的原生实现语法。在 SpEL 中的方法引用是在 Java 8 未出现之前产生的。

范例：【mldnspring-base 模块】实现 SpEL 的方法引用操作。

```
package cn.mldn.mldnspring;
import java.lang.reflect.Method;
import org.springframework.expression.Expression;
import org.springframework.expression.ExpressionParser;
import org.springframework.expression.spel.standard.SpelExpressionParser;
import org.springframework.expression.spel.support.StandardEvaluationContext;
public class SpELVarDemo04 {
    public static void main(String[] args) throws Exception {
        String str = "#myInt('123')";                                   // myInt表示方法引用设置的别名
        Method method = Integer.class.getMethod("parseInt", String.class) ; // 方法对象
        ExpressionParser parser = new SpelExpressionParser();
        Expression exp = parser.parseExpression(str);                  // 设置一个自定义的根变量
```

```
        StandardEvaluationContext context = new StandardEvaluationContext();
        context.registerFunction("myInt", method);                // 为myInt设置一个引用的具体方法
        Integer result = exp.getValue(context,Integer.class) ;
        System.out.println(result * 2);
    }
}
```

程序执行结果	246

本程序实现了 Integer 类中 parseInt 方法的引用，相当于在 Spring 容器中使用 myInt 引用了 parseInt 方法。

对于变量的使用，实际上还包含有实例化对象中的访问变量，类对象在访问属性时肯定不是直接访问，而是间接通过 getter 方法实现调用。下面就利用反射调用 Date 类中的 getTime 方法。

范例：【mldnspring-base 模块】调用对象中的属性（getter 方法）。

```
package cn.mldn.mldnspring;
import java.util.Date;
import org.springframework.expression.EvaluationContext;
import org.springframework.expression.Expression;
import org.springframework.expression.ExpressionParser;
import org.springframework.expression.spel.standard.SpelExpressionParser;
import org.springframework.expression.spel.support.StandardEvaluationContext;
public class SpELVarDemo05 {
    public static void main(String[] args) throws Exception {
        String str = "time";                                      // Date类中的getTime方法
        ExpressionParser parser = new SpelExpressionParser();
        Expression exp = parser.parseExpression(str);             // 设置了一个自定义的根变量
        EvaluationContext context = new StandardEvaluationContext(new Date());
        Long result = exp.getValue(context,Long.class) ;
        System.out.println(result);
    }
}
```

程序执行结果	1517364183173

本程序利用反射机制实现了 Date 类中 getTime 方法的调用。调用时只需要通过表达式定义要调用的属性名称，就可以自动找到 getter 方法。在定义表达式时，也可以使用 Time，Spring 会自动处理大小写问题（仅局限于第一个字母）。

在调用以上代码时，会发现已经设置好了根变量的内容。如果未设置根变量对象（设置对象是 null），执行时将无法找到指定对象中的属性，就会产生异常。为了解决这样的麻烦，SpEL 里继续使用 Groovy 表达式对 null 进行处理。

范例：【mldnspring-base 模块】null 处理。

```
package cn.mldn.mldnspring;
import org.springframework.expression.EvaluationContext;
```

```
import org.springframework.expression.Expression;
import org.springframework.expression.ExpressionParser;
import org.springframework.expression.spel.standard.SpelExpressionParser;
import org.springframework.expression.spel.support.StandardEvaluationContext;
public class SpELVarDemo06 {
    public static void main(String[] args) throws Exception {
        String str = "#root?.time";                                   // Date类中的getTime方法
        ExpressionParser parser = new SpelExpressionParser();
        Expression exp = parser.parseExpression(str);                 // 设置一个自定义的根变量
        EvaluationContext context = new StandardEvaluationContext();   // 不设置对象
        Long result = exp.getValue(context,Long.class) ;
        System.out.println(result);
    }
}
```

程序执行结果	null

本程序并没有设置根变量的内容，所以根变量此时的对象为 null。按照传统做法，程序执行后会出现以下错误提示信息：

```
org.springframework.expression.spel.SpelEvaluationException: EL1007E: Property or field 'time' cannot be
found on null
```

由于这里采用了 Groovy 中的#root?.time 表达式，因此可以避免此异常的发生，只是返回了 null。

5.7　集合表达式

SpEL 对于集合的访问处理也有所支持，利用集合表达式不仅可以定义集合内容，也可以对集合进行查询、修改或转换处理操作。

范例：【mldnspring-base 模块】定义 List 集合。

```
package cn.mldn.mldnspring;
import java.util.List;
import org.springframework.expression.EvaluationContext;
import org.springframework.expression.Expression;
import org.springframework.expression.ExpressionParser;
import org.springframework.expression.spel.standard.SpelExpressionParser;
import org.springframework.expression.spel.support.StandardEvaluationContext;
public class SpELCollectionDemo01 {
    public static void main(String[] args) {
        String str = "{'mldn','jixianit','mldnjava'}";                    // 定义表达式
        ExpressionParser parser = new SpelExpressionParser();
        Expression exp = parser.parseExpression(str);
```

```
        EvaluationContext context = new StandardEvaluationContext();
        List<String> result = exp.getValue(context, List.class);              // 获取集合
        System.out.println(result);
    }
}
```

程序执行结果	[mldn, jixianit, mldnjava]

本程序在定义字符串表达式的时候采用了数组的形式，而这种形式被 SpEL 解析后会将其定义为 List 集合。

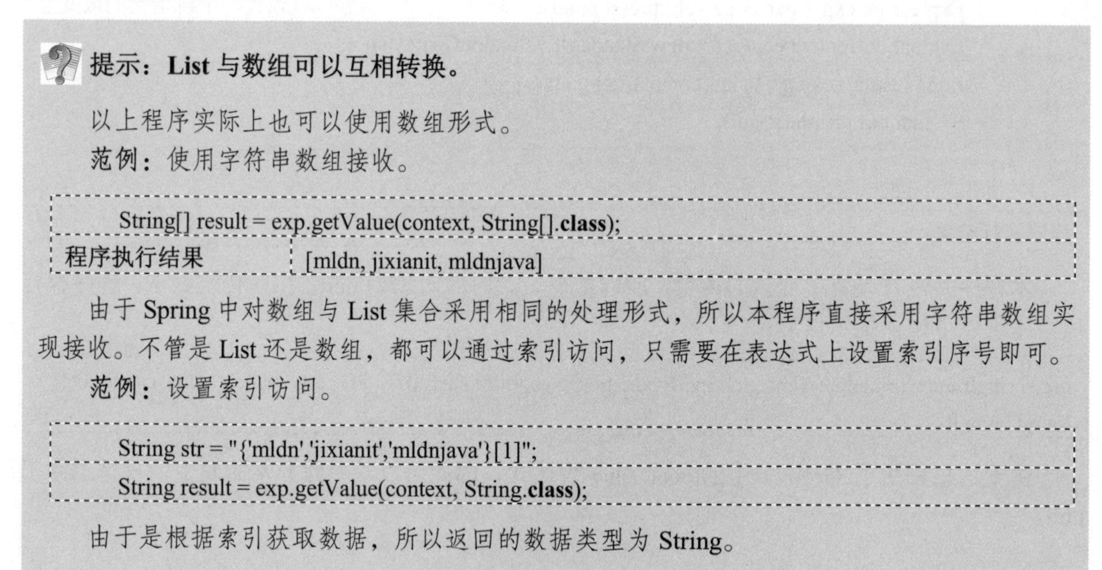

提示：List 与数组可以互相转换。

以上程序实际上也可以使用数组形式。

范例：使用字符串数组接收。

```
    String[] result = exp.getValue(context, String[].class);
```

程序执行结果	[mldn, jixianit, mldnjava]

由于 Spring 中对数组与 List 集合采用相同的处理形式，所以本程序直接采用字符串数组实现接收。不管是 List 还是数组，都可以通过索引访问，只需要在表达式上设置索引序号即可。

范例：设置索引访问。

```
    String str = "{'mldn','jixianit','mldnjava'}[1]";
    String result = exp.getValue(context, String.class);
```

由于是根据索引获取数据，所以返回的数据类型为 String。

以上程序采用的是一个固定集合表达式来定义集合内容，在 SpEL 中也可以通过表达式访问外部集合中的数据。

范例：【mldnspring-base 模块】访问外部集合。

```
package cn.mldn.mldnspring;
import java.util.ArrayList;
import java.util.Collections;
import java.util.List;
import org.springframework.expression.EvaluationContext;
import org.springframework.expression.Expression;
import org.springframework.expression.ExpressionParser;
import org.springframework.expression.spel.standard.SpelExpressionParser;
import org.springframework.expression.spel.support.StandardEvaluationContext;
public class SpELCollectionDemo04 {
    public static void main(String[] args) {
        List<String> all = new ArrayList<String>() ;                          // 定义集合对象
        Collections.addAll(all, "mldn","jixianit","mldnjava") ;              // 增加一组数据
        String str = "#allData[1]";                                          // 定义表达式
```

```
        ExpressionParser parser = new SpelExpressionParser();
        Expression exp = parser.parseExpression(str);
        EvaluationContext context = new StandardEvaluationContext();
        context.setVariable("allData", all);                          // 设置集合
        String result = exp.getValue(context, String.class);          // 获取集合
        System.out.println(result);
    }
}
```

程序执行结果	jixianit

本程序在表达式中定义了一个 allData 变量，该变量的内容将通过外部的 List 集合获取，随后根据索引获取数据。

提示：Set 集合也同样支持。

对于本程序如果在设置 allData 变量内容时，采用 Set 集合也同样可以使用，实际上这一项功能相当于弥补了 Set 集合无法通过索引获取数据的缺陷。

范例：【mldnspring-base 模块】设置 Map 集合。

```
package cn.mldn.mldnspring;
import java.util.HashMap;
import java.util.Map;
import org.springframework.expression.EvaluationContext;
import org.springframework.expression.Expression;
import org.springframework.expression.ExpressionParser;
import org.springframework.expression.spel.standard.SpelExpressionParser;
import org.springframework.expression.spel.support.StandardEvaluationContext;
public class SpELCollectionDemo06 {
    public static void main(String[] args) {
        Map<String,String> map = new HashMap<String,String>() ;
        map.put("mldn", "魔乐科技：www.mldn.cn") ;             // 定义Map集合
        map.put("mldnjava", "魔乐科技软件训练营：www.mldnjava.cn") ;
        map.put("jixianit", "极限IT训练营：www.jixianit.com") ;
        String str = "#allData['jixianit']" ;                  // 设置操作表达式，根据key获取数据
        ExpressionParser parser = new SpelExpressionParser();  // 定义一个Spring表达式解析器
        Expression exp = parser.parseExpression(str);          // 从字符串里面解析出内容
        EvaluationContext context = new StandardEvaluationContext();
        context.setVariable("allData", map);                   // 设置可变集合
        String result = exp.getValue(context, String.class);   // 获取集合
        System.out.println(result);
    }
}
```

程序执行结果	极限IT训练营：www.jixianit.com

本程序通过表达式变量的设置，引入了外部的 Map 集合，并且根据 Map 集合的 key 获取了对应的 value 数据。

对于 SpEL 而言，除了可以获取集合的数据外，还可以利用表达式实现集合数据的修改。

范例：【mldnspring-base 模块】修改 List 集合的一个数据。

```
package cn.mldn.mldnspring;
import java.util.ArrayList;
import java.util.Collections;
import java.util.List;
import org.springframework.expression.EvaluationContext;
import org.springframework.expression.Expression;
import org.springframework.expression.ExpressionParser;
import org.springframework.expression.spel.standard.SpelExpressionParser;
import org.springframework.expression.spel.support.StandardEvaluationContext;
public class SpELCollectionDemo08 {
    public static void main(String[] args) {
        List<String> all = new ArrayList<String>() ;                          // 定义集合对象
        Collections.addAll(all, "mldn","jixianit","mldnjava") ;              // 增加一组数据
        String str = "#allData[1]='www.jixianit.com'";                       // 定义表达式
        ExpressionParser parser = new SpelExpressionParser();
        Expression exp = parser.parseExpression(str);
        EvaluationContext context = new StandardEvaluationContext();
        context.setVariable("allData", all);                                 // 设置集合
        String result = exp.getValue(context, String.class);                 // 获取集合
        System.out.println(result);
        System.out.println(all);
    }
}
```

程序执行结果	www.jixianit.com [mldn, www.jixianit.com, mldnjava]

本程序通过表达式修改了集合中指定索引的数据。可以发现，修改完成后表达式可以返回修改后的数据，同时会影响原始集合数据。

范例：【mldnspring-base 模块】迭代修改 List 集合数据。

```
package cn.mldn.mldnspring;
import java.util.ArrayList;
import java.util.Collections;
import java.util.List;
import org.springframework.expression.EvaluationContext;
import org.springframework.expression.Expression;
import org.springframework.expression.ExpressionParser;
import org.springframework.expression.spel.standard.SpelExpressionParser;
```

```
import org.springframework.expression.spel.support.StandardEvaluationContext;
public class SpELCollectionDemo09 {
    public static void main(String[] args) {
        List<String> all = new ArrayList<String>() ;                    // 定义集合对象
        Collections.addAll(all, "mldn","jixianit","mldnjava") ;         // 增加一组数据
        String str = "#allData.!['学习资源：' + #this]";                 // 定义表达式
        ExpressionParser parser = new SpelExpressionParser();
        Expression exp = parser.parseExpression(str);
        EvaluationContext context = new StandardEvaluationContext();
        context.setVariable("allData", all);                            // 设置集合
        List<String> result = exp.getValue(context, List.class);        // 获取集合
        System.out.println(result);
        System.out.println(all);
    }
}
```

程序执行结果	[学习资源：mldn, 学习资源：jixianit, 学习资源：mldnjava] [mldn, jixianit, mldnjava]

本程序采用迭代表达式实现了集合中的内容修改。在表达式中可以使用#this 表示当前元素，所有修改后的数据会形成一个新的集合返回，对于原始集合并不做修改。

在修改集合时，除了可以修改 List 集合，也可以修改 Map 集合中的内容。对于 Map 集合，可以修改某一个 key 的内容，也可以采用迭代修改全部集合内容。

范例：【mldnspring-base 模块】修改指定 Map 集合数据。

```
package cn.mldn.mldnspring;
import java.util.HashMap;
import java.util.Map;
import org.springframework.expression.EvaluationContext;
import org.springframework.expression.Expression;
import org.springframework.expression.ExpressionParser;
import org.springframework.expression.spel.standard.SpelExpressionParser;
import org.springframework.expression.spel.support.StandardEvaluationContext;
public class SpELCollectionDemo10 {
    public static void main(String[] args) {
        Map<String,String> map = new HashMap<String,String>() ;
        map.put("mldn", "魔乐科技：www.mldn.cn") ;                         // 定义Map集合
        map.put("mldnjava", "魔乐科技软件训练营：www.mldnjava.cn") ;
        map.put("jixianit", "极限IT训练营：www.jixianit.com") ;
        String str = "#allData['jixianit']='极限IT在线学习训练营：www.jixianit.com'" ;
        ExpressionParser parser = new SpelExpressionParser();          // 定义一个Spring表达式解析器
        Expression exp = parser.parseExpression(str);                   // 从字符串里面解析出内容
        EvaluationContext context = new StandardEvaluationContext();
```

```
		context.setVariable("allData", map);				// 设置可变集合
		String result = exp.getValue(context, String.class);		// 获取集合
		System.out.println(result);
		System.out.println(map);
	}
}
```

程序执行结果	极限IT在线学习训练营：www.jixianit.com {mldn=魔乐科技：www.mldn.cn, jixianit=极限IT在线学习训练营：www.jixianit.com, mldnjava=魔乐科技软件训练营：www.mldnjava.cn}

本程序通过表达式修改了 Map 集合中指定 key 的内容。与 List 集合类似，也会影响到原始集合的内容。

范例：【mldnspring-base 模块】对 Map 集合实现过滤，将包含 mldn 字符串的 key 保存到新的集合之中。

```
package cn.mldn.mldnspring;
import java.util.HashMap;
import java.util.Map;
import org.springframework.expression.EvaluationContext;
import org.springframework.expression.Expression;
import org.springframework.expression.ExpressionParser;
import org.springframework.expression.spel.standard.SpelExpressionParser;
import org.springframework.expression.spel.support.StandardEvaluationContext;
public class SpELCollectionDemo12 {
	public static void main(String[] args) {
		Map<String,String> map = new HashMap<String,String>() ;
		map.put("mldn", "魔乐科技：www.mldn.cn") ;			// 定义Map集合
		map.put("mldnjava", "魔乐科技软件训练营：www.mldnjava.cn") ;
		map.put("jixianit", "极限IT训练营：www.jixianit.com") ;
		String str = "#allData.?[#this.key.contains('mldn')]" ;		// key查找
		ExpressionParser parser = new SpelExpressionParser();		// 定义一个Spring表达式解析器
		Expression exp = parser.parseExpression(str);			// 从字符串里面解析出内容
		EvaluationContext context = new StandardEvaluationContext();
		context.setVariable("allData", map);				// 设置可变集合
		Map<String,String> result = exp.getValue(context, Map.class); // 获取集合
		System.out.println(result);
	}
}
```

程序执行结果	{mldn=魔乐科技：www.mldn.cn, mldnjava=魔乐科技软件训练营：www.mldnjava.cn}

本程序对 Map 集合实现了迭代查询，并且使用#this.key 获取了 Map 中 key 的信息，由于 Key 的类型为 String，所以可以利用 contains 方法进行判断。最后，将满足条件的数据保存到新的集合中。

范例：【mldnspring-base 模块】将 Map 集合转换为 List 集合。

```
package cn.mldn.mldnspring;
import java.util.HashMap;
import java.util.List;
import java.util.Map;
import org.springframework.expression.EvaluationContext;
import org.springframework.expression.Expression;
import org.springframework.expression.ExpressionParser;
import org.springframework.expression.spel.standard.SpelExpressionParser;
import org.springframework.expression.spel.support.StandardEvaluationContext;
public class SpELCollectionDemo13 {
    public static void main(String[] args) {
        Map<String,String> map = new HashMap<String,String>() ;
        map.put("mldn", "魔乐科技：www.mldn.cn") ;                    // 定义Map集合
        map.put("mldnjava", "魔乐科技软件训练营：www.mldnjava.cn") ;
        map.put("jixianit", "极限IT训练营：www.jixianit.com") ;
        String str = "#allData.![#this.key + \" - \" + #this.value]" ;   // key查找
        ExpressionParser parser = new SpelExpressionParser();           // 定义Spring表达式解析器
        Expression exp = parser.parseExpression(str);                   // 从字符串里面解析出内容
        EvaluationContext context = new StandardEvaluationContext();
        context.setVariable("allData", map);                            // 设置可变集合
        List<String> result = exp.getValue(context, List.class);        // 获取集合
        System.out.println(result);
    }
}
```

程序执行结果	[mldn - 魔乐科技：www.mldn.cn, jixianit - 极限IT训练营：www.jixianit.com, mldnjava - 魔乐科技软件训练营：www.mldnjava.cn]

本程序将 Map 集合中的 key 与 value 分别取出，按照 key-value 的形式将所有的数据重新保存到了 List 集合中。

5.8　Spring 配置文件与 SpEL

SpEL 为 Spring 增强了字符串的处理能力，在实际的使用之中，开发者可以结合 SpEL 利用配置文件或 Annotation 实现更加强大的注入处理。

5.8.1　基于配置文件使用 SpEL

定义 Spring 配置文件时，会将所有配置的字符串按照指定属性的类型实现自动转型，而有了 SpEL 后就可以将表达式的执行结果注入到属性中，以简化程序逻辑。下面将基于 Spring 配置文件实现一个普通类的对象，并且在注入属性的时候使用 SpEL 表达式处理数据。

1.【mldnspring-base 项目】建立 Item.java 类。

```
package cn.mldn.mldnspring.vo;
import java.util.Date;
public class Item {
    private String title ;
    private Date createDate ;
    // setter、getter、toString略
}
```

2.【mldnspring-base 项目】修改 spring-base.xml 配置文件，使用 SpEL 配置 Item 类。

```
<bean id="item" class="cn.mldn.mldnspring.vo.Item">
    <property name="title" value="#{'www.mldn.cn'.substring(4,8)}"/>          <!-- 字符串处理 -->
    <property name="createDate" value="#{new java.util.Date()}"/>             <!-- 实例化对象 -->
</bean>
```

本程序利用 SpEL 定义了两个表达式，由于是在配置文件中定义，所以为了与 String 类型区分，需要为表达式加上“#{}”边界分隔符。

3.【mldnspring-base 项目】编写测试类，实现 Item 对象注入与输出。

```
package cn.mldn.mldnspring;
import org.junit.Test;
import org.junit.runner.RunWith;
import org.springframework.beans.factory.annotation.Autowired;
import org.springframework.test.context.ContextConfiguration;
import org.springframework.test.context.junit4.SpringJUnit4ClassRunner;
import cn.mldn.mldnspring.vo.Item;
@ContextConfiguration(locations = { "classpath:spring/spring-base.xml" })        // 资源文件定位
@RunWith(SpringJUnit4ClassRunner.class)                                          // 设置测试工具
public class TestItem {
    @Autowired
    private Item item ;
    @Test
    public void testItem() {
        System.out.println(this.item);
    }
}
```

程序执行结果	Item [title=mldn, createDate=Wed Jan 31 15:19:47 CST 2019]

通过结果可以发现，在进行 Item 对象属性设置时会自动执行 SpEL 表达式，并且将处理后的结果保存下来。

5.8.2 基于 Annotation 使用 SpEL

除了可以在配置文件中使用 SpEL，也可以在配置类中结合 Annotation 实现 SpEL，以实现

配置 Bean 的引用或进行表达式处理，此时就需要使用@Value 注解处理。

1.【mldnspring-base 项目】定义 Dept.java 类，并且引入 Item 配置资源进行处理。

```
package cn.mldn.mldnspring.vo;
import org.springframework.beans.factory.annotation.Value;
import org.springframework.stereotype.Component;
@Component
public class Dept {
    @Value("#{item.createDate?.time}")
    private Long deptno ;
    @Value("#{item.title.toUpperCase()}")
    private String dname ;
    @Value("#{'中国北京'.substring(2,4)}")
    private String loc ;
    // setter、getter、toString略
}
```

为了方便读者理解，这里直接使用@Component 注解进行 Bean 的定义，在配置其属性时利用@Value 注解与 SpEL 表达式引用其他 Bean 的配置内容，并利用类属性获取了设置内容，以进行再次处理。

2.【mldnspring-base 项目】编写测试类，实现部门资源的注入。

```
package cn.mldn.mldnspring;
import org.junit.Test;
import org.junit.runner.RunWith;
import org.springframework.beans.factory.annotation.Autowired;
import org.springframework.test.context.ContextConfiguration;
import org.springframework.test.context.junit4.SpringJUnit4ClassRunner;
import cn.mldn.mldnspring.vo.Dept;
@ContextConfiguration(locations = { "classpath:spring/spring-base.xml" })       // 资源文件定位
@RunWith(SpringJUnit4ClassRunner.class)                                          // 设置测试工具
public class TestDept {
    @Autowired
    private Dept dept ;
    @Test
    public void testItem() {
        System.out.println(this.dept);
    }
}
```

程序执行结果	Dept [deptno=1517384012092, dname=MLDN, loc=北京]

本测试类中注入了已经配置好的 Dept 对象，随后实现了内容输出。通过输出结果可以发现，所有在 Dept 类中配置的 SpEL 表达式已经全部自动处理完成。

5.9　本章小结

1．Spring 表达式增强了字符串的处理功能，可以使用字符串实现各种复杂程序的代码功能。

2．SpEL 默认采用的分隔符为“#{}”，也可以手动实例化 ParserContext 接口对象，自定义分隔符。

3．SpEL 除了支持计算表达式、类表达式、集合处理表达式外，还可以在配置文件中使用，也可以通过@Value 注解形式使用。

第 6 章

定时调度

通过本章学习，可以达到以下目标：

1．理解 Timer、TimerTask 处理缺陷。
2．掌握 QuartZ 组件与 Spring 的整合使用。
3．掌握 SpringTask 组件的使用。
4．掌握线程池与 SpringTask 组件的关系。

定时调度是企业项目开发中的重要环节，开发者可以利用定时调度实现自动处理操作支持，以避免人力资源的浪费。本章将为读者分析JDK中的定时调度组件问题以及QuartZ与SpringTask组件的使用。

6.1 传统定时调度组件问题分析

JDK 中提供了专门的定时调度组件——java.util.Timer 和 java.util.TimerTask，两者关系如图 6-1 所示。利用这两个类，可以实现间隔调度处理。

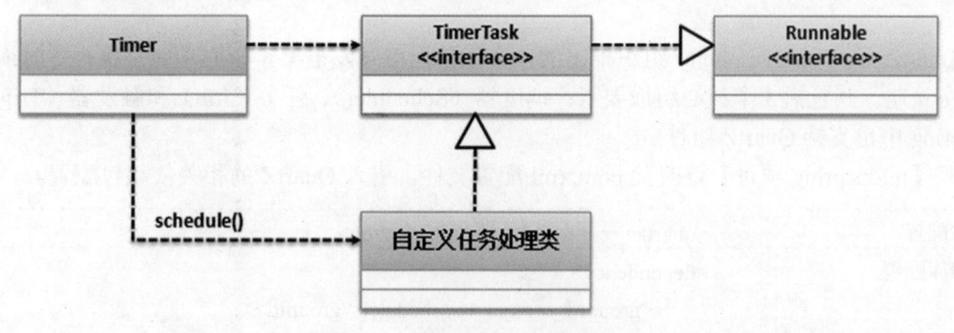

图 6-1 Timer 类与 TimerTask 类

范例：【mldnspring-base 项目】建立一个定时调度类。

```
package cn.mldn.mldnspring.demo;
import java.text.SimpleDateFormat;
import java.util.Timer;
import java.util.TimerTask;
class MyTask extends TimerTask {                    // 定义一个定时调度需要处理的任务程序类
```

```
    @Override
    public void run() {
        System.out.println(
                "【当前日期时间】" + new SimpleDateFormat("yyyy-MM-dd HH:mm:ss.SSS")
                        .format(new java.util.Date()));
    }
}
public class TimeTaskDemo {
    public static void main(String[] args) {
        Timer timer = new Timer();
        timer.schedule(new MyTask(), 0, 1000);                    // 每秒执行一次
    }
}
```

以上程序实现了一个 JDK 支持的间隔调度处理操作，每秒执行一次 TimerTask 定时任务。通过输出结果，可以清晰看到程序的执行间隔。在实际企业项目开发中，仅依靠间隔调度是远远不够的。例如，现在有如下两种需求：

☑ 每年 2 月 13 日对旅馆访客进行备份，因为 2 月 14 日会有很多客人入住。

☑ 每年 12 月 31 日对一年的销售记录进行统计清空。

这时可以发现，如果只依靠间隔调度，那么对于定时任务的执行控制实在是过于复杂，所以传统 JDK 中的定时调度实现根本就无法提供准确时间或日期的调用保障。为了解决这个问题，早期使用的是 QuartZ 组件，而后在 Spring 中又提供了 SpringTask 组件。

6.2 QuartZ 定时调度

QuartZ 是 OpenSymphony 组织推出的一个任务调度开发框架，核心功能是执行定时或定期的任务策略。其包括 3 个核心组成要素：调度器（Scheduler）、任务（Job）和触发器（Trigger）。在 Spring 中也支持 QuartZ 组件。

1.【mldnspring 项目】修改父 pom.xml 配置文件，引入 QuartZ 的相关依赖包配置。

属性配置	<quartz.version>2.3.0</quartz.version>
依赖库配置	<dependency> <groupId>org.quartz-scheduler</groupId> <artifactId>quartz</artifactId> <version>${quartz.version}</version> </dependency> <dependency> <groupId>org.springframework</groupId> <artifactId>spring-tx</artifactId> <version>${spring.version}</version> </dependency>

2．【mldnspring-base 项目】在子项目的 pom.xml 配置文件中引入 Quartz 组件依赖库。

```
<dependency>
    <groupId>org.quartz-scheduler</groupId>
    <artifactId>quartz</artifactId>
</dependency>
<dependency>
    <groupId>org.springframework</groupId>
    <artifactId>spring-tx</artifactId>
</dependency>
```

此时，就可以在项目中使用 QuartZ 组件了。如果要在 Spring 中实现 QuartZ 组件的定时任务，有以下两种实现方式：

☑　直接继承一个定时调度的父类。

☑　通过 Spring 进行定时调度的处理配置。

6.2.1　继承 QuartzJobBean 类实现定时任务

要进行 QuartZ 组件开发，首先需要有一个继承的父类（类似于之前的 TimerTask）。该父类可使用 org.springframework.scheduling.quartz.QuartzJobBean 实现，在此类中提供如下的调度执行方法。

```
protected abstract void executeInternal(JobExecutionContext context)
    throws JobExecutionException;
```

1．【mldnspring-base 项目】建立一个定时的任务处理类。

```
package cn.mldn.mldnspring.task;
import java.text.SimpleDateFormat;
import org.quartz.JobExecutionContext;
import org.quartz.JobExecutionException;
import org.springframework.scheduling.quartz.QuartzJobBean;
public class MyTask extends QuartzJobBean {                          // 定义负责任务处理的程序类
    @Override
    protected void executeInternal(JobExecutionContext context) throws JobExecutionException {
        System.out.println("【当前的日期时间】" + new SimpleDateFormat(
            "yyyy-MM-dd HH:mm:ss.SSS").format(new java.util.Date()));
    }
}
```

为了更好地观察定时调度处理，本程序在执行任务处理时只格式化输出了当前的日期系统时间。

2．【mldnspring-base 项目】要想执行任务，需指明该任务是间隔触发还是定时触发。因此下面在 Spring 配置文件里进行一系列修改，主要是修改 spring-base.xml 配置文件，追加定时任务的相关配置。

☑ 建立任务调度工厂类，所有的任务被工厂类管理。

```
<!-- 要想进行任务处理，需要有一个任务调度的工厂类 -->
<bean id="taskFactory" class="org.springframework.scheduling.quartz.JobDetailFactoryBean">
    <!-- 任务工厂中需明确设置执行本次任务所需要的任务类，要求继承QuartzJobBean父类 -->
    <property name="jobClass" value="cn.mldn.mldnspring.task.MyTask"/>
    <property name="jobDataMap">                    <!-- 定义一些任务的相关配置信息 -->
        <map>
            <entry key="timeout" value="0"/>        <!-- 任务启动后不延迟，立即执行 -->
        </map>
    </property>
</bean>
```

☑ 任务执行有两种处理模式：间隔触发和定时触发（CRON）。这里使用间隔触发操作。

```
<!-- 定义任务触发的工厂类，主要是定义任务的执行类型 -->
<bean id="simpleTrigger"
    class="org.springframework.scheduling.quartz.SimpleTriggerFactoryBean">
    <property name="jobDetail" ref="taskFactory"/>           <!-- 引用任务处理类 -->
    <property name="repeatInterval" value="1000"/>           <!-- 任务间隔时间 -->
</bean>
```

☑ 配置启动任务的调度器，并设置任务执行触发器。

```
<!-- 配置任务的执行调度器 -->
<bean id="scheduleFactory"
    class="org.springframework.scheduling.quartz.SchedulerFactoryBean">
    <property name="triggers">                               <!-- 定义触发器 -->
        <array>
            <ref bean="simpleTrigger" />                     <!-- 设置任务触发器 -->
        </array>
    </property>
</bean>
```

配置完成后，启动 Spring 容器时会自动处理 QuartZ 定时调度任务，随后将按照间隔执行任务。

6.2.2 使用 CRON 实现定时调度

使用 QuartZ 任务调度组件的主要目的就是利用其功能可以实现定期调度，例如，每周五晚上 12 点启动任务，或者每年 12 月 31 日启动统计任务等，而间隔调度的处理需要结合 CRON 表达式来完成。

CRON 原本是一个 Linux 系统下执行定时任务的处理工具，利用如下格式的字符串来安排任务调度。

格式一（7个描述域）	秒 分 时 日 月 周 年（Seconds Minutes Hours DayofMonth Month DayofWeek Year）
格式二（6个描述域）	秒 分 时 日 月 周（Seconds Minutes Hours DayofMonth Month DayofWeek）

对于 CRON 格式中的作用域内容组成可以参考表 6-1 所定义的内容。

表 6-1　CRON 作用域组成

编号	数据域	字符选择范围
1	秒（Seconds）	可出现“,”“-”“*”“/”4 个字符，有效范围为 0～59 的整数
2	分（Minutes）	可出现“,”“-”“*”“/”4 个字符，有效范围为 0～59 的整数
3	时（Hours）	可出现“,”“-”“*”“/”4 个字符，有效范围为 0～23 的整数
4	日（DayofMonth）	可出现“,”“-”“*”“/”“?”“L”“W”“C”8 个字符，有效范围为 1～31 的整数
5	月（Month）	可出现“,”“-”“*”“/”4 个字符，有效范围为 1～12 的整数或 JAN～DEC
6	周（DayofWeek）	可出现“,”“-”“*”“/”“?”“L”“C”“#”8 个字符，有效范围为 1～7 的整数或 SUN～SAT 两个范围。1 表示星期天，2 表示星期一，以此类推
7	年（Year）	可出现“,”“-”“*”“/”4 个字符，有效范围为 1970～2099

如果想将 QuartZ 间隔调度修改为 CRON 定时调度，可直接修改调度触发器。

范例：【mldnspring-base 项目】修改 spring-base.xml 配置文件，采用定时调度（每分钟一执行）。

```
<!-- 定义任务触发的工厂类，使用CRON进行任务的触发操作 -->
<bean id="cronTrigger" class="org.springframework.scheduling.quartz.CronTriggerFactoryBean">
    <property name="jobDetail" ref="taskFactory"/>              <!-- 定义任务详情 -->
    <property name="cronExpression" value="0 * * * * ?"/>       <!-- 设置任务的触发时间 -->
</bean>
<!-- 配置任务的执行调度器 -->
<bean id="scheduleFactory"
    class="org.springframework.scheduling.quartz.SchedulerFactoryBean">
    <property name="triggers">                                  <!-- 定义触发器 -->
        <array>
            <ref bean="cronTrigger" />                          <!-- 设置任务触发器 -->
        </array>
    </property>
</bean>
```

此时执行的是定时调度，所以要将间隔触发器更换为 CronTriggerFactoryBean，同时设置调度时间。这里，只要秒数一变为 0（每分钟只会有一次秒数为 0 的情况），就会自动执行定时任务。

6.2.3 基于 Spring 配置实现 QuartZ 调度

前面已经实现了一个 QuartZ 定时调度处理，但是实际上当前的定时调度从现在的设计角度来讲，并不符合面向对象开发要求了，因为一个定时的任务必须强制性地去继承 QuartzJobBean 父类。为了解决这样的问题，在 Spring 里可以通过配置实现任务的执行处理操作。

1.【mldnspring-base 项目】修改 MyTask 类，此类不做任何的继承处理也没有强制性的方法覆写要求。

```
package cn.mldn.mldnspring.task;
import java.text.SimpleDateFormat;
public class MyTask {                                            // 不再强制性继承任何父类
    protected void runTask() {                                   // 任意定义一个方法名称
        System.out.println("【当前的日期时间】" + new SimpleDateFormat(
                "yyyy-MM-dd HH:mm:ss.SSS").format(new java.util.Date()));
    }
}
```

此时在 MyTask 类中为定时任务的执行定义了一个 runTask 方法，而如果要想让此方法生效则一定要在 Spring 配置文件中进行定义。

2.【mldnspring-base 项目】修改 spring-base.xml 配置文件进行自定义任务处理类与自定义任务处理方法的配置。

☑ 使用 MethodInvokingJobDetailFactoryBean 类实现自定义的任务类与任务执行方法。

```
<!-- 定义了一个新的专门用于自定义任务调度方法的工厂任务类 -->
<bean id="methodTaskFactory"
        class="org.springframework.scheduling.quartz.MethodInvokingJobDetailFactoryBean">
    <property name="targetObject">                          <!-- 定义要执行的任务程序类对象 -->
        <bean class="cn.mldn.mldnspring.task.MyTask"/>  <!-- 任务执行类 -->
    </property>
    <property name="targetMethod" value="runTask"/>         <!-- 设置执行任务的处理方法 -->
</bean>
```

☑ 定义 CRON 触发器，每秒触发执行任务。

```
<!-- 要定义任务触发的工厂类，使用CRON进行任务的触发操作 -->
<bean id="cronTrigger"
        class="org.springframework.scheduling.quartz.CronTriggerFactoryBean">
    <property name="jobDetail" ref="methodTaskFactory"/>         <!-- 定义任务详情 -->
    <property name="cronExpression" value="* * * * * ?"/>         <!-- 每秒一触发 -->
</bean>
```

☑ 在任务调度器中配置 CRON 触发器。

```
<!-- 配置任务的执行调度器 -->
<bean id="scheduleFactory"
    class="org.springframework.scheduling.quartz.SchedulerFactoryBean">
```

```
        <property name="triggers">                              <!-- 定义触发器 -->
            <array>
                <ref bean="cronTrigger" />                      <!-- 设置任务触发器 -->
            </array>
        </property>
    </bean>
```

此时实现了与之前同样效果的 QuartZ 触发器，而这样的处理操作由于没有了强制性的继承要求，也更加符合当前面向对象设计的趋势，用户使用也会更加灵活。

6.3　SpringTask 任务调度

QuartZ 是一个成熟的调度组件，发展时间较长，所以在与 Spring 整合处理时需要进行大量的配置。为了解决 QuartZ 配置复杂的问题，Spring 3.0 后的版本提供了一个新的定时任务工具——SpringTask 组件。可以把它作为一个轻量级的 Quartz，使用起来简单方便，而且支持注解和配置文件两种形式。

6.3.1　基于配置文件实现 SpringTask 任务调度处理

要在配置文件中使用 SpringTask 组件，需要先在配置文件中使用 task 命名空间进行配置，如图 6-2 所示。

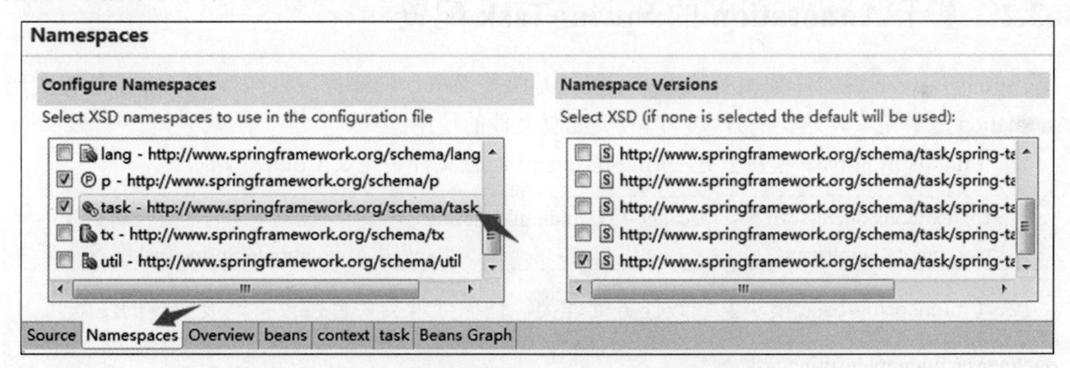

图 6-2　在配置文件中引入 task 命名空间

配置完成后，在 spring-base.xml 配置文件中可发现如下命名空间信息。

```
    xmlns:task="http://www.springframework.org/schema/task"
```

1.【mldnspring-base 模块】建立一个专门的任务执行类，该类不需要做任何继承，只是一个纯粹的类。开发者可以自己定义一个任务的执行方法。

```
package cn.mldn.mldnspring.task;
import java.text.SimpleDateFormat;
public class MyTask {
    public void runTask() {                                     // 任意定义一个方法名称
```

```
        System.out.println("【当前的日期时间】" + new SimpleDateFormat(
                "yyyy-MM-dd HH:mm:ss.SSS").format(new java.util.Date()));
    }
}
```

2.【mldnspring-base 模块】修改 spring-base.xml 配置文件，追加任务配置。

☑ 采用间隔触发形式执行任务调用。

```
<bean id="myTask" class="cn.mldn.mldnspring.task.MyTask"/>      <!-- 任务处理类 -->
<task:scheduled-tasks>                                  <!-- 进行SpringTask任务的调度配置 -->
    <!-- 定义任务类、执行方法与间隔调度的间隔时间 -->
    <task:scheduled ref="myTask" method="runJobTask" fixed-rate="1000"/>
</task:scheduled-tasks>
```

☑ 采用 CRON 定时触发形式执行任务调用。

```
<bean id="myTask" class="cn.mldn.mldnspring.task.MyTask"/>          <!-- 任务处理类 -->
<task:scheduled-tasks>                                  <!-- 进行SpringTask任务的调度配置 -->
    <!-- 定义处理任务的Bean对象与任务处理方法，同时设置调度时间 -->
    <task:scheduled ref="myTask" method="runTask" cron="* * * * * ?"/>
</task:scheduled-tasks>
```

本程序在配置 SpingTask 时，直接使用了 task 命名空间。在此命名空间中可直接配置任务执行类、任务处理方法与调度模式。

6.3.2 基于 Annotation 的 SpringTask 配置

SpringTask 实现的任务调度其优势不仅仅是配置简单，同样也可以结合 Context 配置实现 Annotation 注解定义，这样的开发可以减少配置文件的使用。

1.【mldnspring-base 模块】在 spring-base.xml 配置文件中定义 Context 路径。

```
<context:component-scan base-package="cn.mldn.mldnspring" />  <!-- 此包下所有注解自动生效 -->
<task:annotation-driven/>                               <!-- 启用Annotation的SpringTask配置 -->
```

2.【mldnspring-base 模块】修改任务处理类，任务处理类要直接设置调度安排的注解。

```
package cn.mldn.mldnspring.task;
import java.text.SimpleDateFormat;
import org.springframework.scheduling.annotation.Scheduled;
import org.springframework.stereotype.Component;
@Component
public class MyTask {
    @Scheduled(cron="* * * * * ?")                          // 任务调度设置
    public void runTask() {                                 // 任意定义一个方法名称
        System.out.println("【当前的日期时间】" + new SimpleDateFormat(
                "yyyy-MM-dd HH:mm:ss.SSS").format(new java.util.Date()));
    }
}
```

本程序采用了注解的形式实现了 SpringTask 任务定义，同时在要执行的任务方法上使用了@Scheduled 注解设置了任务的执行时间，该操作可以配置 CRON 定时调用，也可以使用 fixedDelay 实现间隔调用。

6.3.3　SpringTask 任务调度池

实际开发中，有时需要多个定时任务同时执行。默认情况下，Spring 一般会使用单线程池来执行定时任务处理，这样就会出现一个问题：当某一个定时任务执行时间过长时，其他定时任务就不得不延期执行。为了解决这个问题，可采用多线程池形式定义多个线程对象，实现多个定时任务的并行调用。

范例：【mldnspring-base 模块】修改 spring-base.xml 配置文件，追加线程池定义。

```
<task:annotation-driven/>                                   <!-- 启用Annotation的SpringTask配置 -->
<task:scheduler id="schedulePool" pool-size="2"/>           <!-- 定义线程池中的线程个数 -->
```

这里，程序定义了两个大小相同的线程池。当有两个定时任务执行时，将不会彼此影响，从而不会造成任务执行的延迟。

6.4　本章小结

1. 定时任务可随 Spring 容器自动启动，并可在后台自动执行。
2. 可使用的两种定时组件 QuartZ、SpringTask，它们都支持间隔触发与 CRON 定时触发。
3. 多个任务并行执行时，如果不想造成延迟处理，需要增加线程池个数。

第 7 章

AOP 切面编程

通过本章学习，可以达到以下目标：

1．理解代理设计与动态代理设计缺陷。
2．理解 AOP 的基本概念。
3．掌握 AOP 通知处理与配置。
4．掌握 Annotation 的 AOP 配置。

AOP（Aspect Oriented Programming，面向方面编程，也称为面向切面编程），是 Spring 的第二大特征，其本身属于代理功能的加强。利用 AOP 与切面表达式，可以更加方便地实现代理设计，且可以采用灵活的方式实现切面处理的支持。

7.1　AOP 产生动机

在进行实际项目开发时，除了进行核心业务处理外，还需要执行一些与核心业务有关的辅助性操作。以业务层调用数据层为例，除了要调用数据层实现数据处理外，还有可能进行日志记录、事务处理等辅助性操作。这样的辅助性操作最初是借助代理设计模式实现的，结构如图 7-1 所示。

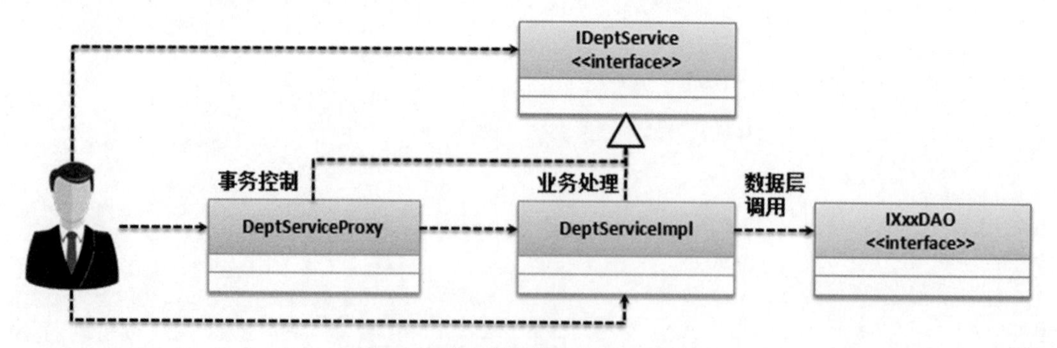

图 7-1　代理设计模式与业务设计

在传统代理设计模式中，需要为每个业务接口设计一个代理子类。代理子类除了要执行真实业务子类定义的业务方法外，还要进行一些辅助设计，如事务控制、日志采集等。

范例：定义部门业务处理，使用伪代码模拟数据层调用。

定义部门业务接口	定义部门业务接口实现子类

package cn.mldn.service;
import cn.mldn.vo.Dept;
public interface IDeptService {
    public boolean add(Dept vo) ;
}
	package cn.mldn.service.impl;
	import cn.mldn.service.IDeptService;
	import cn.mldn.vo.Dept;
	public class DeptServiceImpl implements IDeptService {
	    @Override
	    public boolean add(Dept vo) {
	        // 调用数据层实现一系列的数据处理操作
	        return true;
	    }
	}

范例：定义部门代理设计，实现事务控制与真实业务调用。

```
package cn.mldn.service.proxy;
import cn.mldn.service.IDeptService;
import cn.mldn.vo.Dept;
public class DeptServiceProxy implements IDeptService {
    private IDeptService deptService ;
    public DeptServiceProxy(IDeptService deptService) {
        this.deptService = deptService ;
    }
    @Override
    public boolean add(Dept vo) {
        try {
            // 开启数据库事务处理
            this.deptService.add(vo) ; // 调用真实业务层方法
            // 【COMMIT】执行完毕提交事务
        } catch (Exception e) {
            // 【ROLLBACK】出现异常事务回滚
        }
        return true;
    }
}
```

采用这种静态代理设计模式会产生一个实际问题：真实项目开发过程中一定会存在大量的业务接口，按照这样的处理模式需要编写大量的代理设计类，并且这些代理设计类的功能几乎是相同的。为了简化这样的操作，可以采用动态代理设计模式来解决，程序结构如图 7-2 所示。

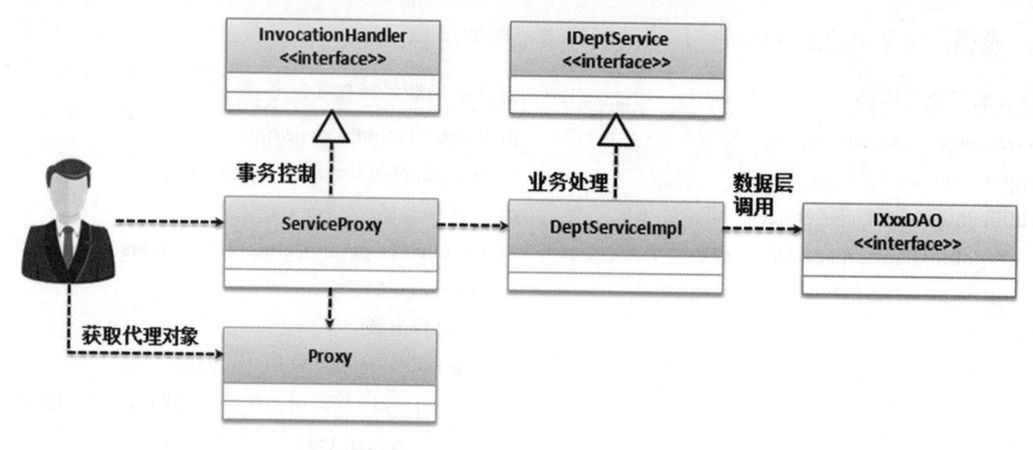

图 7-2　动态代理设计结构

利用动态代理设计结构实现业务层代理操作的最大优势在于：可以采用统一形式实现业务层的代理操作，从而避免代理子类过多的情况。

范例：使用动态代理解决代理类过多问题。

```
package cn.mldn.service.proxy;
import java.lang.reflect.InvocationHandler;
import java.lang.reflect.Method;
import java.lang.reflect.Proxy;
public class ServiceProxy implements InvocationHandler {
    private Object realObject;                                     // 被代理的真实对象
    /**
     * 真实对象的绑定处理，返回一个动态生成的接口类对象
     * @param realObject 真实主题类
     * @return 代理类对象
     */
    public Object bind(Object realObject) {
        this.realObject = realObject;
        return Proxy.newProxyInstance(realObject.getClass().getClassLoader(),
            realObject.getClass().getInterfaces(),this);
    }
    @Override
    public Object invoke(Object proxy, Method method, Object[] args) throws Throwable {
        if (this.checkTransactionMethod(method.getName())) {        // 检测是否需要开启事务
            // 开启数据库事务处理
        }
        Object backResult = method.invoke(this.realObject, args) ;   // 调用真实业务
        try {
            if (this.checkTransactionMethod(method.getName())) {    // 检测是否需要提交事务
```

```
            // 【COMMIT】数据库事务提交
            }
        } catch (Exception e) {
            // 【ROLLBACK】数据库事务回滚
            throw e ;
        }
        return backResult ;
    }
    /**
     * 检测当前方法是否需要开启事务控制处理
     * @param methodName 方法名称
     * @return 如果需要开启，返回true；如果不需要，返回false
     */
    private boolean checkTransactionMethod(String methodName) {
        return methodName.startsWith("add") || methodName.startsWith("edit") ||
            methodName.startsWith("delete") ;
    }
}
```

虽然动态代理设计在一定程度上可以满足开发设计要求，但其最大的缺陷在于：设计之初就必须明确定义出要进行业务处理的标准，且必须与业务操作代码混合在一起进行处理。很明显，这样的配置会造成复杂的代码耦合性问题。AOP 就是为了解决代理设计缺陷而产生的。

7.2　AOP 简介

AOP 是一种编程范式，开发者可从另一个角度考虑程序结构，从而完善面向对象编程。AOP 是基于动态代理的一种更高级的应用，可结合 AspectJ 组件，利用切面表达式将代理类织入到程序组成中，实现组件的解耦合设计。

AOP 主要用于横切关注点分离和织入，因此在 AOP 处理中需要关注如下几个核心概念（见图 7-3）。

- ☑ **关注点**：所关注的任何东西，如业务接口、支付处理、消息发送处理等。
- ☑ **关注点分离**：将业务处理逐步拆分后形成的一个个不可拆分的独立组件。
- ☑ **横切关注点**：实现代理功能，利用代理功能可以将辅助操作在多个关注点上执行，横切点可能包括事务处理、日志记录、角色或权限检测、性能统计等。
- ☑ **织入**：将横切关注点分离后，有可能需要确定关注点的执行位置，可能在业务方法调用前，也可能是调用后，或者是产生异常时。

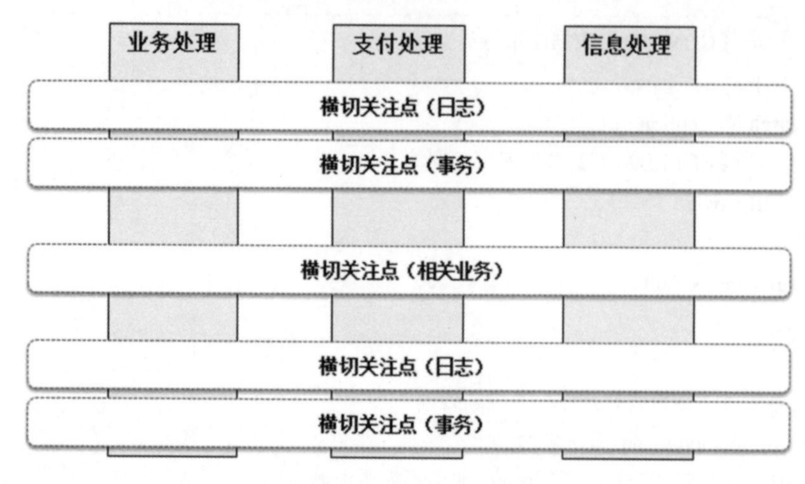

图 7-3　AOP 核心概念

分析图 7-3 可以发现，整个 AOP 设计中，横切关注点的设置与处理是关键所在。可以采用如下通知（Advice）处理形式来实现不同横切点的配置（见图 7-4）。

☑ **前置通知处理（Before Advice）**：在真正的核心功能调用前触发代理操作。

☑ **后置通知（After Advice）**：真正的核心功能调用后进行触发。

|- 后置返回通知（After Returning Advice）：当执行的核心方法返回数据的时候处理。

|- 后置异常通知（After Throwing Advice）：当执行真实处理方法产生异常后通知。

|- 后置最终通知（After Finally Advice）：不管是否出现问题，都执行此操作。

☑ **环绕通知（Round Advice）**：可在方法调用前或后自定义任何行为（包括前置与后置通知处理），且可以决定是否执行连接点处的方法、替换返回值、抛出异常等。

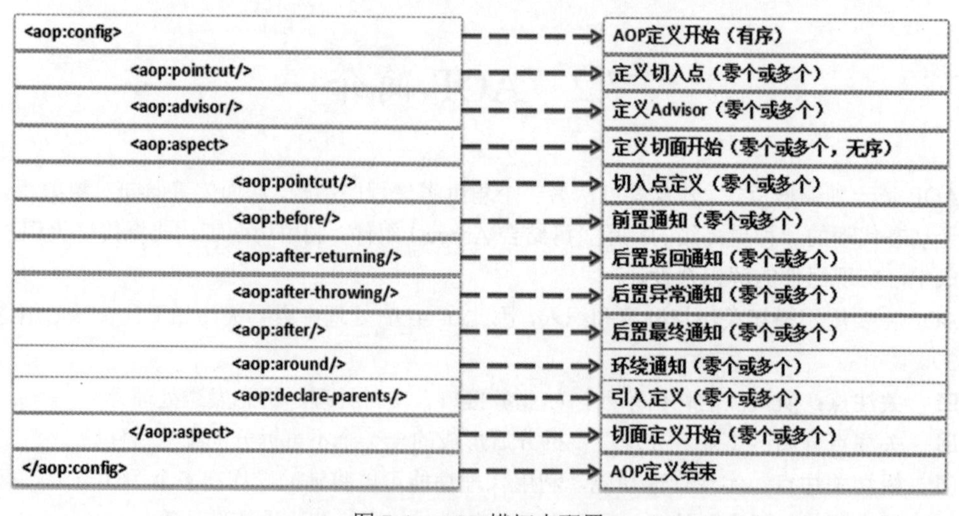

图 7-4　AOP 横切点配置

 **提示：AOP 早期实现方式。**

图 7-4 中的配置操作属于 AOP 的第二代配置形式。这种配置不局限于某一个父接口或父类，而是完全由开发者自行定义通知方法。在 Spring 早期，为了处理的统一，开发者需要定

义 AOP 程序类以及要实现的一系列 Advice 接口，如图 7-5 所示，然后再基于 XML 配置实现不同的通知处理。

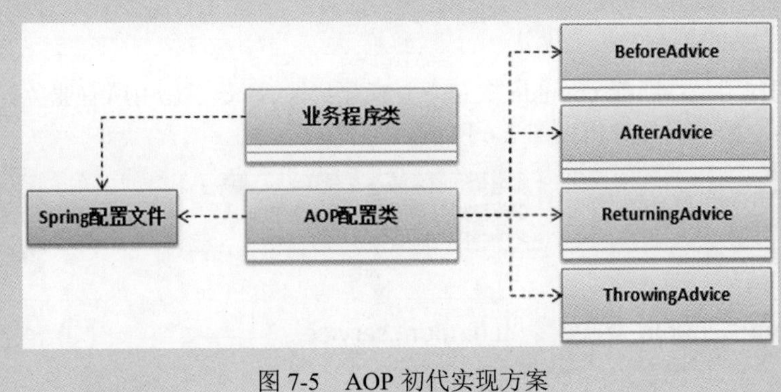

图 7-5　AOP 初代实现方案

7.3　AOP 切入点表达式

在 AOP 执行过程中切入点的处理是最关键的核心步骤，如果切入点配置错误，那么所有的通知处理方法将无法进行正常的织入，Spring AOP 支持的主要的 AspectJ 切入点标识符如下。

☑　execution：定义通知的切入点。

☑　this：用于匹配当前 AOP 代理对象类型的执行方法。

☑　target：用于匹配当前目标对象类型的执行方法。

☑　args：用于匹配当前执行的方法传入的参数为指定类型的执行方法。

在使用 AOP 定义切入点时主要依靠 execution 语句定义的，该语句的定义语法如下。

```
execution(注解匹配? 修饰符匹配? 方法返回值类型 操作类型匹配 方法名称匹配(参数匹配)) 异常匹配?
```

下面介绍一下各参数的含义。

☑　注解匹配【可选】：匹配方法上指定注解的定义，如@Override。

☑　修饰符匹配【可选】：方法修饰符，可以使用 public 或 protected。

☑　方法返回值类型【必填】：可以设置任何类型，可以使用“*”匹配所有返回值类型。

☑　操作类型匹配【必填】：定义方法所在类的名称，可以使用“*”匹配所有类型。

☑　方法名称匹配【必填】：匹配要调用的处理方法，使用“*”匹配所有方法。

☑　参数匹配【必填】：用于匹配切入方法的参数，有 5 种设计方式。

　|- ()：没有参数。

　|- (..)：匹配所有参数。

　|- (..,java.lang.String)：以 String 作为最后一个参数，前面的参数个数可以任意。

　|- (java.lang.String,..)：以 String 作为第一个参数，后面的参数个数可以任意。

　|- (*.java.lang.String)：以 String 作为最后一个参数，前面可以任意设置一个参数。

☑　异常匹配【可选】：定义方法名称中 throws 抛出的异常，可以设置多个，使用“,”分隔。

由于在本次讲解中是以业务层 AOP 设计为主，业务类所在包均在 cn.mldn 的 service 子包中。如果要匹配所有的业务层方法，可以使用如下切面表达式。

```
execution(public * cn.mldn..service..*.*(..))
```

该表达式的作用是：匹配 cn.mldn 父包下任意层级 service 子包中所有业务类的方法，可以设置任意的参数个数，具体分析如图 7-6 所示。

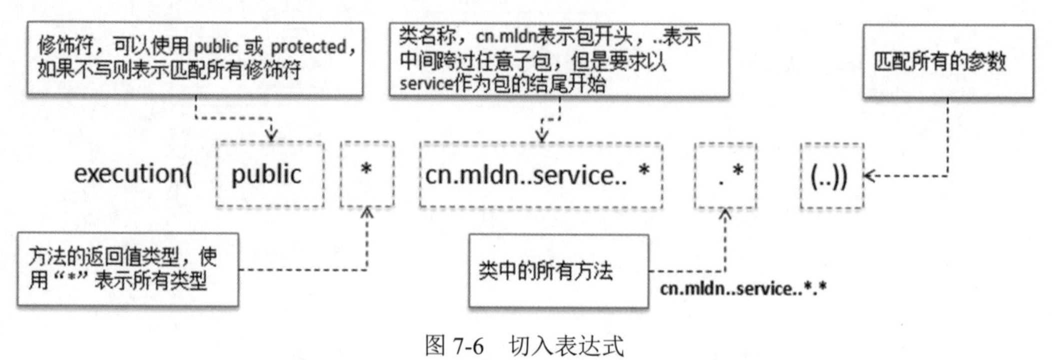

图 7-6　切入表达式

7.4　AOP 基础实现

要在 Spring 中实现 AOP 开发，需要先导入 spring-aop、spring-aspects 依赖库。导入配置完成后，还需要开启 AOP 的命名空间，这样才可以通过配置文件实现 AOP 配置，如图 7-7 所示。

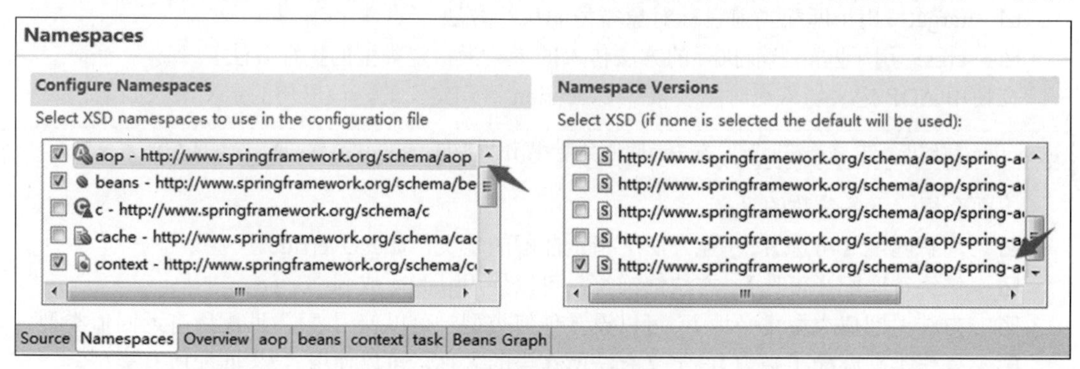

图 7-7　配置 AOP 命名空间

1.【mldnspring 项目】修改父 pom.xml 配置文件，追加 aop 与 aspects 开发包的配置。

```
<dependency>
    <groupId>org.springframework</groupId>
    <artifactId>spring-aop</artifactId>
    <version>${spring.version}</version>
</dependency>
<dependency>
    <groupId>org.springframework</groupId>
    <artifactId>spring-aspects</artifactId>
```

```
        <version>${spring.version}</version>
    </dependency>
```

2.【mldnspring-base 项目】修改 pom.xml 配置文件，追加依赖库。

```
    <dependency>
        <groupId>org.springframework</groupId>
        <artifactId>spring-aop</artifactId>
    </dependency>
    <dependency>
        <groupId>org.springframework</groupId>
        <artifactId>spring-aspects</artifactId>
    </dependency>
```

3.【mldnspring-base 项目】建立一个消息回显的业务接口。

```
package cn.mldn.mldnspring.service;
public interface IMessageService {
    public String echo(String str) ;
}
```

4.【mldnspring-base 项目】建立 MessageServiceImpl 子类，实现 IMessageService 业务接口。

```
package cn.mldn.mldnspring.service.impl;
import org.springframework.stereotype.Service;
import cn.mldn.mldnspring.service.IMessageService;
@Service
public class MessageServiceImpl implements IMessageService {
    @Override
    public String echo(String str) {
        System.out.println("【MessageServiceImpl业务层实现】接收到消息内容：" + str);
        return "【ECHO】msg = " + str;
    }
}
```

本程序模拟了一个消息处理过程，在 echo 方法中实现了提示信息输出。该类将采用注解配置形式，利用 context 扫描配置。

5.【mldnspring-base 项目】建立 AOP 处理程序类 ServiceAdvice。

```
package cn.mldn.mldnspring.advice;
public class ServiceAdvice {                              // 该类不需要继承任何父类，独立存在
    public void handleBefore() {                          // 处理前置操作通知
        System.out.println("【### ServiceAdvice-handleBefore ###】进行业务的前置处理操作。");
    }
    public void handleAfter() {                           // 处理后置操作通知
        System.out.println("【### ServiceAdvice-handleAfter ###】进行业务的后置处理操作。");
    }
}
```

本程序中实现了一个前置通知与后置通知，此类并没有任何附加的处理，方法名称也是由开发者自行定义。要将其织入到业务处理中，需要在 Spring 配置文件中处理。

6.【mldnspring-base 项目】修改 spring-base.xml 配置文件，追加通知织入处理。

```
<bean id="serviceAdvice" class="cn.mldn.mldnspring.advice.ServiceAdvice"/>
<aop:config>            <!-- 进行AOP的配置处理 -->
    <!-- 配置切入点AspectJ表达式，它决定了在哪里执行代理操作 -->
    <aop:pointcut id="myPointcut"
        expression="execution(public * cn.mldn..service..*.*(..))" />
    <aop:aspect ref="serviceAdvice">        <!-- 切入处理操作方法定义，设置程序类名称 -->
        <aop:before method="handleBefore" pointcut-ref="myPointcut"/>    <!-- 引入切入点 -->
        <aop:after method="handleAfter" pointcut-ref="myPointcut"/>      <!-- 引入切入点 -->
    </aop:aspect>
</aop:config>
```

AOP 配置中一定要引入一个执行 AOP 通知的处理类，同时还需要定义通知处理类的切入点，并且为切入点设置不同的通知处理方法。

7.【mldnspring-base 项目】编写测试类，注入 IMessageService 接口对象，观察执行处理。

```
package cn.mldn.mldnspring;
import org.junit.Test;
import org.junit.runner.RunWith;
import org.springframework.beans.factory.annotation.Autowired;
import org.springframework.test.context.ContextConfiguration;
import org.springframework.test.context.junit4.SpringJUnit4ClassRunner;
import cn.mldn.mldnspring.service.IMessageService;
@ContextConfiguration(locations = { "classpath:spring/spring-base.xml" })     // 资源文件定位
@RunWith(SpringJUnit4ClassRunner.class)                                     // 设置要使用的测试工具
public class TestMessageService {
    @Autowired
    private IMessageService messageService ;
    @Test
    public void testEcho() {
        System.out.println(this.messageService.echo("www.mldn.cn"));
    }
}
```

程序执行结果	【### ServiceAdvice-handleBefore ###】进行业务的前置处理操作。 【MessageServiceImpl业务层实现】接收到消息内容：www.mldn.cn 【### ServiceAdvice-handleAfter ###】进行业务的后置处理操作。 【ECHO】msg = www.mldn.cn

通过此时的结果可以发现，在 ServiceAdvice 中定义的通知处理方法都根据切面表达式与通知配置整合到了业务方法调用上，而在整体处理中 AOP 代理与业务层之间并没有明确的代码关联，只是通过配置文件将其整合在一起。

7.5　前置通知参数接收

截至目前，我们实现了一个 AOP 的基本处理操作。为 AOP 定义处理方法时，实际上还可以进行一些参数的接收操作。执行 IMessageService 接口的 echo 方法时，可发现该方法是有参数的，可以实现参数接收。一旦要进行参数接收，切入点就必须明确地追加上参数的具体信息。

1.【mldnspring-base 项目】修改 ServiceAdvice 处理类，为前置通知方法定义参数信息。

```
public void handleBefore(String tempMsg) {          // 处理前置通知
    System.out.println("【### ServiceAdvice-handleBefore ###】进行业务的前置处理操作，参数：" + tempMsg);
}
```

2.【mldnspring-base 项目】修改 spring-base.xml 配置文件，为前置通知定义不同的切面表达式。

```
<aop:config>                                    <!-- 进行AOP的配置处理 -->
    <!-- 配置切入点AspectJ表达式，它决定了在哪里执行代理操作 -->
    <aop:pointcut id="myPointcut"
        expression="execution(public * cn.mldn..service..*.*(..))" />
    <aop:aspect ref="serviceAdvice">            <!-- 切入处理操作方法定义，设置程序类名称 -->
        <aop:before method="handleBefore"
                pointcut="execution(public * cn.mldn..service..*.*(..)) and args(msg)"
                arg-names="msg"/>               <!-- 定义新的切入点 -->
        <aop:after method="handleAfter" pointcut-ref="myPointcut"/> <!-- 引入切入点 -->
    </aop:aspect>
</aop:config>
```

为了在前置通知处理方法中进行参数接收，我们在配置前置通知切入表达式时使用了andargs(msg)参数标记。因此需要在<aop:before>元素中使用 arg-names="msg"描述参数定义。此时再次执行测试程序类，将出现如下提示信息。

```
【### ServiceAdvice-handleBefore ###】进行业务的前置处理操作，参数：www.mldn.cn
【MessageServiceImpl业务层实现】接收到消息内容：www.mldn.cn
【### ServiceAdvice-handleAfter ###】进行业务的后置处理操作。
【ECHO】msg = www.mldn.cn
```

7.6　后置通知

后置通知处理一共有 3 种形式：后置最终通知、后置返回通知和后置异常通知。其中，后置最终通知的作用是不管是否出现问题都会执行，而下面来观察后置返回与后置异常通知处理。

1.【mldnspring-base 项目】修改 MessageServiceImpl 子类，追加异常抛出处理。

```
package cn.mldn.mldnspring.service.impl;
import org.springframework.stereotype.Service;
import cn.mldn.mldnspring.service.IMessageService;
@Service
public class MessageServiceImpl implements IMessageService {
    @Override
    public String echo(String str) {
        if (str == null) {
            throw new RuntimeException("空消息，无法处理！") ;
        }
        System.out.println("【MessageServiceImpl业务层实现】接收到消息内容：" + str);
        return "【ECHO】msg = " + str;
    }
}
```

这里为了更好地观察后置异常通知，抛出了一个 RuntimeException 异常类对象。

2.【mldnspring-base 项目】修改 ServiceAdvice 处理类，追加新的通知处理方法。

```
    public void handleReturn(String retMsg) {                    // 处理后置操作通知
        System.out.println("【### ServiceAdvice-handleReturn ###】业务方法执行完毕：" + retMsg);
    }
    public void handleThrow(Exception exp) {                     // 异常处理通知
        System.out.println("【### ServiceAdvice-handleThrow ###】方法执行产生了异常：" + exp);
    }
```

3.【mldnspring-base 项目】修改 spring-base.xml 配置文件，追加后置返回通知与后置异常通知。

```
    <aop:config>   <!-- 进行AOP配置处理 -->
        <!-- 配置切入点AspectJ表达式，它决定了在哪里执行代理操作 -->
        <aop:pointcut id="myPointcut"
            expression="execution(public * cn.mldn..service..*.*(..))" />
        <aop:aspect ref="serviceAdvice">          <!-- 切入处理操作方法定义，设置程序类名称 -->
            <aop:before method="handleBefore"
                pointcut="execution(public * cn.mldn..service..*.*(..)) and args(msg)"
                arg-names="msg"/>   <!-- 定义新的切入点 -->
            <aop:after method="handleAfter" pointcut-ref="myPointcut"/> <!-- 引入切入点 -->
            <!-- 定义后置返回通知，需要将方法的返回数据传递到方法中 -->
            <aop:after-returning method="handleReturn" pointcut-ref="myPointcut"
                returning="val" arg-names="val"/>
            <!-- 定义后置异常通知，需要将产生的异常对象传递到方法中 -->
            <aop:after-throwing method="handleThrow" pointcut-ref="myPointcut"
                throwing="e" arg-names="e"/>
```

```
    </aop:aspect>
</aop:config>
```

定义了通知处理后，就可以通过测试程序实现代码测试了。测试分为两种：调用 echo 方法传递非空参数，或者传递 null。

☑　调用 echo 方法传递非空参数。返回信息如下。

```
【### ServiceAdvice-handleBefore ###】进行业务的前置处理操作，参数：www.mldn.cn
【MessageServiceImpl业务层实现】接收到消息内容：www.mldn.cn
【### ServiceAdvice-handleAfter ###】进行业务的后置处理操作。
【### ServiceAdvice-handleReturn ###】业务方法执行完毕：【ECHO】msg = www.mldn.cn
【ECHO】msg = www.mldn.cn
```

☑　调用 echo 方法传递 null。返回信息如下。

```
【### ServiceAdvice-handleBefore ###】进行业务的前置处理操作，参数：null
【### ServiceAdvice-handleAfter ###】进行业务的后置处理操作。
【### ServiceAdvice-handleThrow ###】方法执行产生了异常：java.lang.RuntimeException: 空消息，无法处理！
```

对比执行结果可以发现，不管 echo 方法是否出现异常，都会执行后置最终通知处理。

7.7　环绕通知

环绕通知的使用类似于动态代理设计模式的结构，也是 AOP 中处理功能最强大的通知，利用环绕通知可以明确地实现前置调用处理与后置调用处理。

1.【mldnspring-base 项目】修改 ServiceAdvice 类，定义环绕通知。

```
package cn.mldn.mldnspring.advice;
import java.util.Arrays;
import org.aspectj.lang.ProceedingJoinPoint;
public class ServiceAdvice {                                    // 该类不需要继承任何父类，独立存在
    public Object handleRound(ProceedingJoinPoint point) throws Throwable {// 定义环绕通知
        System.out.print("【A、环绕通知 - handleRound】业务方法调用前。参数："
                + Arrays.toString(point.getArgs())) ;
        Object returnValue = null ;                             // 表示的是进行方法返回值的接收处理
        try {
            returnValue = point.proceed(new Object[] {"假的参数我乐意传"}) ;// 修改了真实参数
        } catch (Exception e) {                                 // 异常向上继续抛出
            System.out.print("【C、环绕通知 - handleRound】产生异常。异常：" + e) ;
            throw e ;
        }
        System.out.print("【B、环绕通知 - handleRound】业务方法执行完毕。返回值：" + returnValue) ;
        return returnValue ;
```

```
    }
}
```

本程序实现了环绕通知的定义。环绕通知的处理操作类似于动态代理设计模式，在进行环绕处理的时候一定要在方法上设置 ProceedingJoinPoint 类对象，利用该对象可以找到所有的方法传递参数，也可以继续调用真实的业务方法。

2.【mldnspring-base 项目】修改 spring-base.xml 配置文件，追加环绕通知配置。

```
<aop:config>    <!-- 进行AOP的配置处理 -->
    <!-- 配置切入点AspectJ表达式，它决定了到底在哪里执行代理操作 -->
    <aop:pointcut id="myPointcut"
        expression="execution(public * cn.mldn..service..*.*(..))" />
    <aop:aspect ref="serviceAdvice">        <!-- 切入处理操作方法定义，设置程序类名称 -->
        <aop:around method="handleRound" pointcut-ref="myPointcut"/>    <!-- 环绕通知 -->
    </aop:aspect>
</aop:config>
```

程序编写完后，启动测试程序，可看到如下信息提示。

```
【A、环绕通知 - handleRound】业务方法调用前。参数：[www.mldn.cn]【MessageServiceImpl业务层实现】接收到消息内容：假的参数我乐意传
【B、环绕通知 - handleRound】业务方法执行完毕。返回值：【ECHO】msg = 假的参数我乐意传【ECHO】msg = 假的参数我乐意传
```

可以发现，环绕通知根据 point.proceed 方法调用了前置或后置通知处理，也可以动态修改传递的参数内容。

7.8 基于 Annotation 的 AOP 配置

除了可以采用 XML 的模式实现 AOP 配置外，也可以利用 Annotation 注解的形式进行配置，并且所有的注解需要通过 context 扫描包配置生效。利用注解配置 AOP 时，可以直接在通知处理方法上使用@Around、@Before、@AfterThrowing、@AfterReturning、@After 注解，实现不同通知类型的定义。

1.【mldnspring-base 项目】修改 spring-base.xml 配置文件，启用 AOP 注解配置。

```
<context:component-scan base-package="cn.mldn.mldnspring"/>
<aop:aspectj-autoproxy/>                                    <!-- 启用AOP的Annotation支持 -->
```

2.【mldnspring-base 项目】修改 ServiceAdvice，使用注解进行配置。

```
package cn.mldn.mldnspring.advice;
import java.util.Arrays;
import org.aspectj.lang.ProceedingJoinPoint;
import org.aspectj.lang.annotation.After;
import org.aspectj.lang.annotation.AfterReturning;
```

```
import org.aspectj.lang.annotation.AfterThrowing;
import org.aspectj.lang.annotation.Around;
import org.aspectj.lang.annotation.Aspect;
import org.aspectj.lang.annotation.Before;
import org.springframework.stereotype.Component;
@Component                                                  // Bean注解配置
@Aspect                                                     // 该类为AOP处理类
public class ServiceAdvice {                    // 该类不需要继承任何父类，独立存在
    @Before(value="execution(public * cn.mldn..service..*.*(..)) and args(msg)",
            argNames="msg")
    public void handleBefore(String tempMsg) {              // 处理前置操作通知
        System.out.println("【### ServiceAdvice-handleBefore ###】进行业务的前置处理操作，参数：
" + tempMsg);
    }
    @After(value="execution(public * cn.mldn..service..*.*(..))")
    public void handleAfter() {                             // 处理后置操作通知
        System.out.println("【### ServiceAdvice-handleAfter ###】进行业务的后置处理操作。");
    }
    @AfterReturning(
        value="execution(public * cn.mldn..service..*.*(..))",argNames="r",returning="r")
    public void handleReturn(String retMsg) {               // 处理后置操作通知
        System.out.println("【### ServiceAdvice-handleReturn ###】业务方法执行完毕：" + retMsg);
    }
    @AfterThrowing(
        value="execution(public * cn.mldn..service..*.*(..))",throwing="e",argNames="e")
    public void handleThrow(Exception exp) {                    // 异常处理操作通知
        System.out.println("【### ServiceAdvice-handleThrow ###】方法执行产生了异常：" + exp);
    }
    @Around("execution(public * cn.mldn..service..*.*(..))")
    public Object handleRound(ProceedingJoinPoint point) throws Throwable {   // 定义环绕通知
        System.out.print("【A、环绕通知 - handleRound】业务方法调用前。参数：" +
            Arrays.toString(point.getArgs())) ;
        Object returnValue = null ;                          // 方法返回值的接收处理
        try {
            returnValue = point.proceed(new Object[] {"假的参数我乐意传"}) ;// 修改真实参数
        } catch (Exception e) {                                  // 异常向上继续抛出
            System.out.print("【C、环绕通知 - handleRound】产生异常。异常：" + e) ;
            throw e ;
        }
        System.out.print("【B、环绕通知 - handleRound】业务方法执行完毕。返回值：" + returnValue) ;
        return returnValue ;
    }
}
```

此时的程序处理可以达到与之前 XML 配置同样的效果，并且代码结构更加简单。

7.9 本章小结

1．AOP 实现与代理设计模式相比，最大的特点在于其可以通过切入表达式进行代码整合。

2．AOP 通知分为 3 类：前置通知、后置通知和环绕通知。其中，环绕通知可以实现全部的切面控制。

3．AOP 支持 Annotation 配置，以简化配置文件定义。

第 8 章

Spring 与 JMS 消息组件

通过本章学习，可以达到以下目标：

1. 掌握 ActiveMQ 的配置与使用。
2. 掌握 Spring 与 JMS 整合处理。
3. 可以基于 Bean 实现 JMS 整合配置。

JMS（Java Message Service，Java 消息服务）是 Java 定义的消息服务标准，但 Spring 对其使用进行了重新定义。本章将结合 ActiveMQ 标准组件，为读者讲解 Spring 与 JMS 组件的整合使用。

8.1 JMS 消息组件

JMS 是 Java EE 提出的一个消息数据的传输标准，即根据规定格式传递的消息都可以被标准 JMS 组件处理。在实际项目开发过程中，利用消息组件可以实现数据采集、队列缓冲的处理机制，如图 8-1 所示。

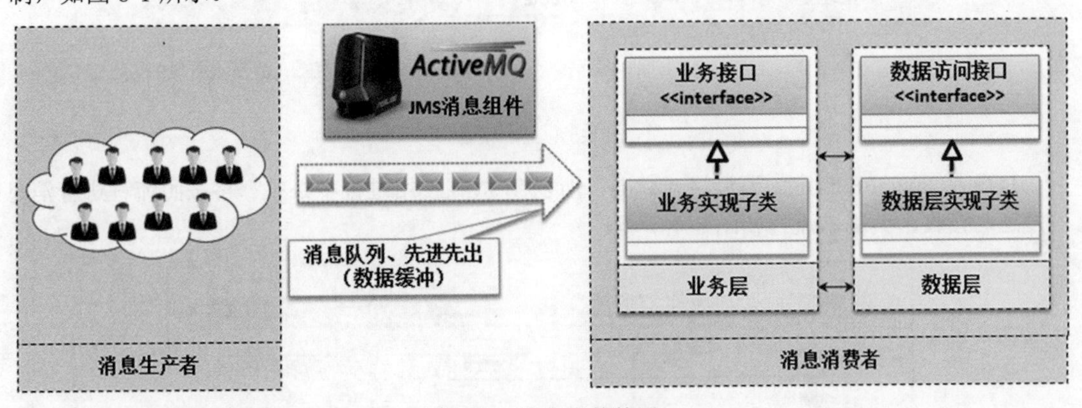

图 8-1　消息组件使用

图 8-1 演示了消息组件的作用。进行消息处理时，分为消息生产者（Provider）与消息消费者（Consumer），消息生产者按照 JMS 消息格式标准发送消息（不需要关注消息处理结果）到消息组件，而后消费者通过消息组件获取消息，进行消费。由于消息组件本身提供一个消息队列，并且按照先进先出的使用原则进行消息处理，因此即便有大量的访问数据出现，也可以实现数据缓冲的处理。在 JMS 实现标准中，最著名的组件就是 Apache ActiveMQ 组件。

提示：ActiveMQ 属于历史悠久的广谱性组件。

严格意义上讲，ActiveMQ 并不是性能最好的消息组件，但却是出现时间最长、使用最广泛、基于 JMS 标准实现的组件。由于 JMS 是传输结构的定义，而不是传输协议的定义，因此需要花费大量时间进行数据处理，造成其性能相对较差。

为了提升消息组件的处理性能，产生了 AMQP 高级消息处理协议，这类组件以 RabbitMQ 为代表。随后又因为大数据时代来临，产生了 Kafka 消息组件，进一步提升了消息处理的性能。对这些消息组件有兴趣的读者，可以翻阅笔者的其他书籍，或者登录 www.mldn.cn 进行视频学习。

在进行 JMS 消息处理时，可以实现队列、主题两类消息的发送与接收。

☑ **队列消息（Queue）**：指的是点对点消息，所有的消费者按照顺序进行接收，如图 8-2 所示。

☑ **主题消息（Topic）**：一个消息发送出去之后，可以有若干个消费者共同处理消息，如图 8-3 所示。

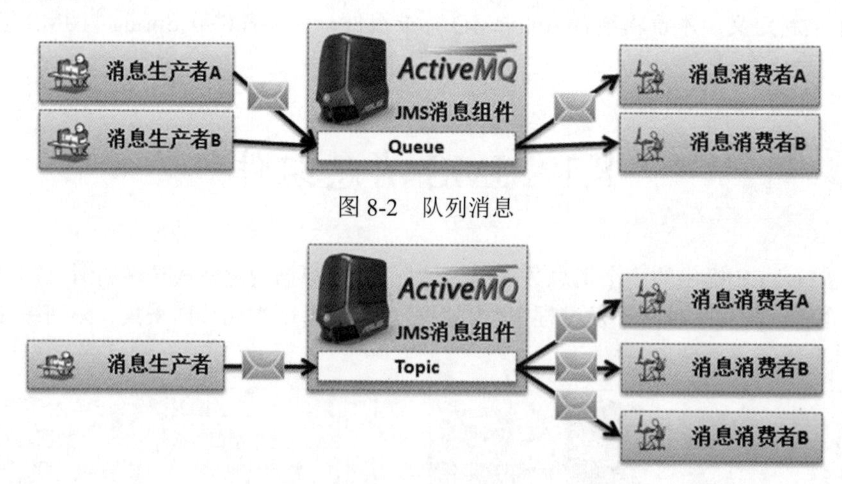

图 8-2　队列消息

图 8-3　主题消息

JMS 是 Java EE 制定的消息格式标准，所以在 JMS 使用过程中必须严格按照流程进行消息的发送与接收，其基本流程如图 8-4 所示。

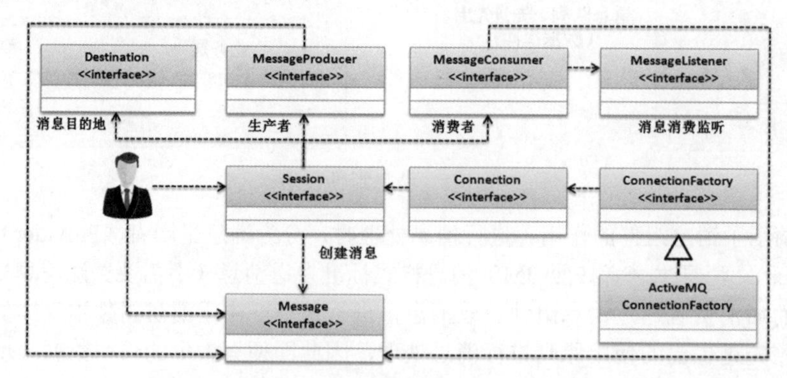

图 8-4　JMS 消息处理流程

在使用 JMS 进行消息处理时，需要进行如下几个核心步骤配置：

☑ JSM 定义了 ConnectionFactory 标准接口，可以利用 ConnectionFactory 创建 Connection。

☑ 获取 Connection 接口对象后如果要进行消息的发送与消费，则必须创建 Session 接口实例。

☑ 利用 Session 可以创建消息的发送者与消息接收者对象，并且可以利用 Session 创建一个消息信息。

|- JMS 中所有的消息都使用 Message 接口描述，并且提供一系列的子类，每种子类可以描述一种消息类型，如文本消息（TextMessage）、对象消息（ObjectMessage）。

☑ 消息发送完毕后，需要释放资源。此时消费端需要定义 MessageListener 子类，实现消息监听。

8.2　配置 ActiveMQ 组件

ActiveMQ 是 Apache 组织推出的一款较为流行的开源消息总线，也是一个完全支持 JMS 1.1 规范标准的消息组件，所有的开发者都可以通过 Apache 的官方网站（http://activemq.apache.org）获取 ActiveMQ 组件包，如图 8-5 所示。

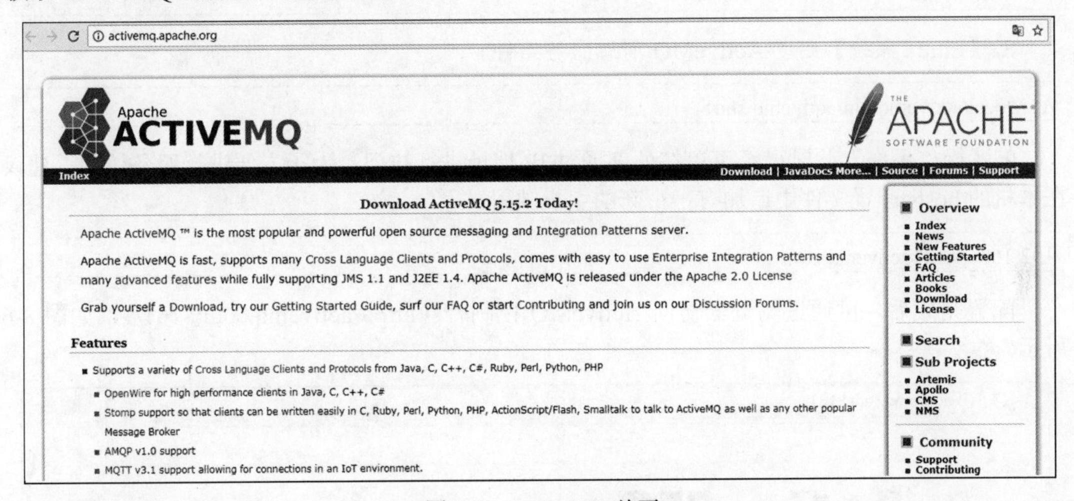

图 8-5　ActiveMQ 首页

本书选用的 ActiveMQ 组件版本为 5.15.2。同时，为了尽可能还原实际项目开发环境，将基于 Linux 系统进行 ActiveMQ 消息组件的配置。

1.【Linux 系统】将下载的 apache-activemq-5.15.2-bin.tar.gz 上传到系统中的/srv/ftp 目录中。

2.【Linux 系统】将上传后的 activemq 开发包解压缩到/usr/local 目录中。

```
tar xzvf /srv/ftp/apache-activemq-5.15.2-bin.tar.gz -C /usr/local/
```

3.【Linux 系统】为了配置方便，修改解压缩后的 activemq 文件夹。

```
mv /usr/local/apache-activemq-5.15.2/ /usr/local/activemq
```

4.【Linux 系统】修改 ActiveMQ 组件的 Web 管理账号配置文件 jetty-realm.properties。

```
vim /usr/local/activemq/conf/jetty-realm.properties
```

5.【Linux 系统】在 jetty-realm.properties 配置文件中追加新的管理员账号（mldnjava/hello）。

```
mldnjava: hello, admin
```

6.【Linux 系统】消息系统可以接收外部发送的任何 JMS 消息数据，因此为了安全，需要修改 activemq.xml 配置文件，追加消息发送与接收的用户信息。打开 activemq.xml 配置文件：

```
vim /usr/local/activemq/conf/activemq.xml
```

7.【Linux 系统】在 activemq.xml 配置文件（<broker>元素中）追加新的认证信息 mldn/jixianit。

```
<plugins>
    <simpleAuthenticationPlugin>
        <users>
            <authenticationUser username="mldn" password="jixianit" groups="admin" />
        </users>
    </simpleAuthenticationPlugin>
</plugins>
```

8.【Linux 系统】启动 ActiveMQ 消息服务组件。

```
/usr/local/activemq/bin/activemq start
```

9.【操作系统】此时已经可以在外部通过 IP 地址进行访问，为了方便进行服务访问，建议在本机的 hosts 配置文件中追加有一个新的主机映射。

```
192.168.79.131activemq.com
```

配置完成后，可通过浏览器访问 ActiveMQ 控制台（http://activemq.com:8161/），如图 8-6 所示。

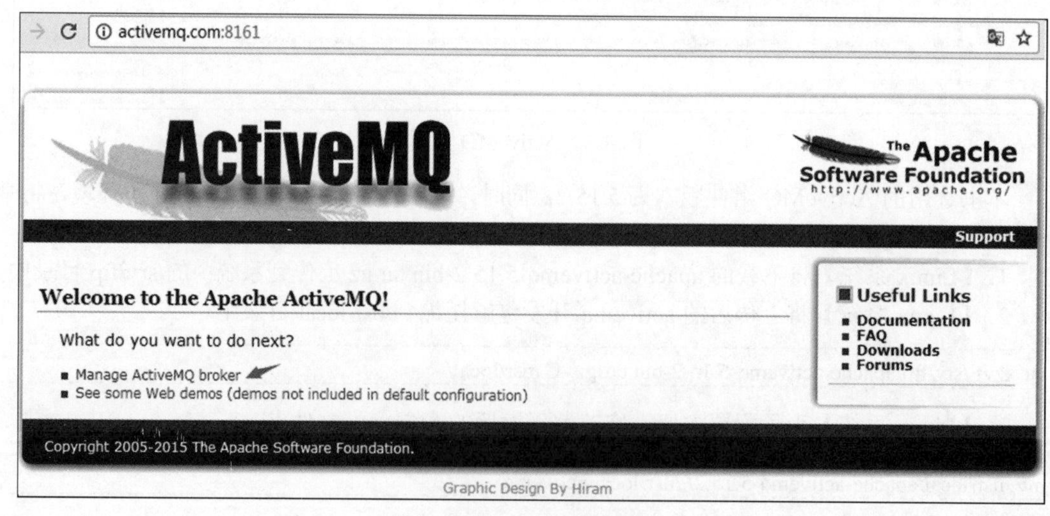

图 8-6　ActiveMQ 控制台

选择 Manage ActiveMQ broker 配置 ActiveMQ 主机，随后输入配置的 Web 管理账号（mldnjava/hello），登录之后可以看见如图 8-7 所示的界面信息。

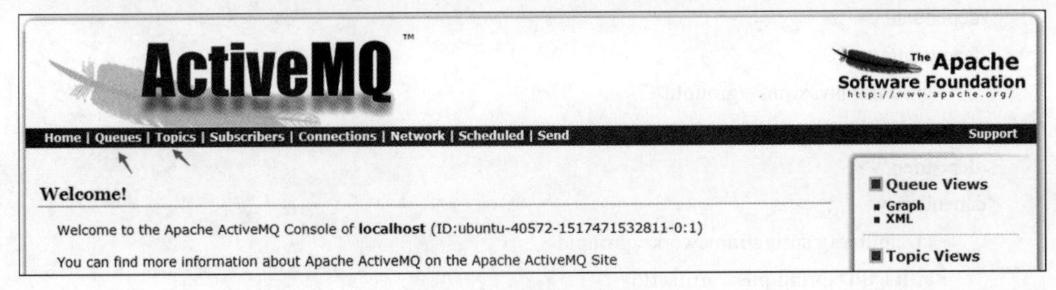

图 8-7　控制台登录首页

此时可以直接选择 Queues 与 Topics，建立要使用的队列名称与主题名称，本次创建的队列名称为 mldn.queue.msg，主题名称为 mldn.topic.msg。

8.3　使用 ActiveMQ 实现消息处理

消息组件在进行消息处理时一定需要消息的生产者与消息的消费者，所以在本程序中将在 mldnspring 项目中创建两个子模块：消息生产者（mldnspring-message-provider）、消息消费者（mldnspring-message-consumer）。

1.【mldnspring 项目】修改父 pom.xml 配置文件，引入 ActiveMQ 的相关依赖库配置。

属性配置	<activemq.version>5.15.2</activemq.version> <jms.version>1.1</jms.version>
依赖库配置	<dependency> <groupId>org.apache.activemq</groupId> <artifactId>activemq-all</artifactId> <version>${activemq.version}</version> </dependency> <dependency> <groupId>javax.jms</groupId> <artifactId>jms</artifactId> <version>${jms.version}</version> </dependency> <dependency> <groupId>org.springframework</groupId> <artifactId>spring-jms</artifactId> <version>${spring.version}</version> </dependency>

2.【mldnspring-message-*项目】在子模块中进行依赖库配置。

```
<dependency>
```

```
        <groupId>org.apache.activemq</groupId>
        <artifactId>activemq-all</artifactId>
</dependency>
<dependency>
        <groupId>javax.jms</groupId>
        <artifactId>jms</artifactId>
</dependency>
<dependency>
        <groupId>org.springframework</groupId>
        <artifactId>spring-jms</artifactId>
</dependency>
```

在本次配置依赖库中，为了方便直接引入了 activemq-all 依赖库，由于 ActiveMQ 出现较早，所以这个 jar 文件的体积非常庞大，开发者也可以根据需要自行导入子模块。另外需要注意的是，spring-jms 提供了 Spring 的 ActiveMQ 解决方案，可以帮助用户减少配置。

8.3.1 处理 Queue 消息

使用 Spring-JMS 依赖包实现 ActiveMQ 整合处理操作，只需要根据 JMS 的处理标准，在配置文件中进行配置即可。下面为了处理方便，将在 src/main/profiles/dev 目录中创建 config/activemq.properties 配置文件，而后在 Spring 配置文件中将通过此资源文件定义的资源信息进行 ActiveMQ 配置。

1.【mldnspring-message-*项目】由于生产者与消费者需要使用统一的服务信息，所以在两个项目模块中分别创建 src/main/profiles/dev/config/activemq.properties 配置文件。文件内容如下：

activemq.broker.url=tcp://activemq.com:61616	# 定义要处理的ActiveMQ连接地址信息
activemq.user.name=mldn	# ActiveMQ连接用户名称
activemq.user.password=jixianit	# ActiveMQ连接密码
activemq.queue.name=mldn.msg.queue	# 定义要处理的消息队列信息

2.【mldnspring-message-provider 项目】配置消息发送端的 Spring 配置文件 spring-base.xml。

```
<context:component-scan base-package="cn.mldn.mldnspring"/>
<context:property-placeholder location="classpath:config/*.properties"/>          <!-- 加载资源 -->
<!-- Spring中需通过ActiveMQ工厂操作 -->
<bean id="targetConnectionFactory" class="org.apache.activemq.ActiveMQConnectionFactory">
        <!-- 定义连接工厂负责连接处理的程序网络路径 -->
        <property name="brokerURL" value="${activemq.broker.url}"/>           <!-- 连接地址 -->
        <property name="userName" value="${activemq.user.name}"/>             <!-- 用户名 -->
        <property name="password" value="${activemq.user.password}"/>         <!-- 密码 -->
</bean>
<!-- 发送前，需将ActiveMQ处理交由JMS控制 -->
<bean id="connectionFactory"
        class="org.springframework.jms.connection.SingleConnectionFactory">
```

```
        <!-- 配置要进行连接处理的ActiveMQ的连接工厂类 -->
        <property name="targetConnectionFactory" ref="targetConnectionFactory"/>
    </bean>
    <!-- 定义消息发送的目的地信息 -->
    <bean id="destination" class="org.apache.activemq.command.ActiveMQQueue">
        <constructor-arg index="0" value="${activemq.queue.name}"/>
    </bean>
    <!-- 定义JmsTemplate对象类，进行消息发送 -->
    <bean id="jmsTemplate" class="org.springframework.jms.core.JmsTemplate">
        <property name="connectionFactory" ref="connectionFactory"/>
    </bean>
```

在本配置文件中，最为重要的就是 JmsTemplate 消息处理模板类的配置。此类由 Spring 提供，开发者可以利用该类轻松实现消息的发送处理。

3.【mldnspring-message-provider 项目】为了进一步规范配置管理，直接定义一个消息发送的业务接口。

```
package cn.mldn.mldnspring.service;
public interface IMessageService {
    public void send(String msg) ;    // 消息发送
}
```

4.【mldnspring-message-provider 项目】建立 IMessageService 接口子类，并将 JmsTemplate（发送消息）与 Destination（消息目的地）对象注入。

```
package cn.mldn.mldnspring.service.impl;
import javax.jms.Destination;
import javax.jms.JMSException;
import javax.jms.Message;
import javax.jms.Session;
import org.springframework.beans.factory.annotation.Autowired;
import org.springframework.jms.core.JmsTemplate;
import org.springframework.jms.core.MessageCreator;
import org.springframework.stereotype.Component;
import cn.mldn.mldnspring.service.IMessageService;
@Component
public class MessageServiceImpl implements IMessageService {
    @Autowired
    private JmsTemplate jmsTemplate ;                    // Spring提供的JMS消息操作模板
    @Autowired
    private Destination destination ;                    // 消息发送的目的地
    @Override
    public void send(String msg) {
        this.jmsTemplate.send(this.destination, new MessageCreator(){
```

```
            @Override
            public Message createMessage(Session session) throws JMSException {
                return session.createTextMessage(msg);                    // 发送文本消息
            }
        });
    }
}
```

5.【mldnspring-message-provider 项目】编写测试类，注入消息发送业务类，并使用 send 发送消息。

```
package cn.mldn.mldnspring;
import org.junit.Test;
import org.junit.runner.RunWith;
import org.springframework.beans.factory.annotation.Autowired;
import org.springframework.test.context.ContextConfiguration;
import org.springframework.test.context.junit4.SpringJUnit4ClassRunner;
import cn.mldn.mldnspring.service.IMessageService;
@ContextConfiguration(locations = { "classpath:spring/spring-base.xml" })     // 资源文件定位
@RunWith(SpringJUnit4ClassRunner.class)                                       // 设置要使用的测试工具
public class TestMessageService {
    @Autowired
    private IMessageService messageService ;
    @Test
    public void testSend() {
        this.messageService.send("www.mldn.cn");                              // 发送消息
    }
}
```

本测试程序启动后，会向 ActiveMQ 服务主机发送消息，读者可通过 ActiveMQ 控制台观察到微消费信息，如图 8-8 所示。

Queues:

Name	Number Of Pending Messages	Number Of Consumers	Messages Enqueued	Messages Dequeued	Views	Operations
mldn.msg.queue	1	0	1	0	Browse Active Consumers Active Producers atom rss	Send To Purge Delete

图 8-8　ActiveMQ 保存消息

6.【mldnspring-message-consumer 项目】消息消费处理主要是通过监听程序实现的，即监听程序打开后将一直监听消息组件，一旦有消息发送就可以直接进行消费处理，该消息监听类需要实现 MessageListener 接口。

```
package cn.mldn.mldnspring.listener;
import javax.jms.JMSException;
import javax.jms.Message;
```

```
import javax.jms.MessageListener;
import javax.jms.TextMessage;
import org.springframework.stereotype.Component;
@Component
public class ConsumerMessageListener implements MessageListener {
    @Override
    public void onMessage(Message message) {                        // 接收消息
        if (message instanceof TextMessage) {                       // 判断发送的是否为文本信息
            TextMessage textMsg = (TextMessage) message ;           // 接收发送的消息信息
            try {
                System.out.println("***  【接收到新消息】" + textMsg.getText());
            } catch (JMSException e) {
                e.printStackTrace();
            }
        }
    }
}
```

这里为了配置方便，直接使用@Component 注解实现自动配置处理，相当于定义了一个名称为 consumerMessageListener 的 Bean，并且直接在 Spring 配置文件中进行了注入。

7.【mldnspring-message-consumer 项目】修改 spring-base.xml 配置文件，实现消息消费端配置。此时对于连接工厂与连接的配置与生产端完全一样，唯一不同的是需要配置消息监听处理类。

```
<!-- 定义消息监听处理程序类 -->
<bean id="jmsContainer" class="org.springframework.jms.listener.DefaultMessageListenerContainer">
    <!-- 消息监听连接工厂类 -->
    <property name="connectionFactory" ref="connectionFactory"/>
    <!-- 消息监听目的地 -->
    <property name="destination" ref="destination"/>
    <!-- 消息监听处理程序类 -->
    <property name="messageListener" ref="consumerMessageListener"/>
</bean>
```

再次启动 Spring 容器后，消费端将会接收消息。

8.3.2　处理 Topic 消息

要想将队列消息修改为主题消息，可直接修改项目中消息配置的类型，将其由 ActiveMQQueue 修改为 ActiveMQTopic。

1.【mldnspring-message-*项目】修改生产者与消费者的 activemq.properties 配置文件，定义消息主题名称。

```
# 定义要处理的消息主题信息
```

```
activemq.topic.name=mldn.msg.topic
```

2.【mldnspring-message-*项目】修改 Spring 配置文件中的消息目标配置。

```
<bean id="destination" class="org.apache.activemq.command.ActiveMQTopic">
    <constructor-arg index="0" value="${activemq.topic.name}"/>
</bean>
```

此时可以实现主题订阅消息，即多个消费端将同时获取同一个消息内容。

8.3.3 基于 Bean 配置

对于资源的访问处理，除了可以基于配置文件配置之外，也可以利用@Configuration 注解实现基于类的配置处理，这样可以减少 XML 配置文件的数量，便于对多个项目模块进行统一管理。

1.【mldnspring-message-*项目】由于前面基于 properties 资源文件实现了访问信息的定义，所以此时修改生产者与消费者程序，依然要读取 activemq.properties 配置。

```
<context:component-scan base-package="cn.mldn.mldnspring"/>
<context:property-placeholder location="classpath:config/*.properties"/>  <!-- 加载资源 -->
```

在定义配置 Bean 的时候，需要将其定义在 cn.mldn.mldnspring 子包之中。

2.【mldnspring-message-provider 项目】定义 MessageProviderConfig 配置类。

```
package cn.mldn.mldnspring.config;
import javax.jms.Destination;
import org.apache.activemq.ActiveMQConnectionFactory;
import org.apache.activemq.command.ActiveMQTopic;
import org.springframework.beans.factory.annotation.Value;
import org.springframework.context.annotation.Bean;
import org.springframework.context.annotation.Configuration;
import org.springframework.jms.connection.SingleConnectionFactory;
import org.springframework.jms.core.JmsTemplate;
@Configuration //本类是一个配置程序类
//以下配置等价于<context:property-placeholder location="classpath:config/*.properties"/>
//@PropertySource("classpath:config/activemq.properties")
public class MessageProviderConfig {
    @Value("${activemq.broker.url}")      // 读取指定资源文件内容，格式为${key}
    private String brokerURL ;
    @Value("${activemq.topic.name}")
    private String topicName ;
    @Value("${activemq.user.name}")
    private String username ;
    @Value("${activemq.user.password}")
    private String password ;
    // 定义JmsTemplate对象类，进行消息发送
    // <bean id="jmsTemplate" class="org.springframework.jms.core.JmsTemplate">
```

```
    @Bean(name="jmsTemplate")
    public JmsTemplate getJmsTemplate(SingleConnectionFactory connectionFactory) {
        JmsTemplate jmsTemplate = new JmsTemplate() ;
        // <property name="connectionFactory" ref="connectionFactory"/>
        jmsTemplate.setConnectionFactory(connectionFactory);
        return jmsTemplate ;
    }
    // 定义消息发送的目的地信息
    // <bean id="destination" class="org.apache.activemq.command.ActiveMQQueue">
    @Bean(name="destination")
    public Destination createDestination() {
        ActiveMQTopic amq = new ActiveMQTopic(this.topicName) ;
        return amq ;
    }
    // 发送前，需将ActiveMQ处理交由JMS控制
    // <bean id="connectionFactory"
class="org.springframework.jms.connection.SingleConnectionFactory">
    @Bean(name="connectionFactory")
    public SingleConnectionFactory createConnectionFactory(ActiveMQConnectionFactory
targetConnectionFactory) {
        SingleConnectionFactory scf = new SingleConnectionFactory() ;
        // <property name="targetConnectionFactory" ref="targetConnectionFactory"/>
        scf.setTargetConnectionFactory(targetConnectionFactory);
        return scf ;
    }
    // 明确给出连接地址信息，需通过ActiveMQ工厂操作
    // <bean id="targetConnectionFactory" class="org.apache.activemq.ActiveMQConnectionFactory">
    @Bean(name="targetConnectionFactory")
    public ActiveMQConnectionFactory createTargetConnectionFactory() {
        ActiveMQConnectionFactory acf = new ActiveMQConnectionFactory() ;
        acf.setBrokerURL(this.brokerURL);
        acf.setUserName(this.username);
        acf.setPassword(this.password);
        return acf ;
    }
}
```

3.【mldnspring-message-consumer 项目】定义 MessageConsumerConfig 配置类，在进行消费端配置时实际上核心处理与消息生产端的配置差别不大，唯一的区别在于消息消费端只需要配置 MessageListener，而不需要使用 JmsTemplate 进行消息发送。

```
    // <bean id="jmsContainer"
class="org.springframework.jms.listener.DefaultMessageListenerContainer">
```

```
@Bean(name = "jmsContainer")  // 定义JMS消息容器
public DefaultMessageListenerContainer createJmsContainer(SingleConnectionFactory
connectionFactory, Destination destination, ConsumerMessageListener consumerMessageListener) {
    DefaultMessageListenerContainer container = new DefaultMessageListenerContainer();
    container.setConnectionFactory(connectionFactory);                  // 设置连接工厂
    container.setDestination(destination);                              // 设置消息目的地
    container.setMessageListener(consumerMessageListener);              // 设置监听对象
    return container ;
}
```

本程序实现了基于 Bean 的配置管理，实际上在使用 Bean 进行配置时会牵扯到更多的内容，如果不是开发中明确要求，建议在传统的 Spring 开发中还是使用配置文件的形式处理。

8.4　本章小结

1．JMS 是 Java EE 定义的一套数据传输规范，ActiveMQ 是其开源实现。

2．消息分为两种类型：队列消息（Queue）和主题消息（Topic）。

3．基于 Spring 管理的 ActiveMQ 开发者只需要配置好消息的生产者与消费者处理信息项，就可以轻松实现消息处理。

第 9 章

Spring 与 WebService

通过本章学习，可以达到以下目标：

1. 理解 WebService 的主要作用以及操作流程。
2. 可以使用 CXF 实现 WebService 基础开发。
3. 使用 Spring 整合 CXF 实现 WebService 开发。

WebService 是一种早期的异构系统通信规范，利用 WebService 可实现不同语言的远程 RPC 接口。本章将为读者讲解如何基于 Spring 实现 WebService 的开发。

9.1 WebService 简介

WebService 是一种早期的异构系统整合标准，可以直接在已有的 Web 服务器上进行服务发布。其基于 SOAP（Simple Object Access Protocol，简单对象访问协议）实现数据 XML 交换处理，同时使用 WSDL（Web Services Description Language，Web 服务描述语言）作为业务接口描述，以方便客户端调用，如图 9-1 所示。

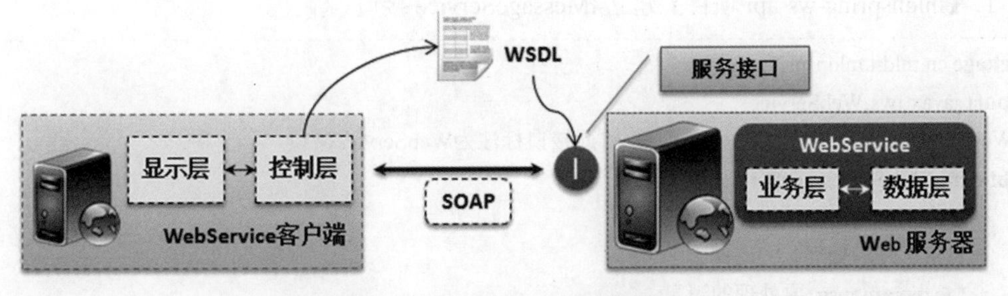

图 9-1　WebService 实现架构

提示：WebService 历史发展。

WebService 是一个技术标准，最早应用在微软.NET 平台与 SUN 公司 J2EE 平台的整合上，而后陆续有许多编程语言也开始支持 WebService 标准。以 Java 语言为例，出现了许多 WebService 开发技术，如 Axis、XFire、CXF 等。现在基本上都采用 CXF 组件实现。需要注意的是，WebService 作为技术标准，使用了 XML 作为接口描述，所以其性能相对较差。目前，一些新项目逐渐开始使用 Restful 技术标准，以替代 WebService。有兴趣的读者可通过笔者的

《Java 微服务架构实战》一书进行学习。

9.2 WebService 基础开发

WebService 是以远程业务接口为操作标准实现的 RPC 项目开发，所以在本章将创建 3 个 Maven 项目模块：mldnspring-ws-api、mldnspring-ws-provider 和 mldnspring-ws-consumer，如图 9-2 所示。

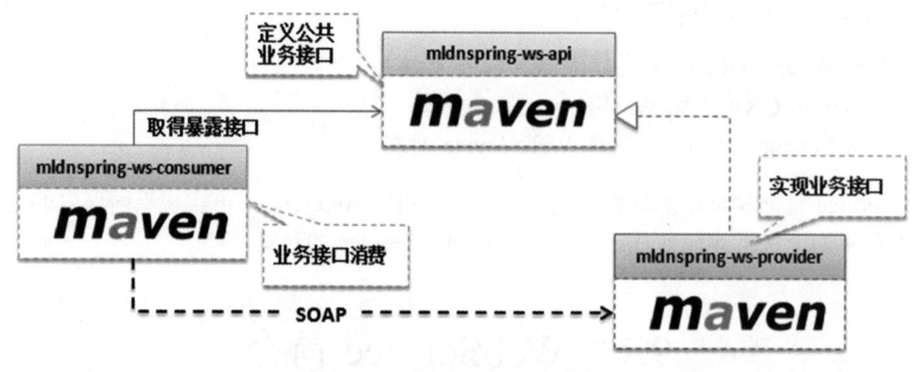

图 9-2　WebService 项目结构

9.2.1 创建公共接口项目

创建一个 Maven 模块 mldnspring-ws-api，其中不需要引入任何依赖库，只需提供整体项目中的核心业务接口。

1.【mldnspring-ws-api 项目】建立 IMessageService 接口。

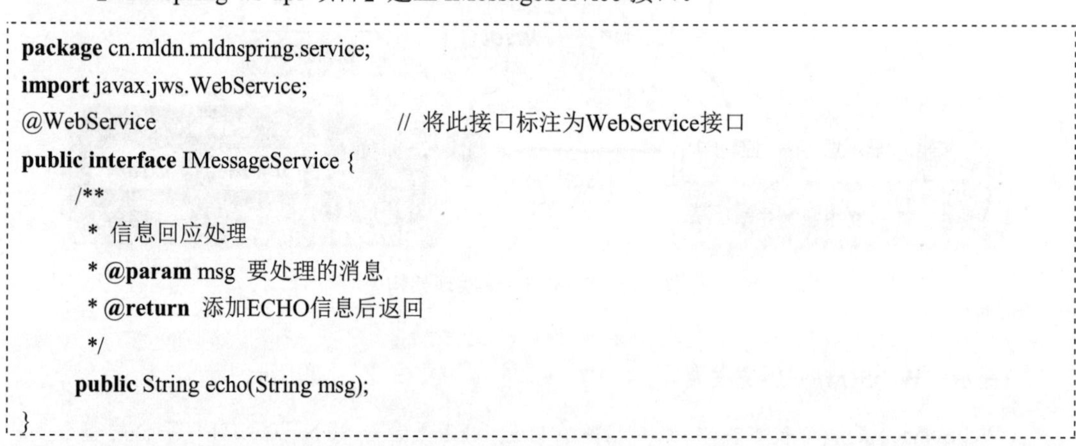

```
package cn.mldn.mldnspring.service;
import javax.jws.WebService;
@WebService                          // 将此接口标注为WebService接口
public interface IMessageService {
    /**
     * 信息回应处理
     * @param msg 要处理的消息
     * @return 添加ECHO信息后返回
     */
    public String echo(String msg);
}
```

由于 WebService 需要通过远程接口的映射形式来进行处理，所以在定义时使用了 @WebService 注解。

2.【mldnspring 项目】对于 WebService 的生产者与消费者项目模块，都需要通过 IMessageService 接口提供服务与处理消费，为了方便操作，可以直接修改父 pom.xml 配置文件，

引入 mldnspring-ws-api 模块。

属性配置	<mldnspring.version>0.0.1</mldnspring.version>
依赖库配置	<dependency> <groupId>cn.mldn</groupId> <artifactId>mldnspring-ws-api</artifactId> <version>${mldnspring.version}</version> </dependency>

配置完成后对于 mldnspring-ws-provider、mldnspring-ws-consumer 两个子模块只需要修改 pom.xml 配置文件引入 mldnspring-ws-api 模块即可使用公共业务接口。

9.2.2　创建 WebService 服务提供者

WebService 服务提供者主要功能是定义业务接口的实现子类，并且需要将业务接口的实现子类通过 Web 服务器进行项目的发布。

1.【mldnspring-ws-provider 项目】建立 IMessageService 接口实现子类。

```
package cn.mldn.mldnspring.service.impl;
import javax.jws.WebService;
import cn.mldn.mldnspring.service.IMessageService;
@WebService(                                                        // 定义Web服务接口
        endpointInterface="cn.mldn.mldnspring.service.IMessageService",    // 定义终端业务接口
        serviceName="messageService")                               // 定义服务名称
public class MessageServiceImpl implements IMessageService {
    @Override
    public String echo(String str) {
        if (str == null) {
            throw new RuntimeException("空消息，无法处理！") ;
        }
        return "【ECHO】msg = " + str;
    }
}
```

本程序作为 WebService 业务接口的实现子类，定义时通过@WebService 注解定义了该业务接口的发布名称。

2.【mldnspring-ws-provider 项目】编写 WebService 服务启动类，并注册 WebService 接口实例。

```
package cn.mldn.mldnspring;
import javax.xml.ws.Endpoint;
import cn.mldn.mldnspring.service.IMessageService;
import cn.mldn.mldnspring.service.impl.MessageServiceImpl;
public class StartWebServiceApplication {
    public static void main(String[] args) {
```

```
        IMessageService msgObj = new MessageServiceImpl() ;                    // 实例化接口对象
        Endpoint.publish("http://192.168.1.101:7777/message", msgObj) ;
    }
}
```

3.【mldnspring-ws-provider 项目】启动 WebService 提供者项目模块，通过浏览器访问地址 http://192.168.1.101:7777/message?wsdl，可看到 WebService 定义描述结构，如图 9-3 所示。

```
192.168.1.101:7777/message?wsdl
xmlns="http://schemas.xmlsoap.org/wsdl/" targetNamespace="http://impl.service.mldnspring.mldn.cn/" name="messageService">
  <import namespace="http://service.mldnspring.mldn.cn/" location="http://192.168.1.101:7777/message?wsdl=1"/>
 ▼<binding xmlns:ns1="http://service.mldnspring.mldn.cn/" name="MessageServiceImplPortBinding" type="ns1:IMessageService">
    <soap:binding transport="http://schemas.xmlsoap.org/soap/http" style="document"/>
   ▼<operation name="echo">
      <soap:operation soapAction=""/>
     ▼<input>
        <soap:body use="literal"/>
      </input>
     ▼<output>
        <soap:body use="literal"/>
      </output>
    </operation>
  </binding>
 ▼<service name="messageService">
   ▼<port name="MessageServiceImplPort" binding="tns:MessageServiceImplPortBinding">
      <soap:address location="http://192.168.1.101:7777/message"/>
    </port>
  </service>
</definitions>
```

图 9-3　WebService 接口描述

9.2.3　创建 WebService 服务消费者

WebService 对于接口的描述主要通过 WSDL 进行业务访问，但是如果要想实现这样的访问，则必须通过相关工具生成 WebService 辅助代理程序代码后，才可以实现远程接口消费处理，开发者可以直接使用 Eclipse 来生成此辅助代码。

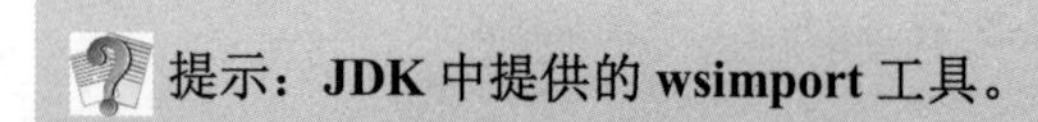

提示：JDK 中提供的 wsimport 工具。

如果要进行 WebService 辅助代码的生成处理操作，开发者可以不使用 Eclipse 这样的开发工具，直接通过 JDK 提供的命令来实现远程 WebService 接口调用转换。

```
wsimport -keep http://192.168.1.101:7777/message?wsdl
```

执行后会在当前目录中生成相关辅助代码，开发者可以将其导入到项目中直接使用，但是这样的配置过于烦琐，所以本书将直接利用 Eclipse 生成。

1.【mldnspring-ws-consumer 项目】创建 WebService 客户端程序，如图 9-4 所示。

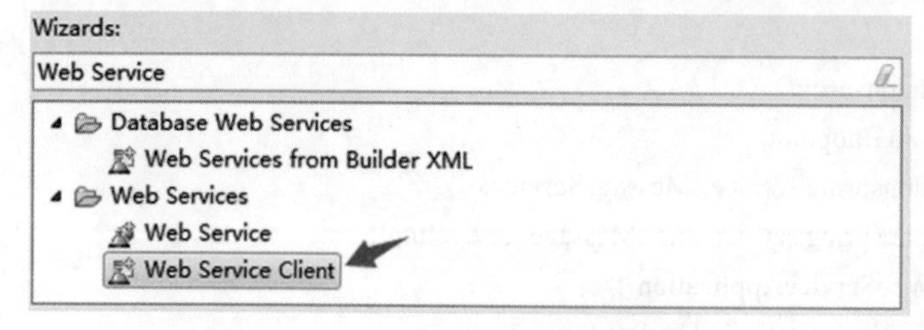

图 9-4　创建 WebServiceClient

2.【mldnspring-ws-consumer 项目】输入 WSDL 访问地址，如图 9-5 所示。

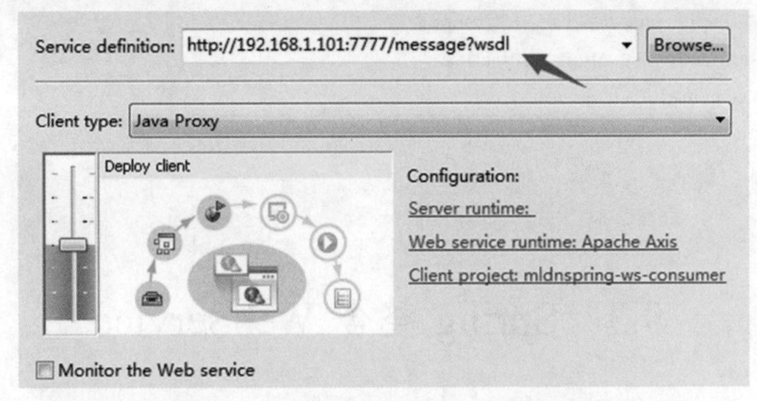

图 9-5　生成 WebService 客户端辅助代码

3.【mldnspring-ws-consumer 项目】编写客户端程序，进行业务调用。

```
package cn.mldn.mldnspring.client;
import cn.mldn.mldnspring.service.IMessageService;
import cn.mldn.mldnspring.service.impl.MessageServiceLocator;
public class TestMessageClient {
	public static void main(String[] args) throws Exception {
		IMessageService messageService = new
				MessageServiceLocator().getMessageServiceImplPort();
		System.out.println(messageService.echo("www.mldn.cn"));
	}
}
```

程序执行结果	【ECHO】msg = www.mldn.cn

WebService 消费端通过辅助代码实现了远程服务接口的调用，整体调用就仿佛在本地执行业务接口一样简单。

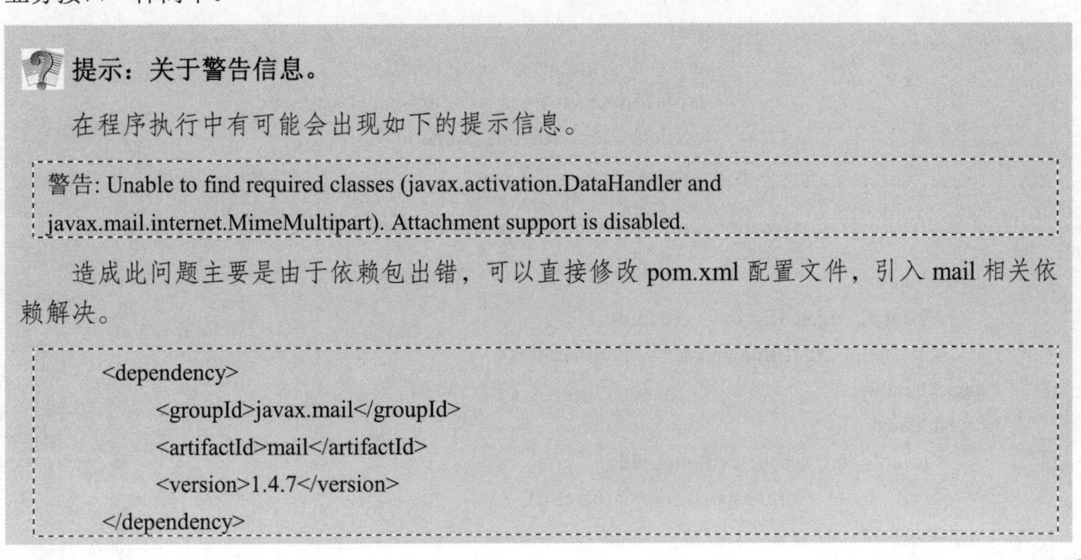

提示：关于警告信息。

在程序执行中有可能会出现如下的提示信息。

```
警告: Unable to find required classes (javax.activation.DataHandler and
javax.mail.internet.MimeMultipart). Attachment support is disabled.
```

造成此问题主要是由于依赖包出错，可以直接修改 pom.xml 配置文件，引入 mail 相关依赖解决。

```
<dependency>
	<groupId>javax.mail</groupId>
	<artifactId>mail</artifactId>
	<version>1.4.7</version>
</dependency>
```

```
<dependency>
    <groupId>javax.activation</groupId>
    <artifactId>activation</artifactId>
    <version>1.1.1</version>
</dependency>
```

引入后进行消费端调用时，将不会再出现错误提示信息。

9.3 Spring 整合 WebService

通过完整的 WebService 程序可以发现，WebService 是基于 Web 服务器的应用，进行消费端调用时需要先生成一系列的辅助代码，而后才可以调用。由于 WebService 中不只需要发布服务接口，还有可能牵扯到一些复杂的业务逻辑与其他开发框架的整合，所以为了开发方便，可将 WebService 与 Spring 进行整合处理，此时需要使用 CXF 依赖库支持。

1.【mldnspring 项目】修改 pom.xml 配置文件，引入 CXF 相关依赖库。

属性配置	`<cxf.version>3.2.0</cxf.version>`

依赖库配置	`<dependency><groupId>org.apache.cxf</groupId><artifactId>cxf-rt-frontend-jaxws</artifactId><version>${cxf.version}</version></dependency>`
	`<dependency><groupId>org.apache.cxf</groupId><artifactId>cxf-rt-transports-http</artifactId><version>${cxf.version}</version></dependency>`
	`<dependency><groupId>org.apache.cxf</groupId><artifactId>cxf-rt-transports-http-jetty</artifactId><version>${cxf.version}</version></dependency>`

2.【mldnspring-ws-provider、mldnspring-ws-consumer 项目】导入 CXF 相关依赖库。

```
<dependency>
    <groupId>org.apache.cxf</groupId>
    <artifactId>cxf-rt-frontend-jaxws</artifactId>
</dependency>
<dependency>
    <groupId>org.apache.cxf</groupId>
    <artifactId>cxf-rt-transports-http</artifactId>
</dependency>
```

```
<dependency>
        <groupId>org.apache.cxf</groupId>
        <artifactId>cxf-rt-transports-http-jetty</artifactId>
</dependency>
```

3.【mldnspring-ws-provider 项目】建立 src/main/resources/spring/spring-webservice-provider.xml 配置文件，引入 jaxws 命名空间，进行 WebService 配置，如图 9-6 所示。

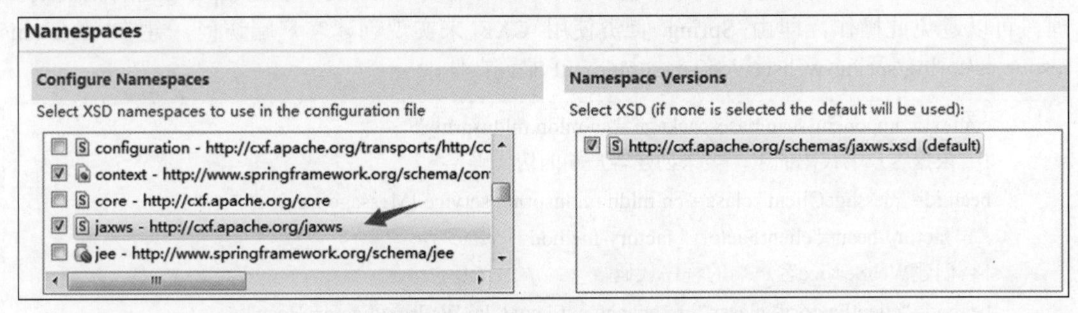

图 9-6　配置 jaxws 命名空间

4.【mldnspring-ws-provider 项目】编辑 spring-webservice-provider.xml 配置文件。

```
<?xml version="1.0" encoding="UTF-8"?>
<beans xmlns="http://www.springframework.org/schema/beans"
    xmlns:xsi="http://www.w3.org/2001/XMLSchema-instance"
    xmlns:context="http://www.springframework.org/schema/context"
    xmlns:jaxws="http://cxf.apache.org/jaxws"
    xsi:schemaLocation="
        http://cxf.apache.org/jaxws
        http://cxf.apache.org/schemas/jaxws.xsd
        http://www.springframework.org/schema/beans
        http://www.springframework.org/schema/beans/spring-beans-4.3.xsd
        http://www.springframework.org/schema/context
        http://www.springframework.org/schema/context/spring-context-4.3.xsd">
    <context:component-scan base-package="cn.mldn.mldnspring"/>
    <!-- 发布WebService终端接口，同时绑定服务地址 -->
    <jaxws:endpoint implementor="cn.mldn.mldnspring.service.impl.MessageServiceImpl"
        id="messageService" address="http://192.168.1.101:7777/message" />
</beans>
```

5.【mldnspring-ws-provider 项目】编写程序启动类，该类只需要加载 spring-webservice-provider.xml 配置文件，并且启动 Spring 容器。

```
package cn.mldn.mldnspring;
import org.springframework.context.ApplicationContext;
import org.springframework.context.support.ClassPathXmlApplicationContext;
public class StartWebServiceApplication {
    public static void main(String[] args) {
        ApplicationContext context = new
```

```
            ClassPathXmlApplicationContext("classpath:spring/spring-*.xml") ;
    }
}
```

容器启动后，开发者可以通过 http://192.168.1.101:7777/message?wsdl 获得此接口描述信息。

6.【mldnspring-ws-consumer 项目】在原始 WebService 实现过程之中，消费端如果想通过接口调用 WebService 定义的接口，必须先生成一系列的辅助代理程序。基于 Spring 进行消费端管理，可以避免此操作，即由 Spring 负责使用 CXF 来实现创建客户端访问，建立 src/main/resources/spring/spring-webservice-consumer.xml 配置文件。

```
<context:component-scan base-package="cn.mldn.mldnspring"/>
<!-- 依据客户端代码的工厂类来创建客户端的伪代码 -->
<bean id="messageClient" class="cn.mldn.mldnspring.service.IMessageService"
      factory-bean="clientFactory" factory-method="create" />
<!-- 创建WebService客户端的程序代码 -->
<bean id="clientFactory" class="org.apache.cxf.jaxws.JaxWsProxyFactoryBean">
      <property name="serviceClass"
            value="cn.mldn.mldnspring.service.IMessageService" />          <!-- 定义远程接口 -->
      <property name="address"
            value="http://192.168.1.101:7777/message" />                   <!-- 定义服务地址 -->
</bean>
```

7.【mldnspring-ws-consumer 项目】通过 Spring 的依赖配置管理注入 IMessageService 远程对象，编写测试类，调用远程业务方法。

```
@ContextConfiguration(locations = { "classpath:spring/spring-*.xml" })
@RunWith(SpringJUnit4ClassRunner.class)                          // 设置要使用的测试工具
public class TestMessageClient {
    @Autowired
    private IMessageService messageService;                      // 注入业务对象
    @Test
    public void testEcho() throws Exception {
        System.out.println(this.messageService.echo("www.mldn.cn"));
    }
}
```

此时可以实现远程 WebService 服务调用，不再需要基于开发工具生成的辅助代理访问程序。

9.4 本章小结

1. WebService 是一种异构系统的整合技术，利用 WebService 标准可以使用 XML 实现远程服务描述，并且基于 SOAP 协议实现服务调用，其性能相对较低。

2. 利用 Spring 整合 WebService，可以简化消费端调用处理，方便整合其他组件服务。

第 10 章

Spring 与 Redis 数据库

通过本章学习，可以达到以下目标：

1. 掌握 Redis 数据库的主要作用。
2. 掌握 Redis 数据库的安装与配置。
3. 掌握 Redis 数据库的操作命令。
4. 掌握 Redis 常用数据类型的作用与操作命令。
5. 掌握 SpringDataRedis 与 Jedis 操作。
6. 掌握 Redis 哨兵机制与 Spring 访问。
7. 掌握 RedisCluster 配置与 Spring 访问。

缓存是实际项目开发中的主要工作，在进行分布式缓存的处理上，Redis（Remote Dictionary Server，远程数据服务）是使用最广泛的缓存数据库。本章将为读者讲解 Redis 的安装配置方法以及 SpringDataRedis 技术的使用。

10.1 Redis 简介

Redis 是使用最广泛的 NoSQL 数据库，由意大利人 Salvatore Sanfilippo（网名为 Antirez）开发。这是一款开源的内存高速缓存数据库，采用 key-value 存储模式，可构建高性能、可扩展的 Web 应用程序解决方案。

提示：NoSQL 数据库。

传统数据库开发由于需要考虑 ACID（即原子性、一致性、隔离性和持久性）原则，数据库访问性能会受到影响。NoSQL（Not Only SQL，不仅仅是 SQL）数据库不受 ACID 原则控制，不使用 SQL 标准处理，因此具有采用数据集存储、动态结构定义、存储精简、可扩展性强、适合云计算等特点。

Redis 每秒可以执行大约 110000 个设置操作、81000 个读取操作。它支持丰富的数据类型（如 string、list、set、zset、hash、GEO 等），可以实现数据的持久化存储，能有效防止主机故障时数据丢失问题。如果想获得 NoSQL 数据库，读者可以登录 Redis 官网（https://redis.io/）下载，如图 10-1 所示。

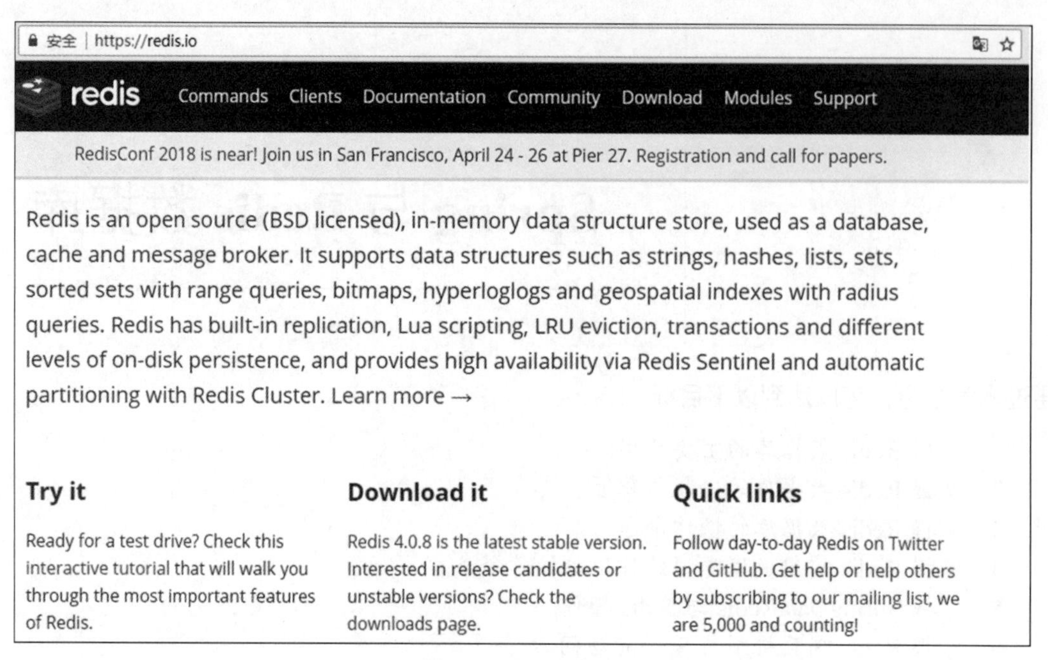

图 10-1　Redis 官网

10.2　Redis 安装与配置

Redis 数据库由 C 语言编写，官方提供的安装包为程序源代码，开发者需要手动进行编译和安装。本书使用 Redis-4.0.8 版本，并基于 Ubuntu（Linux）系统进行安装与配置。具体的安装与配置过程如下。

1.【redis-single 主机】Redis 需要进行 C 程序编译，为了可以正常编译程序源代码，需确保系统中已经配置了如下程序支持库与应用程序：

```
apt-get -y install g++ gcc libpcre3 libpcrecpp* libpcre3-dev libssl-dev autoconf automake libtool
nfs-kernel-server libncurses5-dev make
```

2.【redis-single 主机】将 Redis 程序包（redis-4.0.8.tar.gz）通过 FTP 上传到 Linux 系统中（/srv/ftp 目录下），随后将其解压缩到/usr/local/src 目录下。

```
tar xzvf /srv/ftp/redis-4.0.8.tar.gz -C /usr/local/src/
```

3.【redis-single 主机】源代码解压缩后，可得到/usr/local/src/redis-4.0.8 目录，按照如下步骤进行编译。

（1）进入到源代码目录 cd /usr/local/src/redis-4.0.8/。

（2）对源代码进行编译：make。

（3）安装 Redis：make install。

4.【redis-single 主机】Redis 的主要功能是进行数据的缓存，一旦接触到缓存，往往要将主机所有的可用资源都交给 Redis 控制，所以建议做一个内存的分配策略配置。

☑　将所有内存交由应用程序管理：echo "vm.overcommit_memory=1" >> /etc/sysctl.conf。

☑　将配置写入到 Linux 内核中：/sbin/sysctl -p。

5.【redis-single 主机】创建/usr/local/redis 目录，在其中设置 bin 子目录，以保存编译后的可执行程序。同时，建立 conf 子目录，用于保存 Redis 配置文件。

```
mkdir -p /usr/local/redis/{bin,conf}
```

6.【redis-single 主机】通过 Redis 源代码目录，复制 Redis 的相关命令与配置文件模板。

```
cp /usr/local/src/redis-4.0.8/src/redis-server /usr/local/redis/bin/
cp /usr/local/src/redis-4.0.8/src/redis-cli /usr/local/redis/bin/
cp /usr/local/src/redis-4.0.8/src/redis-benchmark /usr/local/redis/bin/
cp /usr/local/src/redis-4.0.8/redis.conf /usr/local/redis/conf/
```

7.【redis-single 主机】进行 Redis 服务配置，创建相关的数据文件保存目录。

```
mkdir -p /usr/data/redis/{run,logs,dbcache}
```

run 子目录存放的是进程 id，logs 子目录存放的是日志信息，dbcache 子目录存放的是数据信息。

8.【redis-single 主机】利用 vi 打开 Redis 配置文件。

```
vim /usr/local/redis/conf/redis.conf
```

9.【redis-single 主机】修改 redis.conf 配置文件，修改内容如下。

1	设置Redis运行端口	port 6379
2	让Redis在后台运行	daemonize yes
3	Redis运行的进程编号保存路径	pidfile /usr/data/redis/run/redis_6379.pid
4	Redis日志文件	logfile "/usr/data/redis/logs/redis_6379.log"
5	数据文件保存目录	dir /usr/data/redis/dbcache

本配置将 Redis 设置为后台运行。如果不设置此项，则启动时会出现 Redis 启动界面信息。

10.【redis-single 主机】启动 Redis 服务，设置启动的配置文件路径。

```
/usr/local/redis/bin/redis-server /usr/local/redis/conf/redis.conf
```

11.【redis-single 主机】通过 Redis 客户端连接 Redis 服务器。

默认连接本机的Redis	/usr/local/redis/bin/redis-cli
设置Redis连接主机信息	/usr/local/redis/bin/redis-cli -h 127.0.0.1 -p 6379

12.【redis-single 主机】Redis 客户端处理。

设置数据	set mldn java
取得数据	get mldn
查看所有数据key	keys *

13.【redis-single 主机】除了可使用 Redis 客户端进行数据库测试外，也可以利用 Redis 提供的性能测试工具 redis-benchmark 测试 Redis 的使用。

```
/usr/local/redis/bin/redis-benchmark -h 127.0.0.1 -p 6379 -n 10000 -d 30 -c 2000
```

本测试中，各项配置参数说明如下。

☑ -n：各连接用户发出的请求数量。

☑ -c：模拟的访问客户端。

☑ -d：每次请求发出的数据长度大小。

10.3 Redis 数据操作

Redis 作为最常用的高速缓存数据库，除了其访问性能高之外，还支持有各种常见数据类型：字符串（string）、哈希（hash）、列表（list）、无重复集合（set）、有序集合（zset）、地理位置（GEO）。

10.3.1 string 数据类型

string 是 Redis 中常用的基本数据类型，可以包含任何二进制数据（图片、对象等），一个 string 存储单元最大可以存储 512MB 的数据。数据操作命令如表 10-1 所示。

表 10-1 string 型数据

编号	命令	描述
1	set key VALUE	设置一组数据，重复设置相同key会覆盖
2	get key	根据key获取对应数据
3	setnx key VALUE	不覆盖设置数据
4	setex key TIME VALUE	设置数据有效时间（单位：s）
5	mset key1 VALUE1 key2 VALUE2 …	设置多个数据
6	msetnx key1 VALUE1 key2 VALUE2 …	不覆盖设置多个数据
7	append key 追加内容	为指定key追加数据内容
8	strlen key	获取指定key对应的内容长度
9	del key key key	删除指定key的信息

Redis 是 key-value 集合的存储结构，主要使用 set 指令设置数据。如果数据设置成功，则返回 OK。

范例：设置数据。

```
set mldn hello
```

如果设置的 key 相同，则会用新数据内容替换掉旧数据内容。对于要设置的字符串，可以使用 JSON 结构进行复杂数据定义。

范例：设置 JSON 数据。

```
set user-mldn {name:smith,age:10}
```

要获取设置数据的内容，可以利用 get 命令，根据指定的 key 进行获取。

范例：取得一个 key 对应的数据内容。

```
get mldn
```

在实际项目开发中，使用 keys 命令可以查看 Redis 数据库保存的全部 key 信息。该操作可以利用通配符“*”进行匹配。

范例：使用 keys 获取全部的 key 数据信息。

获取全部key	keys *
获取指定关键字的全部key	keys user-*

为了方便数据进行分类存储，Redis 中一共设置了 16 个数据库，使用“select 数据库序号”指令可以进行切换。要进行数据库清空操作，需要考虑以下两种情况。

☑ 清空当前数据库中的数据：flushdb。

☑ 清空全部数据库中的数据：flushall。

默认状态下，如果设置的 key 相同，一定会发生数据覆盖。为了防止误覆盖操作发生，Redis 里提供了 setnx 命令。该操作会返回如下两个信息：

☑ 返回数字 1，表示操作成功。C 语言里，非 0 是 true。

☑ 返回数字 0，表示操作失败。0 是 false。

范例：不覆盖设置数据。

```
setnx user-mldn lixinghua
```

Redis 数据库还有一个重要功能，就是可以设置 key 的保存时间。时间一到，自动将此信息删除。实际项目中经常出现的短信验证码功能，就可以利用 Redis 这一特点来实现。

范例：设置 key 的保存时间。

```
setex user-mldn 10 jixianit
```

本程序设置的 user-mldn 的 key 信息要求在 10s 内进行读取，时间一到将会自动删除。

范例：设置多个数据信息。

```
mset user-mldn-name java user-mldn-address beijing
```

本命令设置了两个 key：user-mldn-name 和 user-mldn-addreess。如果设置的 key 重复，则会发生覆盖情况。如果不希望产生数据覆盖问题，可以使用 msetnx 命令处理。

范例：不覆盖，设置多个数据信息。

```
msetnx user-mldn-name hello user-mldn-sex male
```

如果此时某个 key 产生了重复，那么所有的数据都不会再进行设置。

对于已经保存的数据，可以使用 append 指令进行数据追加。

范例：追加指定 key 的数据。

```
append user-mldn-name world
```

由于 user-mldn-name 的 key 已经存在，因此这里会在已有内容上追加新数据。如果 key 不存在，则相当于使用 set 命令设置新数据。

范例：获取数据长度。

```
strlen user-mldn-name
```

如果不再需要某些数据，则可以使用 del 命令进行删除。del 命令可以删除多个 key。

范例：删除数据。

```
del user-mldn-name user-mldn-sex
```

本程序将删除 user-mldn-name、user-mldn-sex 两个 key 的信息。

10.3.2 hash 数据类型

使用字符串数据类型时，要存储一个对象信息，通常需要设置许多 key 信息才能够完成。针对一组数据，Redis 提供了 hash 数据类型，可以将一组信息统一保存在一个 key 中。每个 hash 可以存储 $2^{32}-1$ 键值对（40 多亿）的数据量。

hash 数据类型的操作命令如表 10-2 所示。

表 10-2　hash 数据类型

编号	命令	描述
1	hset 对象 key 属性 key 内容	存放 hash 数据
2	hget 对象 key 属性 key	取得 hash 数据
3	hsetnx 对象 key 属性 key 内容	不覆盖设置
4	hmset 对象 key 属性 key1 内容 1 属性 key2 内容 2 …	批量设置
5	hmget 对象 key 属性 key1 属性 key2 …	获取对象的所有 key
6	hexists 对象 key 属性 key	判断某个数据是否存在
7	hlen key	取得全部内容数量
8	hdel 对象 key 属性 key …	删除指定 key 信息
9	hkeys 对象 key	取得所有 key
10	hvals 对象 key	取得 hash 中所有内容
11	hgetall 对象 key	获得全部的 key 与 value

hash 数据类型的最大特点是：可以在一个 key 中设置若干个属性 key 数据，因此可保存丰富的数据内容。

范例：设置 hash 数据。

```
hset user-mldn name hello
hset user-mldn addr beijing
hset user-mldn age 16
```

本程序在 user-mldn 的 key 中设置了 3 个属性 key 内容：name、addr 和 age。

范例：获取指定 key 的数据信息。

```
hget user-mldn name
```

正常情况下，直接使用 hset 进行内容设置，将产生数据覆盖问题。下面看看如何不出现数据覆盖问题。

范例：不覆盖处理。

```
hsetnx user-mldn name world
```

由于 name 的属性 key 已经存在，所以此时返回了数据 0（表示设置失败）。

范例：批量设置数据。

```
hmset user-admin name adminstrator age 10
```

本程序在 user-admin 的 key 中设置了 name、age 两个属性 key，获取数据时也同样需要设置 key 与属性 key 的名称。

范例：获取指定 key 的信息。

```
hmget user-mldn name age addr
```

本程序获取了 user-mldn 中的 3 个属性 key 内容——name、age 和 addr，因此返回结果会包含 3 个数据内容。如果获取时指定属性 key 不存在，将返回 nil（null）。

范例：判断某一个 key 是否存在。

```
hexists user-mldn name
```

如果此时指定的 key 存在，则返回 1。如果指定的 key 不存在，则返回 0。

在 hash 类型中一个 key 往往对应若干个属性 key，如果想知道全部的属性 key 数量，可以使用 hlen 命令获取。

范例：获取指定 key 中属性 key 的数量。

```
hlen user-mldn
```

范例：删除 key 中指定的属性 key。

```
hdel user-mldn name
```

范例：获取指定 hash 数据中的全部属性 key 信息。

```
hkeys user-mldn
```

范例：获取指定 hash 数据中的全部数据。

```
hvals user-mldn
```

范例：获取指定 hash 数据中的全部属性 key 与 value。

```
hgetall user-mldn
```

此时数据将按照“属性 key”与“属性 value”的形式顺序返回。

10.3.3　数字操作

在 string 或 hash 中都可以实现数字内容的保存，也可以实现数字的处理操作，支持的命令如表 10-3 所示。

表 10-3　数字操作命令

编号	命令	数据类型	描述
1	incr key	string	自增处理

续表

编号	命令	数据类型	描述
2	incrby key 数字	string	自增指定数据
3	decr key	string	自减处理
4	decrby key	string	自减指定数据
5	hincrby 对象 key 数字	hash	自增指定数据

范例：自增数据。

```
incr num
```

如果 Redis 中已经存在名称为 num 的 key，则会在已有数据上实现自增处理，并返回自增后的结果。如果 key 不存在，则会自动创建指定的 key，并返回 1。重复操作，会继续在已有 num 上实现自增。

默认状态下使用 incr 增长，步长为 1。如果用户需要自定义增长步长，可以使用 incrby 命令完成。

范例：自定义自增步长。

```
incrby num 20
```

本程序将在 num 已有数据的基础上增长 20 个内容，同时返回增长后的数据结果。

范例：自减处理。

```
decr num
```

本程序在已有 num 上实现了自减处理操作，默认每一次自减为 1。如果 key 不存在，则会创建一个新的数据，并且内容设置为 0，处理的结果为-1。

范例：自减指定步长。

```
decrby num 50
```

以上都是在 string 数据类型上实现的自增与自减处理操作。如果存放数据的类型为 hash，也可以实现自增处理。

范例：在 hash 类型上实现自增处理。

```
hincrby shopcar-mldn goods-a 1
```

本程序设置了一个 shopcar-mldn 的 key，针对属性 key goods-a 实现了自增处理。如果设置的数据为负数，则实现自减处理。

提示：可在购物车中使用 hash 数据处理。

高并发程序访问处理下，想要实现购物车操作，可以利用 Redis 缓存数据库来实现。实现方式是采用 hash 类型。

☑ hash 数据 key：shopcar-用户名（如用户名为 mldn，则 key 为 shopcar-mldn）。

☑ 可在指定 key 中保存若干个属性 key，每个属性 key 保存一个商品编号，而后实现指定商品购买数量的控制。

10.3.4　list 数据类型

list 数据实现的是双端数据队列的概念，所保存的数据可从队头（右边）或队尾（左边）实现数据的设置与取得。Redis 中，每一个 list 集合可以保存 $2^{32}-1$ 个元素（超过 40 亿个元素），常见的数据操作命令如表 10-4 所示。

表 10-4　list 数据操作命令

编号	命令	描述
1	lpush list 集合 key 内容 ...	向 list 队尾存放数据
2	lrange 集合 key 头部索引 尾部索引	取得指定索引位置的内容。如果设置为-1，表示获取全部
3	rpush 集合 key 内容 ...	向 list 队头存放元素
4	linsert 集合 key before 内容 1 增加内容	在指定元素前追加内容
5	lset 集合 key 索引 内容	修改指定索引的内容
6	lrem 集合 key 内容 重复个数 删除内容	从队尾删除数据
7	ltrim 集合 key 开始索引 结束索引	保留指定 key 值范围内的数据
8	lpop 集合 key	从指定集合的尾部删除元素，并返回删除元素
9	rpoplpush 集合 key 集合 key	将移除的元素添加到指定的集合中
10	lindex 集合 key 索引	取得元素指定索引的内容
11	llen 集合 key	返回集合中的元素个数

操作 list 集合时，实际上维持的是一个双端队列处理，队列分为队尾（左边 left，命令以 l 开头）与对头（右边 right，命令以 r 开头）。

范例：向队列中保存数据。

```
lpush mldn-query user-a user-b user-c user-d user-e
```

此时向 mldn-query 集合中保存了 5 个数据，并且按照保存顺序从队尾依次追加。命令执行后，会返回保存的数据个数。此时的 list 数据结构如图 10-2 所示。

图 10-2　list 集合

范例：向队头追加数据。

```
rpush mldn-query new-a new-b new-c
```

这里向 mldn-query 集合的头部追加了 3 个数据，因此集合中存放了 8 个数据，list 结构如图 10-3 所示。

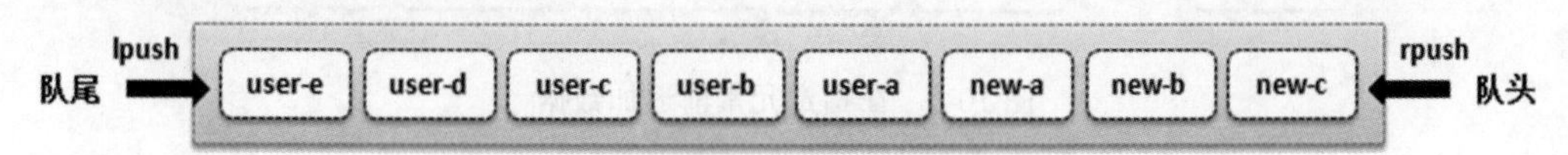

图 10-3　list 集合

设置完成的 list 集合数据可以直接利用 lrange 命令获取，获取时需要设置范围索引。

范例：获取指定范围的数据。

```
lrange mldn-query 0 -1
```

在 mldn-query 中一共保存了 8 个数据，所以范围设置为 0～7。为了方便起见，这里直接将尾部索引设置为了-1，这样就可以实现全部数据列出。

范例：从 list 中弹出数据。

队尾弹出	lpop mldn-query，弹出后返回队尾弹出的数据user-e，此时队列如图10-4所示
队头弹出	rpop mldn-query，弹出后返回对头弹出的数据new-c，此时队列如图10-5所示

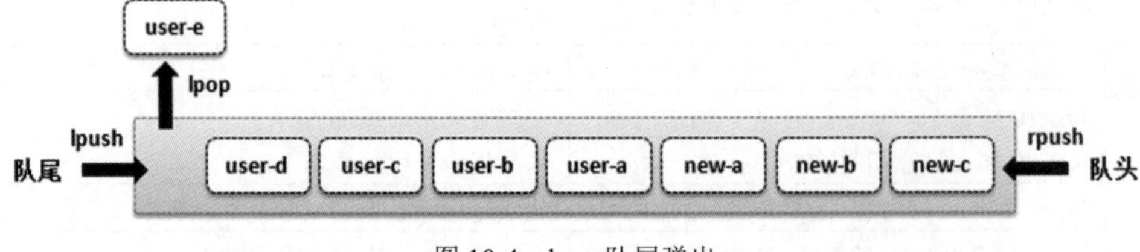

图 10-4　lpop 队尾弹出

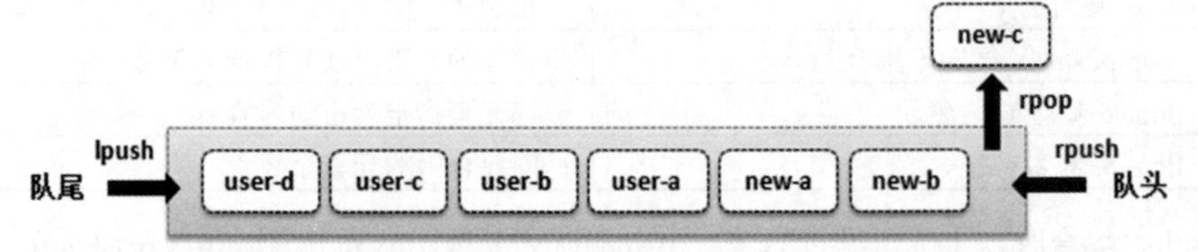

图 10-5　rpop 队头弹出

提示：list 集合也被称为栈操作。

list 集合的特点非常类似于传统的双端队列，很多书中将其解释为栈操作。例如，lpush 表示栈底保存，rpush 表示栈顶保存。本书为了讲解方便，以队列的形式介绍。

范例：在指定元素前追加数据。

```
linsert mldn-query before user-c user-admin
```

本程序在 user-c 数据前追加了 user-admin 数据，追加完成后的 list 集合数据如图 10-6 所示。

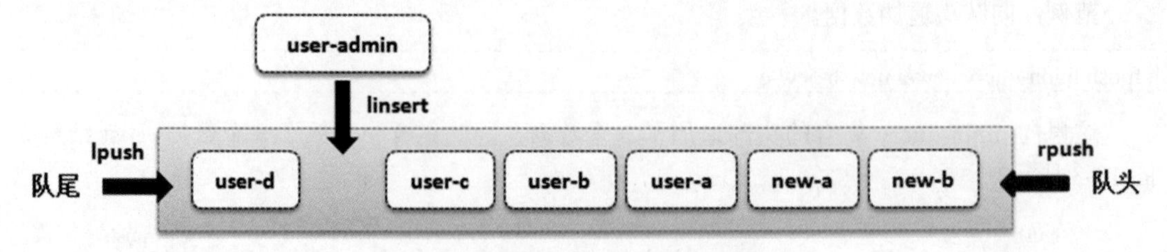

图 10-6　在指定元素前追加数据

范例：在指定元素后追加数据。

```
linsert mldn-query after new-a new-guest
```

本程序在 new-a 元素后追加了 new-guest 数据，追加完成后的 list 集合如图 10-7 所示。

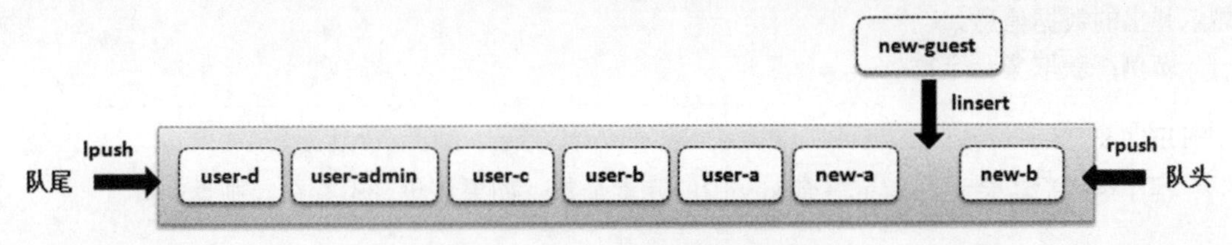

图 10-7　在指定元素后追加数据

范例：修改指定索引数据。

```
lset mldn-query 0 user-lee
```

本程序修改了索引为 0 的元素（即队尾第一个元素），而后保存数据。此时的 list 集合如图 10-8 所示。

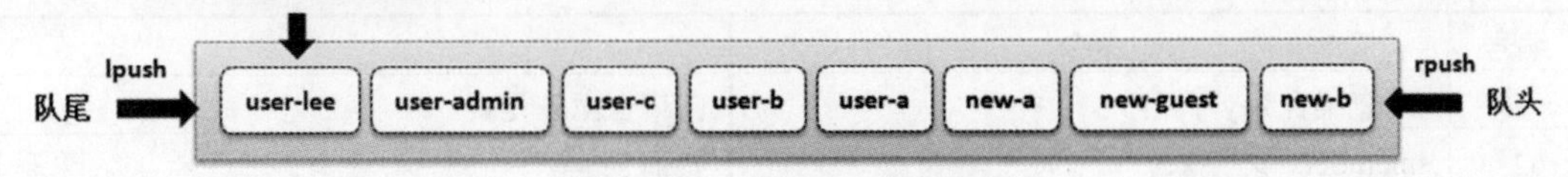

图 10-8　修改指定索引的数据

list 集合中允许存放重复元素。

范例：设置重复元素。

```
lpush mldn-query user-test user-test user-test
```

本程序为 mldn-query 设置了 3 个重复的数据内容，并按照顺序依次添加到了队尾。

如果某些数据不再需要，可以使用 lrem 命令删除，同时可指明删除次数。

范例：删除重复的 user-test 数据 3 次。

```
lrem mldn-query 3 user-test
```

除了可以根据数据内容删除之外，也可以设置数据的保留索引范围（不删除）。

范例：保留集合索引的 3～6 数据信息。

```
ltrim mldn-query 3 6
```

本程序执行后，list 集合中仅保存了索引 3～6 的数据信息，如图 10-9 所示。

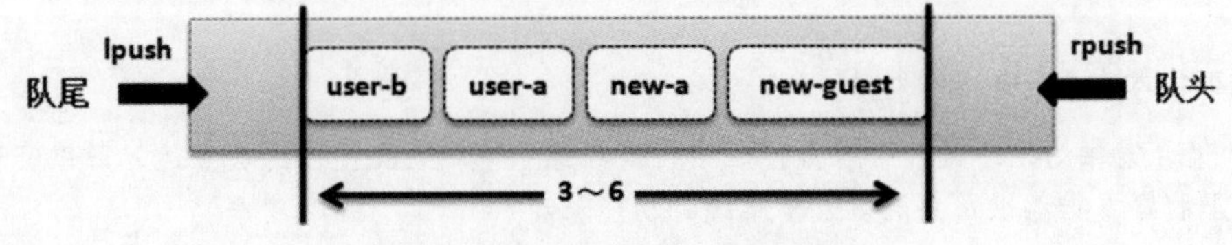

图 10-9　保留索引范围数据

范例：弹出数据，并将其保留到其他集合。

```
rpoplpush mldn-query mldn-pop
```

本程序会从队头实现数据弹出，并且将弹出的数据保存到 mldn-pop 数据集中。执行后，会显示弹出的数据信息。

范例：获取集合个数。

```
llen mldn-query
```

程序执行后，会返回指定集合 key 中的元素个数。如果该集合不存在，则返回 0。

10.3.5 set 数据类型

set 数据类型的主要特点是可以保存无重复数据集合，且属于无序存储。每个集合中最多可保存 $2^{32}-1$ 个元素（超过 40 亿个元素）。set 集合还提供了数据集合计算功能，可以进行数据比较处理。

set 数据的主要操作命令如表 10-5 所示。

表 10-5 set 数据操作命令

编号	命令	描述
1	sadd 集合 key 内容	向集合添加元素
2	smembers 集合 key	查询 set 集合
3	srem 集合 key 内容	删除集合元素
4	spop 集合 key	从集合中随机弹出一个元素
5	sdiff 集合 key1 集合 key2	返回两个集合的差集
6	sdiffstore 存储集合 key 集合 key1 集合 key2	将差集保存到另外一个集合中
7	sinter 集合 key1 集合 key2	交集计算
8	sinterstore 存储集合 集合 key1 集合 key2	将交集保存到新的集合中
9	sunion 集合 key1 集合 key2	并集计算，将两个集合合并在一起
10	sunionstore 存储集合 集合 key1 集合 key2	将并集进行存储
11	smove 集合 key1 集合 key2 第一个集合内容	从一个 key 对应的 set 中移除并添加到另外一个集合中
12	scard 集合 key	返回名称为 key 的集合个数
13	sismember 集合 key 内容	测试 member 是否是名称为 key 的 set 的元素
14	srandmember 集合 key	随机返回名称为 key 的一个集合的元素

通过表 10-5 定义的命令可以发现，Redis 提供的 set 数据类型可以实现集合的交、差、并等数据计算，也可以对集合计算的结果进行存储。

提示：关于 set 集合的使用。

set 集合提供的无序集合可以直接应用在抢红包系统中。set 提供的集合运算可以实现微博或 QQ 上的好友推荐以及共同好友信息统计功能。

范例：定义两个 set 集合，分别实现数据保存。

保存第一个set集合	sadd user-mldn a b c d e f f
保存第二个set集合	sadd user-admin a c d x y z e e

本程序保存了两个 set 数据，且相互有重复内容。由于 set 集合不存放重复数据，所以这些重复数据只会被保存一次。

范例：查看集合中的数据。

操作功能	操作命令	返回结果
查看**user-mldn**全部数据	smembers user-mldn	c、d、e、f、a、b
查看**user-admin**全部数据	smembers user-admin	c、d、e、z、y、x、a

通过返回结果可以发现，set 数据类型是无序保存的，且没有重复数据。

范例：随机弹出数据。

```
spop user-admin
```

执行命令后，将会从 user-admin 集合中随机弹出一个数据，并显示弹出的数据内容。同时，该数据会自动从原始集合中删除。如果不希望删除返回，可以使用 srandmember 命令。

范例：不删除随机返回集合数据。

```
srandmember user-mldn
```

set 数据类型的最大特点在于，可以实现数据内容的集合处理操作（交、差、并）。下面通过具体程序代码进行演示。

范例：求两个集合间的差集。

```
sdiff user-admin user-mldn
```

此时将显示两个集合中的不同内容。如果有需要，也可以将运算结果保存到一个新集合中。

范例：将两个差集的计算结果保存到新集合中。

```
sdiffstore diff-admin-mldn user-admin user-mldn
```

范例：求两个集合间的交集。

```
sinter user-admin user-mldn
```

范例：求两个集合间的并集（合并，不显示重复数据）。

```
sunion user-admin user-mldn
```

set 数据集合还支持数据的移动处理，即可将一个集合的指定数据保存到另外一个集合中。

范例：集合数据移动。

```
smove user-admin user-mldn a
```

本程序将 user-admin 集合中的数据 a 保存到了 user-mldn 集合中，如果原 user-mldn 中包含指定内容，则不会保存。

10.3.6　zset（sorted set）数据类型

set 数据本身是无序的，但实际开发中经常需要对数据进行排序处理。例如，要实现一个新闻点击量排名或全球网站排名，就需要对用户点击量进行实时统计，而后进行排序处理。Redis 中提供的 zset 数据类型，可以实现此操作。

zset 数据类型的操作命令如表 10-6 所示。

表 10-6　zset 数据操作命令

编号	命令	描述
1	zadd 集合 key 分数 内容	追加有序集合数据
2	zrange 集合 key 开始索引 结束索引 withscores	取出有序集合内容，结束索引设置为-1 表示全部取出
3	zrem 集合 key 内容	删除集合数据
4	zincrby 集合 key 分数 内容	数据增长（针对于顺序增加）
5	zrank 集合 key 内容	返回集合中指定元素的索引数值
6	zrevrank 集合 key 内容	反转数据索引
7	zrevrange 集合 开始分数 结束分数 withscores	反转后取得数据
8	zrangebyscore 集合 开始分数 结束分数 withscores	根据分数取得指定范围的数据
9	zcount 集合 key 开始分数 结束分数	取得集合中指定分数范围的数量
10	zcard 集合 key	取得指定集合中的元素个数
11	zremrangebyrank 集合 key 索引开始 结束索引	根据下标排序，删除指定范围中的数据

范例：设置某日的热门词汇信息。

```
zadd keyword-20191010 5.0 mldn
zadd keyword-20191010 2.0 springboot
zadd keyword-20191010 3.0 kafka
```

本命令实现了一个简单的搜索关键词数据定义，每个关键词都有一个分数，这个分数决定了集合的顺序。

范例：取出所有热门关键词。

```
zrange keyword-20191010 0 -1
```

此时返回的数据将按照分数保存的升序进行排列，只显示关键词信息。如果想连同分数一起显示，可在命令的最后使用 withscores 选项。

范例：取出所有热门关键词，并显示分数。

```
zrange keyword-20191010 0 -1 withscores
```

在实际的关键词处理中，需要实时进行分数增长处理，为此可以使用 zincrby 命令完成增长。

范例：将 mldn 关键词的分数增长 5.0。

```
zincrby keyword-20191010 5.0 mldn
```

本程序命令执行后，将返回指定关键字增长后的分数。如果设置的数值不存在，则创建一个新的数据与分数。

默认情况下，所有的数据都会按照升序进行排列显示。使用 zrevrange 命令，可以实现数据的降序排列。

范例：降序排列关键词，取出搜索率最高的前三个关键词（索引从 0 开始）。

```
zrevrange keyword-20191010 0 2 withscores
```

本程序将会把分数排名在前三的关键词信息全部取出，然后按照降序进行排列。

10.3.7　GEO 数据类型

Redis 从 3.2 开始，支持 GEO（地理位置）数据类型的操作。利用地理数据，可以实现标志物数据存储或者以当前坐标为主查询周围建筑物的功能，操作命令如表 10-7 所示。

表 10-7　GEO 操作命令

编号	命令	描述
1	geoadd key 经度 纬度 名称 [经度 纬度 名称]	设置坐标与标记信息
2	geodist key 名称 1 名称 2 [距离单位]	计算两个地理名称间的位置。如果某个位置不存在，将返回 nil（null）。距离单位可以是 m（米）、km（千米）、mi（英里）、ft（英尺）
3	geopos key 名称 ...	获取指定名称的坐标信息
4	geohash key 名称 ...	将指定名称坐标转换为 GEOHash 编码
5	georadius key 经度 纬度 半径 [单位] [WITHDIST] [WITHCOORD] [WITHHASH] [COUNT 个数] [ASC | DESC]	根据指定坐标，找出半径范围内的标志物。 ① WITHDIST：返回坐标之间的距离。 ② WITHCOORD：返回标志物的经纬度。 ③ WITHHASH：以 GEOHash 格式返回。 ④ COUNT：设置返回记录个数。 ⑤ ASC | DESC：数据排序模式
6	georadiusbymember key 名称信息半径 [单位] [WITHDIST] [WITHCOORD] [WITHHASH] [COUNT 个数] [ASC | DESC]	以指定标志物名称为中心，查找指定半径范围内的其他标志物信息

为了方便理解，本次将采用如下几个坐标，其在地图上的位置如图 10-10 所示。

- ☑ 天安门城楼坐标：经度（116.403963）、纬度（39.915119）。
- ☑ 王府井坐标：经度（116.417876）、纬度（39.915411）。
- ☑ 前门大街坐标：经度（116.404354）、纬度（39.904748）。
- ☑ 我的坐标：经度（116.415901）、纬度（39.914805）

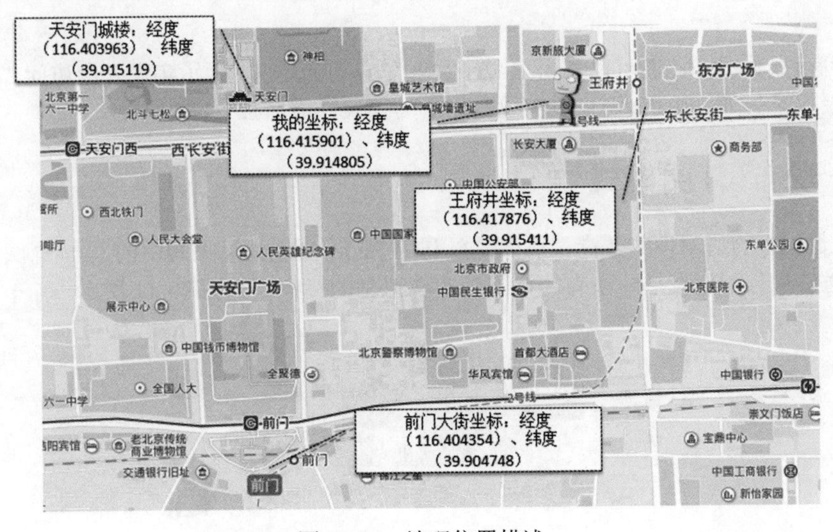

图 10-10　地理位置描述

范例：向数据库中保存 3 个坐标位置。

```
geoadd point 116.403963 39.915119 tiananmen 116.417876 39.915411 wangfujing 116.404354 39.904748
qianmen
```

此时在数据库中保存了 3 个坐标位置信息，后面这些数据可以作为基础查询使用。

范例：取得某一个标志位的坐标。

```
geopos point tiananmen
```

在坐标操作中，可以通过两个标记名称信息，计算出两者间的距离。

范例：计算两个坐标间的位置。

```
geodist point tiananmen qianmen m
```

在实际地理位置查询中，用户更关注的是能否根据当前手机定位，获取周边的一些关键信息。为此，GEO 操作中提供了 georadius 命令，使用此命令可以设置查询半径，显示此范围内的全部数据。

范例：根据当前坐标，获取半径 1000m 内的标志物，并且按照距离降序返回，每次获取 2 个数据信息。

```
georadius point 116.415901 39.914805 1000 m withdist desc count 2
```

本程序根据当前位置查询了周围 1000m 内的标志名称信息，并按照距离降序排列返回了两个查询结果。

除了可以使用坐标位置外，也可以利用已有的标记名称实现半径查询。

范例：以指定标志物为参照，获取半径范围内的其他标志物信息。

```
georadiusbymember point tiananmen 1500 m withdist asc
```

本程序实现了 tiananmen 名称半径 1500m 周围的标志物名称查询。

10.4 Redis 高级配置

除了可存放数据外，还可以利用 Redis 实现发布-订阅处理。为了实现安全访问，Redis 可以进行认证授权。另外，Redis 还可以实现数据库的相关性能监控。

10.4.1 发布-订阅模式

Redis 数据库的访问性能较高，所以可以利用 Redis 实现消息系统中的发布-订阅处理模式，即一个发送者进行消息发送，所有的订阅者实现消息的接收，如图 10-11 所示。

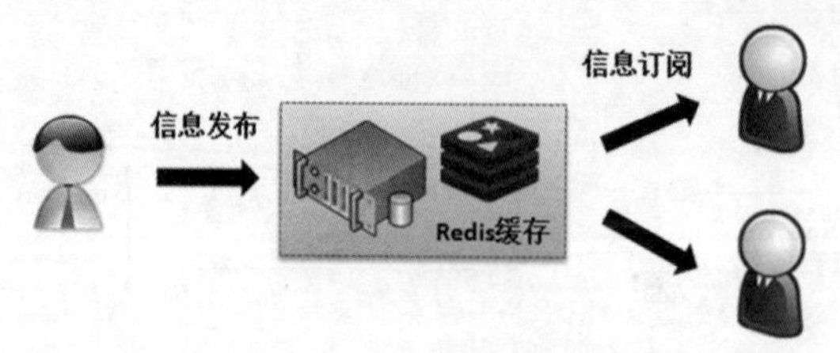

图 10-11　发布-订阅模式

1.【Redis 客户端-A】开启订阅模式，同时设置要订阅数据的通道（可以设置多个通道）。

```
subscribe mldn-channel
```

命令执行后将进入到数据监听状态，等待消息发送。

2.【Redis 客户端-B】开启发布者模式，将消息发布到指定的通道中。

```
publish mldn-channel www.mldn.cn
```

由于发布者与订阅者都通过 mldn-channel 发送和接收信息，所以此时可以实现消息传输。

10.4.2　事务处理

Redis 虽然属于 NoSQL 数据库，但是考虑到自身数据更新的问题也支持事务处理。Redis 提供 3 个用于进行事务控制的命令：multi（开启事务支持）、discard（取消事务支持）、exec（提交事务），可以按照如下步骤操作。

1.【Redis 客户端】在 Redis 之中随意设置一个数据。

```
set age 16
```

2.【Redis 客户端】开启事务支持。

```
multi
```

3.【Redis 客户端】修改 age 属性内容。

```
set age 18
```

此时并不会立即执行此命令，而是将此命令追加到了执行队列中，所以会返回 QUEUED 的信息。

4.【Redis 客户端】提交执行事务。

```
exec
```

执行完本命令后，在事务中所做出的修改将会被执行。如果现在发现修改有问题，则可以执行 discard 命令取消本次事务的更新处理操作。

提示：Redis 事务支持问题。

开启事务后，里面进行的所有操作都是不做检查的。也就是说，所有的数据修改过程都是暂时将执行命令加入到队列中，在提交事务时才会一次性执行。如果此时按照如下操作，将会出现问题。

【Redis 客户端】随意设置一个数据：set hotkey mldn。

【Redis 客户端】开启事务支持：multi。

【Redis 客户端】自增长 hotkey 的内容：incr hotkey，字符串无法增长，此处有错误。

【Redis 客户端】设置一个正确的修改：set age 18。

【Redis 客户端】提交事务：exec。

追加执行队列时即便有操作错误，也不会做任何语法检查，但是在执行时会报错。如果现在某一个更新出现了错误，那么其余的更新不会受到影响。

10.4.3 乐观锁

乐观锁是保证数据并发访问时能正确更新的一种技术手段。其操作是基于版本号进行的，即取得数据时会获取一个版本号，更新数据时，当前数据的版本号必须与取得数据的版本号相一致，方可正常执行。如果版本号不一致，则无法进行更新。Redis 支持乐观锁处理，下面通过具体操作进行演示。

1.【Redis 客户端-A】执行如下数据操作：

☑ 设置一个新的数据。

```
set age 16
```

☑ 对该数据进行监听（数据要进行修改）。

```
watch age
```

☑ 开启事务支持。

```
multi
```

2.【Redis 客户端-B】在第一个客户端未进行数据修改前，利用第二个客户端进行更新操作。

```
set age 30000
```

数据更新后，age 的版本号已经发生改变，则第一个 Session 获取的版本号与当前版本号不相符。

3.【Redis 客户端-A】此时第一个客户端还处于事务中，修改 age 数据并且进行事务提交。

☑ 修改 age 数据内容，此时不会立即修改，而是会追加到修改队列中。

```
set age 12
```

☑ 提交事务。

```
exec
```

此时返回的结果为(nil)（null），说明更新操作并没有执行。因为现在的数据版本编号和之前读取时的版本编号不同（Redis 自动实现处理），所以无法进行正确的更新处理。

10.4.4 安全认证

数据库中保存的都是重要的资源信息，为了保证资源的安全，一定要为数据库提供安全认证机制。在 Redis 数据库里面并没有涉及复杂认证处理，开发者只需要配置一个认证密码即可实现安全访问。

1.【redis-single 主机】利用 vi 打开 Redis 配置文件。

```
vim /usr/local/redis/conf/redis.conf
```

2.【redis-single 主机】配置文件修改如下。

设置密码	requirepass mldnjava
开启远程访问	# bind 127.0.0.1

如果不注释掉 bind 配置，则 Redis 服务只允许本机进行访问。实际开发中，Redis 肯定要通过远程模式进行访问，所以不注释掉就将其修改为当前主机的 IP 地址或者是当前主机名称。

3.【redis-single 主机】结束当前 Redis 服务进程。

```
killall redis-server
```

4.【redis-single 主机】重新启动 Redis 进程。

```
/usr/local/redis/bin/redis-server /usr/local/redis/conf/redis.conf
```

5.【redis-single 主机】通过客户端连接 Redis。

```
/usr/local/redis/bin/redis-cli -h redis-single -p 6379
```

6.【redis-single 主机】此时的客户端连接操作与之前相同，可以直接登录，但在进行数据操作时会出现 (error) NOAUTH Authentication required.错误提示信息，需要认证后才可以访问。

```
auth mldnjava
```

7.【redis-single 主机】如果觉得登录后再进行认证比较麻烦，也可以在客户端连接时直接输入认证密码。

```
/usr/local/redis/bin/redis-cli -h redis-single -p 6379 -a mldnjava
```

这样就可以直接以认证形式连接 Redis 数据库，并且可以直接进行数据操作。

10.4.5　Redis 性能监控

Redis 负责的是内存缓存处理，而关于缓存处理一个很关键的问题就是——用户需要知道当前计算机的运行性能如何。Redis 官方并没有提供有关性能监测的功能，但许多第三方爱好者开发了一些监测工具。

下面将介绍如何使用 redis-stat 工具监测计算机性能。redis-stat 是一个开源软件，可以直接通过 GitHub 下载。

1.【GitHub】下载 redis-stat 工具，下载地址为 https://github.com/junegunn/redis-stat。

2.【redis-single 主机】redis-stat 工具是用 Ruby 编写的，所以想要使用它，必须下载 Ruby 的 gems 管理程序。

```
apt-get -y install ruby ruby-dev rubygems
```

3.【redis-single 主机】此时代码保存在了 GitHub 上，所以应在本机配置好了 git 工具包。

```
apt-get -y install git
```

4.【redis-single 主机】下载 redis-stat 的工具包，下载路径为/usr/local/src 目录。

☑　进入到指定目录。

```
cd /usr/local/src
```

☑　进行 git 仓库克隆。

```
git clone https://github.com/junegunn/redis-stat.git
```

工具下载完成后，实际上这里面只有一个核心命令：/usr/local/src/redis-stat/bin/redis-stat。

5.【redis-single 主机】安装 redis-stat 程序。

☑ 进入 redis-stat 所在目录。

```
cd /usr/local/src/redis-stat/bin/
```

☑ 进行程序安装。

```
gem install redis-stat
```

6.【redis-single 主机】此时，监控程序已经安装完成。在启用前首先把 Redis 服务先启动好。

```
/usr/local/redis/bin/redis-server /usr/local/redis/conf/redis.conf
```

7.【redis-single 主机】启动监控程序。

```
/usr/local/src/redis-stat/bin/redis-stat 192.168.68.139:6379 -a mldnjava
```

此时会出现如图 10-12 所示的监控界面。

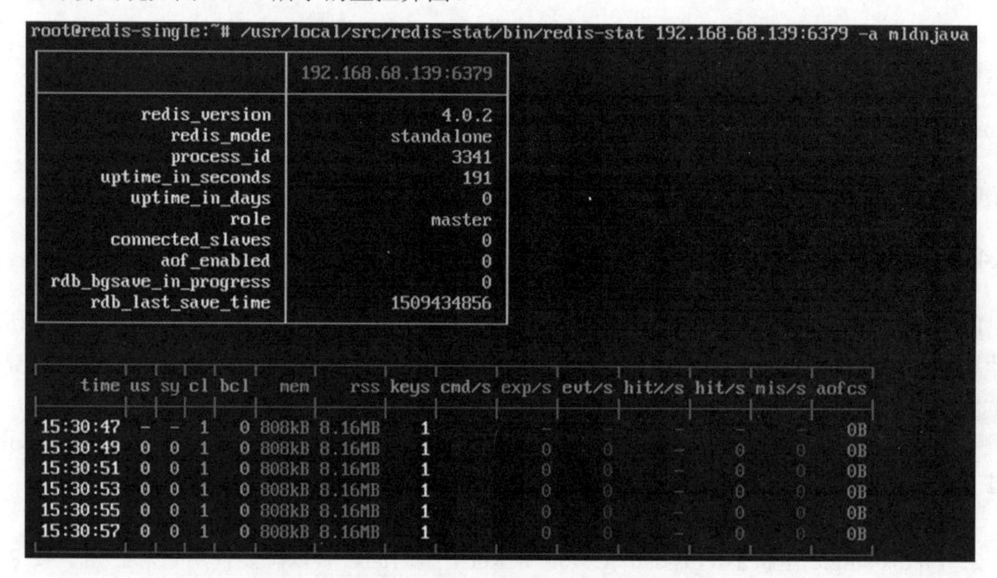

图 10-12 Redis 命令行监控界面

8.【redis-single 主机】Redis 有一个默认的测试程序。

```
/usr/local/redis/bin/redis-benchmark -h redis-single -p 6379 -n 10000 -d 30 -c 200
```

9.【redis-single 主机】redis-stat 里有一个专门的 Web 服务器，可以通过页面进行浏览。

```
/usr/local/src/redis-stat/bin/redis-stat redis-single:6379 -a mldnjava --server=80 --daemon --verbose
```

本命令将在后台启动一个 Web 服务进程，并在 80 端口上进行监听。

10.【Windows 主机】为了方便访问，修改 hosts 配置文件，追加一个主机映射。

```
192.168.68.139 redis-single
```

11.【Windows 主机】配置完成后，可通过浏览器打开 Web 监控页面（http://redis-single/），如图 10-13 所示。

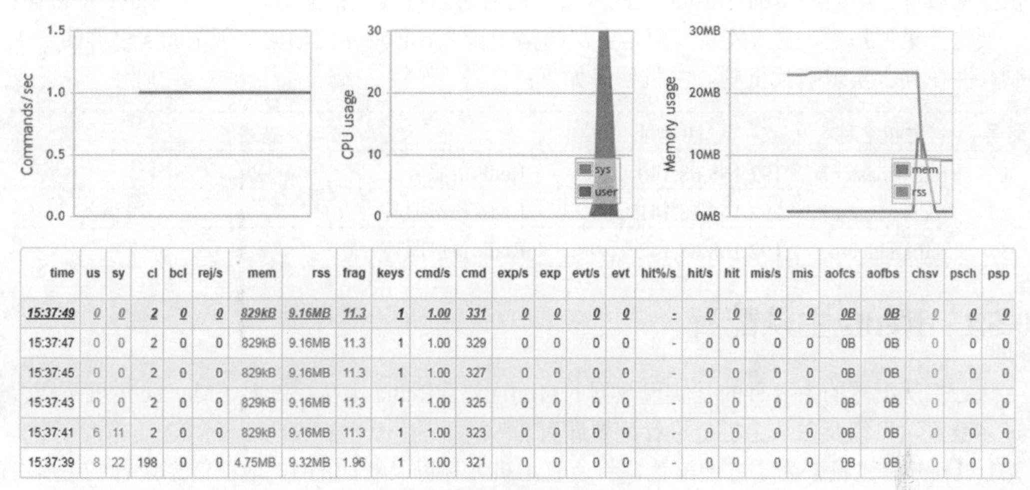

time	us	sy	cl	bcl	rej/s	mem	rss	frag	keys	cmd/s	cmd	exp/s	exp	evt/s	evt	hit%/s	hit/s	hit	mis/s	mis	aofcs	aofbs	chsv	psch	psp
15:37:49	0	0	2	0	0	829kB	9.16MB	11.3	1	1.00	331	0	0	0	0	-	0	0	0	0	0B	0B	0	0	0
15:37:47	0	0	2	0	0	829kB	9.16MB	11.3	1	1.00	329	0	0	0	0	-	0	0	0	0	0B	0B	0	0	0
15:37:45	0	0	2	0	0	829kB	9.16MB	11.3	1	1.00	327	0	0	0	0	-	0	0	0	0	0B	0B	0	0	0
15:37:43	0	0	2	0	0	829kB	9.16MB	11.3	1	1.00	325	0	0	0	0	-	0	0	0	0	0B	0B	0	0	0
15:37:41	6	11	2	0	0	829kB	9.16MB	11.3	1	1.00	323	0	0	0	0	-	0	0	0	0	0B	0B	0	0	0
15:37:39	8	22	198	0	0	4.75MB	9.32MB	1.96	1	1.00	321	0	0	0	0	-	0	0	0	0	0B	0B	0	0	0

图 10-13　Redis 性能监控

10.5　Redis 哨兵机制

Redis 数据库处理性能高，并且可进行数据的持久化处理，所以在当下的企业项目开发设计中应用的越来越多。同时，开发者必须考虑 Redis 主机出错后的数据恢复问题。Redis 默认提供有主备支持能力，并且允许自动同步，这样就可以防止单主机损坏后的数据丢失问题。

所谓“主备”，指的是一台主机可以有 N 台备份同步数据处理主机，当设置了 A 主机的内容之后，那么该内容可以自动同步到 B 主机或其他主机上。图 10-14 显示了一主二仆式的主从配置关系。

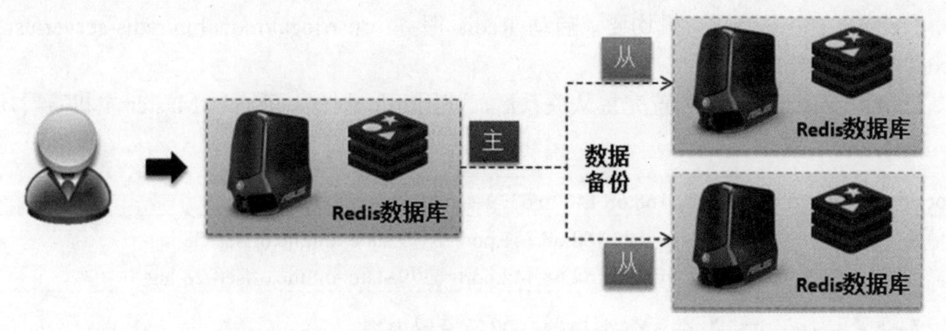

图 10-14　主从关联

利用主从机制可以实现数据的自动备份同步操作。通过 Master 主机操作时，系统会自动将所有的增加、修改、删除操作作用于全部 Slave 主机中。

使用主从配置处理后，存在一个严重的隐患：如果 Master 主机挂掉了，Redis 剩余的两台从主机也将无法正常提供服务。为了解决这个问题，Redis 提出了哨兵（Sentinel）解决方案，即设置哨兵监控进程，让若干个哨兵进程监控整体的 Redis 运行状态，当它们共同发现 Master 主机挂掉后，会根据已有的主从配置关系，在剩余的 Slave 主机中选举出新的 Master 主机，并正常提供主从服务。如果原本的 Master 主机重新启动后可以继续提供服务，将作为 Slave 主机存在。

要进行集群搭建，主从关系匹配是要做的第一步。下面将介绍如何实现 Redis 集群的一主二仆制，并在其上实现哨兵机制。主机信息如下：

编号	主机名称	IP地址	描述
1	redis-master	192.168.68.140	Redis主服务
2	redis-slave-a	192.168.68.141	Redis备份服务
3	redis-slave-b	192.168.68.142	Redis备份服务

10.5.1 Redis 主从配置

进行主从配置的主要目的是实现数据的自动同步。Redis 主从控制时，Master 主机不需要进行过多配置，只需要在从主机上进行配置即可。

1.【redis-*主机】为了方便配置，将所有主机名称以及对应的 IP 地址信息设置到了所有的主机上：vim /etc/hosts。

```
redis-master     192.168.68.140
redis-slave-a     192.168.68.141
redis-slave-b     192.168.68.142
```

2.【redis-slave-*主机】进行主从的配置处理，只需要处理从主机即可，即从主机上一定要保留主从的关系配置。例如，当前环境中有两台从主机，那么修改这两台主机的配置文件。

☑ 打开 Redis 配置文件 vim /usr/local/redis/conf/redis.conf。

☑ 在配置文件中进行如下配置。

设置跟随的Master主机地址	slaveof 192.168.68.140 6379
设置Master主机认证信息	masterauth mldnjava

3.【redis-*主机】所有主机均要求启动 Redis 服务：/usr/local/redis/bin/redis-server/usr/local/redis/conf/redis.conf。

4.【redis-master 主机】配置完主从关系后，使用 info 命令，可通过 Master 主机查看所有的副本信息。

/usr/local/redis/bin/redis-cli -h 192.168.68.140 -p 6379 -a mldnjava info replication	
副本信息	slave0:ip=192.168.68.141,port=6379,state=online,offset=28,lag=0 slave1:ip=192.168.68.142,port=6379,state=online,offset=28,lag=0

5.【redis-master 主机】进入到客户端，随后设置内容。

```
/usr/local/redis/bin/redis-cli -h 192.168.68.140 -p 6379 -a mldnjava set mldn hello
```

此时虽然是在 redis-master 主机上设置的数据，但由于存在主从关系，会自动进行数据同步。用户可以在三台主机上任意获得此数据信息。

6.【redis-slave-a 主机】获取数据信息。

```
/usr/local/redis/bin/redis-cli -h 192.168.68.141 -p 6379 -a mldnjava get mldn
```

7.【redis-master 主机】删除指定的数据。

```
/usr/local/redis/bin/redis-cli -h 192.168.68.140 -p 6379 -a mldnjava del mldn
```

Master 主机数据删除后，会影响所有的从主机数据。主从操作的特征就是以 Master 主机为核心操作用户的接入点，用户通过 Master 操作，本身不关注 Slave 的存在，它们之间自动同步处理。

10.5.2　哨兵机制

在主从关系配置的基础上，为了保证 Master 出现问题后依然可以正常提供服务支持，需要配置哨兵机制，以实现新的 Master 主机选举处理。集群环境下，可以为每一个 Redis 主机配置哨兵进程。

1.【redis-*主机】哨兵的配置文件整体是相同的，所以首先需复制哨兵配置文件的模板。在 Redis 源代码里面提供了一个哨兵配置模板，将其复制到 Redis 工作目录中。

```
cp /usr/local/src/redis-4.0.8/sentinel.conf /usr/local/redis/conf/
```

2.【redis-*主机】打开哨兵配置文件。

```
vim /usr/local/redis/conf/sentinel.conf
```

3.【redis-*主机】修改哨兵配置文件，重点进行如下几项配置。

编号	配置项	配置内容
1	关闭Redis保护模式，如不关闭，无法重新选举	protected-mode no
2	哨兵进程监听的端口号	port 26379
3	设置哨兵目录	dir /usr/data/redis/sentinel
4	设置哨兵监控	sentinel monitor mymaster 192.168.68.140 6379 2 mymaster是一个内部使用的名称，而2表示如果有2个哨兵认为Master无法使用，则在剩余主机中重新选出新的Master
5	设置Master认证信息	sentinel auth-pass mymaster mldnjava
6	设置Master不活跃的时间	sentinel down-after-milliseconds mymaster 30000
7	要同步的Master数量	sentinel parallel-syncs mymaster 1
8	选举失败的超时时间	sentinel failover-timeout mymaster 180000

4.【redis-*主机】建立哨兵配置文件的目录。

```
mkdir -p /usr/data/redis/sentinel
```

5.【redis-*主机】通过 Redis 源代码目录，复制哨兵的启动文件。

```
cp /usr/local/src/redis-4.0.8/src/redis-sentinel /usr/local/redis/bin/
```

6.【redis-*主机】启动所有的哨兵进程（前提是 Redis 进程是正常的）。

```
/usr/local/redis/bin/redis-sentinel /usr/local/redis/conf/sentinel.conf
```

通过 Master 可以获得所有的 Slave 信息，那么此时连接的就是 Master 主机，通过这台主机获得 Slave 数据。

7.【redis-*主机】下面看一下哨兵的配置信息。

☑ +sentinel sentinel：启动其他哨兵后会出现此提示信息。

☑ +slave slave：表示可通过 Master 的 info 信息获得所有的 Slave 信息。

☑ 随后直接杀死 redis-master 主机上的 Redis 服务进程：killall redis-server。

☑ 超过了 Master 的超时存活时间，哨兵会发现 Master 无法连接了，此时就需要重新投票选举：

|- +sdown master mymaster 192.168.68.140 6379：Master 主机出现问题。

|- +vote-for-leader：发出了投票新的 leader 选举。

|- +config-update-from：更新配置文件，此时 sentinel 内容会被自动修改。

|- +switch-master mymaster 192.168.68.140 6379 192.168.68.141 6379：切换了 Master。

此时就实现了新的 Master 选举，而集群状态也变为了一主一从制。虽然哨兵机制能很好地解决主从关系问题，但由于其会自动修改哨兵配置文件，并且 Master 主机也会随机切换，所以维护起来会比较麻烦。

10.6 RedisCluster 集群

RedisCluster 是从 Redis 3.0 版本之后提供的官方 Redis 集群解决方案，其可以避免哨兵机制复杂的 Master 监控与选举操作，也可以方便地实现数据的分片处理，发挥集群主机的性能优势，提供更加高效的 Redis 解决方案。

提示：Redis 集群方案。

Redis 的发展时间较长，实际上也出现过许多第三方的集群方案，如 twemproxy、codis 等，都很常见，有兴趣的读者可登录 www.mldn.cn 自行学习。本书将以官方集群实现方案为主进行讲解。

RedisCluster 在设计的时候，就考虑到了去中心化的架构模式。也就是说，集群中的每个节点都是平等的关系，保存着各自的数据和整个集群的状态。每个节点都和其他所有节点连接，因此开发者只需要连接集群中的一个节点，就可以获取到其他节点的数据，如图 10-15 所示。

Redis 集群会自动根据已有的主机分配主从关系，将所保存的数据平均分配到每一台 Master 主机中。如果某台 Master 主机因出现问题无法提供服务（例如，cluster-node-timeout 超时时间一到，则会认为此 Master 宕机），就会由其他 Master 主机参与选举，最终选举出一台新的 Master 主机继续提供服务，如图 10-16 所示。

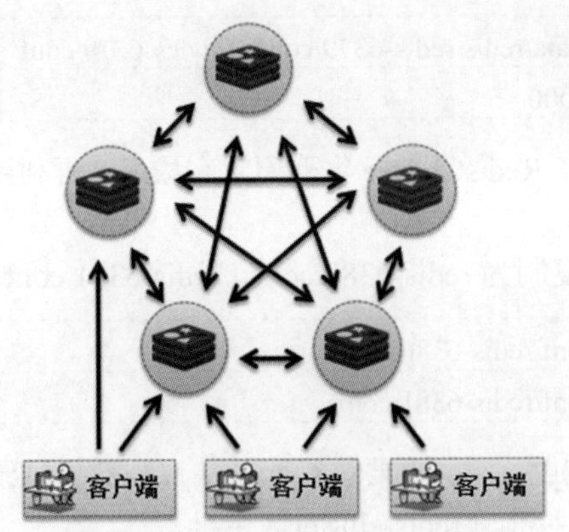

图 10-15　RedisCluster 架构

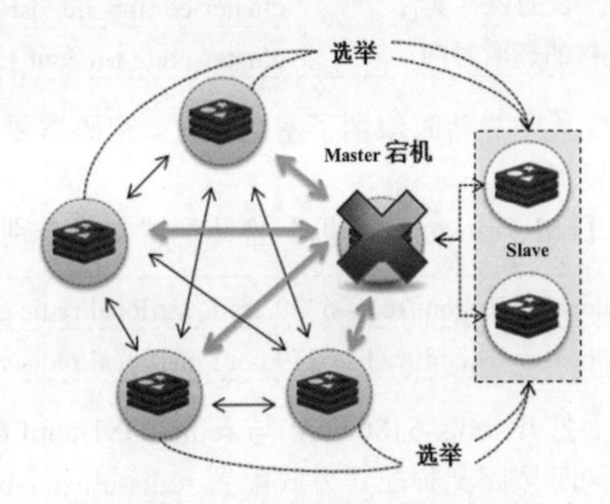

图 10-16　RedisCluster 重新选举

进行集群搭建，肯定要有多台主机。本例将采用一主二从的主备关系，使用 9 台 Redis 数据库服务器（建立 3 台，而后每台服务器上运行 3 个 Redis 进程，模拟 3 个 Redis 进程），主机列表如下。

编号	主机名称	IP地址	描述
1	redis-cluster-a	192.168.68.143	Redis主服务×3（6379、6380、6381）
2	redis-cluster-b	192.168.68.144	Redis主服务×3（6379、6380、6381）
3	redis-cluster-c	192.168.68.145	Redis主服务×3（6379、6380、6381）

1.【redis-cluster-a 主机】要进行 RedisCluster 搭建，必须提供 Ruby 的开发环境支持。首先，在系统里进行相关开发包的配置。

```
apt-get install ruby ruby-dev rubygems
```

2.【redis-cluster-a 主机】要在一台主机上运行 3 个 Redis 实例，需要建立 3 套数据目录。

```
mkdir -p /usr/data/redis/{redis-6379,redis-6380,redis-6381}/{logs,run,dbcache,config}
```

3.【redis-cluster-a 主机】将已有的 Redis 配置文件复制为 3 份，分别为 redis-6379、redis-6380 和 redis-6381。下面以 redis-6379.conf 为例，复制配置文件。

```
cp /usr/local/redis/conf/redis.conf /usr/local/redis/conf/redis-6379.conf
```

4.【redis-cluster-a 主机】利用 vi 打开配置文件。

```
vim /usr/local/redis/conf/redis-6379.conf
```

5.【redis-cluster-a 主机】配置 redis-6379.conf 配置文件。

关闭受保护模式	protected-mode no
监听端口	port 6379
pid保存目录	pidfile /usr/data/redis/redis-6379/run/redis_6379.pid
数据目录	dir /usr/data/redis/redis-6379/dbcache
日志目录	logfile "/usr/data/redis/redis-6379/logs/redis_6379.log"
取消密码配置	# requirepass mldnjava
配置开启**cluster**集群	cluster-enabled yes

定义**cluster**的保存文件	cluster-config-file /usr/data/redis/redis-6379/config/nodes-6379.conf
定义连接的超时时间	cluster-node-timeout 15000

此配置中先暂时取消了密码配置。密码需要在 RedisCluster 集群配置完成后，再利用命令重新进行设置。

6.【redis-cluster-a 主机】将此配置文件分别复制为 redis-6380.conf、redis-6381.conf。

```
cp /usr/local/redis/conf/redis-6379.conf /usr/local/redis/conf/redis-6380.conf
cp /usr/local/redis/conf/redis-6379.conf /usr/local/redis/conf/redis-6381.conf
```

而后打开 redis-6380.conf 与 redis-6381.conf 配置文件，主要进行服务端口与数据目录名称的替换。同时按照此种配置方式配置 redis-cluster-b 主机与 redis-cluster-c 主机。

7.【redis-cluster-*主机】启动每台主机的 3 个 Redis 进程。

```
/usr/local/redis/bin/redis-server /usr/local/redis/conf/redis-6379.conf
/usr/local/redis/bin/redis-server /usr/local/redis/conf/redis-6380.conf
/usr/local/redis/bin/redis-server /usr/local/redis/conf/redis-6381.conf
```

8.【redis-cluster-*主机】查看当前主机的 Redis 进程信息。

ps -ef | grep redis	
查询结果	/usr/local/redis/bin/redis-server *:6379 [cluster] /usr/local/redis/bin/redis-server *:6380 [cluster] /usr/local/redis/bin/redis-server *:6381 [cluster]

可以发现此时的 Redis 将以 cluster 集群模式运行。

9.【redis-cluster-a 主机】要使用 Ruby 管理控制 Redis，首先要安装 Redis 的相关依赖包。

```
gem install redis
```

10.【redis-cluster-a 主机】通过 Redis 源代码目录，复制集群配置程序。

```
cp /usr/local/src/redis-4.0.8/src/redis-trib.rb /usr/local/redis/bin/
```

11.【redis-cluster-a 主机】所有的 Redis 主机加入到集群配置中，采用一主二仆的主从配置模式（副本数为 2）。

```
/usr/local/redis/bin/redis-trib.rb create --replicas 2 192.168.68.143:6379 192.168.68.143:6380
192.168.68.143:6381 192.168.68.144:6379 192.168.68.144:6380 192.168.68.144:6381
192.168.68.145:6379 192.168.68.145:6380 192.168.68.145:6381
```

此时会由 redis-trib.rb 命令自动根据主机状态创建主从配置关系。

12.【redis-cluster-b 主机】如果要进行集群配置，则必须通过客户端登录 Redis 集群中的任意一台主机，本次将通过客户端登录 redis-cluster-a 节点。

```
/usr/local/redis/bin/redis-cli -h 192.168.68.143 -p 6380
```

13.【redis-cluster-b 主机】现在的 Redis 集群是脆弱的，没有密码的保护支持，所以需要对其进行配置。利用已经登录过的操作来进行配置处理，在命令行里面如果要定义 Redis 的配置文件，那么主要通过 config 命令完成。

打开受保护模式	config set protected-mode yes
设置认证密码	config set requirepass mldnjava
由于设置了密码所以要登录	auth mldnjava
整个集群需要一个统一密码	config set masterauth mldnjava
将修改的配置写入到配置文件	config rewrite
关闭当前的**Redis**进程	shutdown

而后在所有 Redis 服务主机上都需要执行以上处理操作，并保证密码相同。

14.【redis-cluster-*主机】在每台主机上重新启动 Redis 进程。

```
/usr/local/redis/bin/redis-server /usr/local/redis/conf/redis-6379.conf
/usr/local/redis/bin/redis-server /usr/local/redis/conf/redis-6380.conf
/usr/local/redis/bin/redis-server /usr/local/redis/conf/redis-6381.conf
```

15.【redis-cluster-a 主机】由于通过配置命令实现了密码处理，所以此时需要手动修改 Ruby 配置文件，追加 Redis 的连接密码，才可以正常使用 RedisCluster 集群。打开配置文件。

```
vim /var/lib/gems/2.3.0/gems/redis-4.0.1/lib/redis/client.rb
```

16.【redis-cluster-a 主机】修改 redis/client.rb 配置文件。

```
class Redis
  class Client
    DEFAULTS = {
      :url => lambda { ENV["REDIS_URL"] },
      :scheme => "redis",
      :host => "127.0.0.1",
      :port => 6379,
      :path => nil,
      :timeout => 5.0,
      :password => "mldnjava",
      :db => 0,
      :driver => nil,
      :id => nil,
      :tcp_keepalive => 0,
      :reconnect_attempts => 1,
      :inherit_socket => false
    }
```

17.【redis-cluster-a 主机】利用 redis-trib.rb 命令检测集群状态。

```
/usr/local/redis/bin/redis-trib.rb check 192.168.68.145:6380
```

如果可以正常连接，将返回当前 RedisCluster 中的相关主机信息。

18.【redis-cluster-c 主机】随意登录一台 Master 主机，使用集群方式登录，使用-c 进行集群查询。

```
/usr/local/redis/bin/redis-cli -h 192.168.68.145 -p 6379 -a mldnjava -c
```

19.【redis-cluster-c 主机】向集群进行数据设置。

```
set mldn helloworld
```

20.【redis-cluster-c 主机】查询数据。

get mldn	
提示信息	-> Redirected to slot [1240] located at 192.168.68.145:6380

通过提示信息可以发现，用户可以随意登录集群中的任意一台主机，系统会自动根据用户查询的数据进行主机重定向处理。

10.7 使用 Java 操作 Redis 数据库

Redis 提供了对常见编程语言的操作支持。如果使用 Java 进行 Redis 数据库操作，则官方提供有 Jedis 支持包，以实现 Java 客户端的处理。如果要在项目中使用 Jedis，则需要修改 Maven 配置文件。

1.【mldnspring 项目】修改父 pom.xml 配置文件，追加 Jedis 依赖库配置。

属性配置	`<jedis.version>2.9.0</jedis.version>`

依赖库配置	`<dependency>`
	`<groupId>redis.clients</groupId>`
	`<artifactId>jedis</artifactId>`
	`<version>${jedis.version}</version>`
	`</dependency>`

2.【mldnspring-redis 项目】在子项目中配置 Jedis 依赖库。

```
<dependency>
	<groupId>redis.clients</groupId>
	<artifactId>jedis</artifactId>
</dependency>
```

此时可以在项目中使用 Jedis 提供的程序类，实现 Redis 数据库的连接与数据处理。

10.7.1 连接 Redis 数据库

如果 Redis 数据库需要连接数据库地址、数据库端口、认证信息 3 个核心信息，可利用 Jedis 程序类实现连接处理。

范例：【mldnspring-redis 项目】建立 Redis 连接工具类。

```
package cn.mldn.mldnspring.util.dbc;
import redis.clients.jedis.Jedis;
public class RedisConnectionUtil {
```

```
    private static final String REDIS_HOST = "redis-single" ;          // 主机名称
    private static final int REDIS_PORT = 6379 ;                       // 端口名称
    private static final String REDIS_AUTH = "mldnjava" ;              // 认证信息
    private Jedis jedis ;                                              // 这个对象主要是连接对象信息
    public RedisConnectionUtil() {                                     // 构造方法中进行数据库的连接
        this.jedis = new Jedis(REDIS_HOST,REDIS_PORT) ;                // 输入主机和端口
        this.jedis.auth(REDIS_AUTH) ;                                  // 认证信息
    }
    public void close() {                                              // 关闭连接
        this.jedis.close();
    }
    public Jedis getConnection() {
        return this.jedis ;
    }
}
```

在本程序中，利用 Jedis 提供的构造方法设置了 Redis 主机名称与连接端口号，由于数据库有安全认证，所以还需要通过 Jedis 中的 auth 方法设置认证信息。

范例：【mldnspring-redis 项目】编写测试类，进行 Redis 连接测试。

```
package cn.mldn.mldnspring.test;
import org.junit.Test;
import cn.mldn.mldnspring.util.dbc.RedisConnectionUtil;
import redis.clients.jedis.Jedis;
public class TestRedisConnection {
    @Test
    public void testConnection() {
        RedisConnectionUtil rcu = new RedisConnectionUtil() ;
        Jedis jedis = rcu.getConnection() ;                 // 获得连接信息
        System.out.println(jedis);                          // 输出对象，不为null表示连接成功
        jedis.close();
    }
}
```

本程序主要测试RedisConnectionUtil工具类是否可以正常连接，如果可以连接则取得的Jedis对象不为 null。

10.7.2　Jedis 数据操作

Jedis 提供了 Redis 全部数据类型操作的方法，为了方便使用，方法名称与 Redis 命令相同。下面将通过具体的实例演示数据操作。

范例：【mldnspring-redis 项目】操作字符串数据。

```
package cn.mldn.mldnspring.test;
import java.util.concurrent.TimeUnit;
```

```
import org.junit.Test;
import cn.mldn.mldnspring.util.dbc.RedisConnectionUtil;
import redis.clients.jedis.Jedis;
public class TestRedisDataDemo {
    public static Jedis jedis = null ;                                // 保存Jedis连接
    static {
        RedisConnectionUtil rcu = new RedisConnectionUtil() ;
        jedis = rcu.getConnection() ;                                 // 获取Jedis连接对象
    }
    @Test
    public void testStringData() throws Exception {
        jedis.set("mldn", "Java") ;                                   // 设置数据
        jedis.setex("mldn-message",3, "helloworld") ;                 // 设置数据，3s后失效
        TimeUnit.SECONDS.sleep(4);                                    // 延迟4s执行
        System.out.println(jedis.get("mldn"));                        // 可以获取数据
        System.out.println(jedis.get("mldn-message"));                // 无法获取数据
    }
}
```

程序执行结果	Java null

本程序通过 Jedis 对象实现了 string 数据操作，使用 set 方法设置一个永久数据，使用 setex 设置数据的过期时间。

范例：【mldnspring-redis 项目】操作 hash 数据。

```
    @Test
    public void testHashData() throws Exception {
        jedis.hset("user-mldn", "name", "李兴华") ;                    // 设置hash数据与属性key
        jedis.hset("user-mldn", "age", String.valueOf(18)) ;          // 设置hash数据与属性key
        jedis.hset("user-mldn", "sex", "男") ;                        // 设置hash数据与属性key
        System.out.println(jedis.hget("user-mldn", "name"));          // 获取指定属性key数据
    }
```

程序执行结果	李兴华

由于一个 hash 数据可以设置多个属性 key 信息，所以本程序为 user-mldn 设置了 3 个属性内容，并输出了其中一个属性 key 的数据。在进行数据设置时，需要将设置的数字转为字符串类型后才可以使用。

范例：【mldnspring-redis 项目】操作 list 数据。

```
    @Test
    public void testlistData() throws Exception {
        jedis.flushDB() ;                                             // 清空数据库
        jedis.lpush("user-mldn", "mldnjava","jixianit") ;             // 设置数据
        jedis.rpush("user-mldn", "hello","world") ;                   // 设置数据
```

```
        System.out.print(jedis.rpop("user-mldn") + "、");          // 从队列头部弹出一个数据
        System.out.print(jedis.rpop("user-mldn") + "、");          // 从队列头部弹出一个数据
        System.out.print(jedis.lpop("user-mldn") + "、");          // 从队列尾部弹出一个数据
        System.out.print(jedis.lpop("user-mldn") + "、");          // 从队列尾部弹出一个数据
        System.out.print(jedis.lpop("user-mldn"));                 // 没有数据，返回null
}
```

程序执行结果	world、hello、jixianit、mldnjava、null

本程序为了防止数据操作混乱，首先使用 flushDB 清空了当前数据库，随后设置了一个 list 数据，并且利用 lpush 与 rpush 实现了数据的追加处理。在获取数据时，可以使用 lpop 或 rpop 弹出数据。如果对 list 集合的数据不使用弹出处理，也可以使用 lrange 方法获取全部内容，该方法将返回一个 list 集合。

范例：【mldnspring-redis 项目】获取全部 list 数据并输出。

```
@Test
public void testlistDataGet() throws Exception {
        jedis.flushDB() ;                                           // 清空数据库
        jedis.lpush("user-mldn", "mldnjava","jixianit") ;           // 设置数据
        jedis.rpush("user-mldn", "hello","world") ;                 // 设置数据
        list<String> all = jedis.lrange("user-mldn", 0, -1) ;       // 获取全部数据
        all.forEach((data)->{
                System.out.print(data + "、") ;});                  // 迭代输出
}
```

程序执行结果	jixianit、mldnjava、hello、world、

本程序利用 lrange 方法获取了 list 集合的全部数据，而后使用 list 接口中的 forEach 方法实现内容输出。

范例：【mldnspring-redis 项目】操作 set 集合。

```
@Test
public void testsetData() throws Exception {
        jedis.flushDB() ;                                           //清空数据库
        jedis.sadd("user-admin", "a", "b", "c", "d", "e");          // 设置数据
        jedis.sadd("user-mldn", "a", "c", "e", "x", "y", "z");      // 设置数据
        set<String> all = jedis.sinter("user-admin","user-mldn") ;  // 交集计算
        all.forEach((data)->{
                System.out.print(data + "、") ;});                  // 迭代输出
}
```

程序执行结果	a、c、e、

本程序实现了两个 set 集合的交集运算，并将计算结果保存在 set 集合中，最后使用迭代输出 set 集合数据。

范例：【mldnspring-redis 项目】操作 zset（sorted set）集合。

```
@Test
public void testzsetData() throws Exception {
```

```
        jedis.flushDB() ;                                                  // 清空数据库
        Map<String,Double> map = new HashMap<String,Double>() ;            // 设置Map集合保存数据
        map.put("pid-1-1", 2.0) ;                                          // 保存数据与分数
        map.put("pid-1-2", 1.0) ;                                          // 保存数据与分数
        map.put("pid-2-1", 5.0) ;                                          // 保存数据与分数
        jedis.zadd("user-mldn", map) ;                                     // 将数据保存到Redis中
        // 根据分数范围获取全部数据内容与分数，此时利用Tuple保存每一组结果
        set<Tuple> all = jedis.zrangeByScoreWithScores("user-mldn", 1.0, 5.0) ;
        all.forEach((data) -> {
            System.out.println("元素内容：" + data.getElement()
                    + "、分数：" + data.getScore());                        // 输出数据
        });                                                                // 迭代输出
    }
```

程序执行结果	元素内容：pid-1-2、分数：1.0 元素内容：pid-1-1、分数：2.0 元素内容：pid-2-1、分数：5.0

由于 zset 数据中包含了内容与分数信息，所以在 Jedis 中使用 Tuple 来描述每一组数据。取得数据时可以利用 Tuple 获取元素内容与分数信息。

范例：【mldnspring-redis 项目】操作 GEO 数据。

```
    @Test
    public void testGEOData() throws Exception {
        jedis.flushDB() ;                                                        // 清空数据库
        Map<String,GeoCoordinate> pointsMap = new HashMap<String,GeoCoordinate>() ; // 坐标
        pointsMap.put("天安门",new GeoCoordinate(116.403963, 39.915119)) ;   // 添加坐标
        pointsMap.put("王府井",new GeoCoordinate(116.417876, 39.915411)) ;   // 添加坐标
        pointsMap.put("前门大街",new GeoCoordinate(116.404354, 39.904748)) ;// 添加坐标
        jedis.geoadd("point", pointsMap) ;                                       // 保存坐标信息
        // 查找距离当前坐标周围1000m的建筑物信息
        list<GeoRadiusResponse> georadius = jedis.georadius("point", 116.415901, 39.914805,
            1000, GeoUnit.M,GeoRadiusParam.geoRadiusParam().withDist());
        georadius.forEach((geoData)->{
            System.out.println("建筑物名称：" + geoData.getMemberByString()
                + "、距离：" + geoData.getDistance());
        });                                                                      // 迭代输出
    }
```

程序执行结果	建筑物名称：王府井、距离：181.6811

本程序先在 Jedis 中保存了 3 个建筑物的坐标信息，而后利用范围查找距当前位置 1000m 的标志物信息，返回结果以 list 集合形式出现。随后可以使用 GeoRadiusResponse 获取坐标的相关信息以及与参考坐标之间的距离（如果要获取距离，必须使用 GeoRadiusParam.geoRadiusParam().withDist()设置查询参数）。

10.7.3　Jedis 连接池

进行数据库操作时，如果每次处理都要打开和关闭数据库，一定会造成严重的性能下降。为了提高数据库的操作性能，Jedis 中提供了数据库连接池处理支持。

范例：【mldnspring-redis 项目】使用连接池配置 Redis 连接。

```
package cn.mldn.mldnspring.util.dbc;
import redis.clients.jedis.Jedis;
import redis.clients.jedis.JedisPool;
import redis.clients.jedis.JedisPoolConfig;
public class RedisConnectionUtil {
    private static final String REDIS_HOST = "redis-single" ;          // 主机名称
    private static final int REDIS_PORT = 6379 ;                       // 端口名称
    private static final String REDIS_AUTH = "mldnjava" ;              // 认证信息
    private static final int TIMEOUT = 2000 ;                          // 连接超时时间
    private static final int MAX_TOTAL = 200 ;                         // 最多允许200个的连接
    private static final int MAX_IDLE = 20 ;                           // 没有访问时的最小维持数量
    private static final int MAX_WAIT_MILLIS = 1000 ;                  // 最大等待时间
    private static final boolean TEST_ON_BORROW = true ;               // 是否要进行连接测试
    private JedisPool pool = null ;                                    // 连接池对象
    public RedisConnectionUtil() {                                     // 构造方法连接数据库
        // 如果要想使用连接池进行控制，那么一定需要进行连接池的相关配置
        JedisPoolConfig config = new JedisPoolConfig() ;               // 进行连接池配置
        config.setMaxTotal(MAX_TOTAL);                                 // 最大连接数
        config.setMaxIdle(MAX_IDLE);                                   // 最小维持连接数
        config.setMaxWaitMillis(MAX_WAIT_MILLIS);                      // 最大等待时间
        config.setTestOnBorrow(TEST_ON_BORROW);                        // 测试通过后返回可用连接
        this.pool = new JedisPool(config,REDIS_HOST,REDIS_PORT,TIMEOUT,REDIS_AUTH) ;
    }
    public Jedis getConnection() {
        return this.pool.getResource() ;                               // 连接池获取连接
    }
    public void close() {
        this.pool.close();                                             // 连接池关闭
    }
}
```

本程序不再直接使用 Jedis 对象实现数据库连接，所有的连接都交由 JedisPool 统一管理。这样可以由 Jedis 管理连接池中的连接信息，从而提高数据库操作性能。

10.7.4　Jedis 访问哨兵机制

进行哨兵访问时，所有的数据库连接信息都是通过哨兵进程获取的。Jedis 开发包中提供了

JedisSentinelPool 类，以实现哨兵连接池配置，并可利用此连接池创建 Jedis 连接。

范例：【mldnspring-redis 项目】。

```
package cn.mldn.mldnspring.util.dbc;
import java.util.Hashset;
import java.util.set;
import redis.clients.jedis.Jedis;
import redis.clients.jedis.JedisPoolConfig;
import redis.clients.jedis.JedisSentinelPool;
public class RedisConnectionUtil {
    public static final String MASTER_NAME = "mymaster" ;              // 定义哨兵的Master配置名称
    private static final String REDIS_AUTH = "mldnjava" ;              // 认证信息
    private static final int MAX_TOTAL = 200 ;                         // 最多允许200个连接
    private static final int MAX_IDLE = 20 ;                           // 没有访问时的最小维持数量
    private static final int MAX_WAIT_MILLIS = 1000 ;                  // 最大等待时间
    private static final boolean TEST_ON_BORROW = true ;               // 是否要进行连接测试
    private JedisSentinelPool pool = null ;                            // 连接池对象
    public RedisConnectionUtil() {                                     // 构造方法连接数据库
        // 要通过哨兵机制进行Redis访问，必须明确设置出所有可使用的哨兵地址与端口
        set<String> sentinels = new Hashset<String>() ;                // 设置所有的哨兵处理地址信息
        sentinels.add("redis-master:26379") ;                          // 哨兵的地址
        sentinels.add("redis-slave-a:26380") ;                         // 哨兵的地址
        sentinels.add("redis-slave-b:26381") ;                         // 哨兵的地址
        JedisPoolConfig config = new JedisPoolConfig() ;               // 进行连接池配置
        config.setMaxTotal(MAX_TOTAL);                                 // 最大连接数
        config.setMaxIdle(MAX_IDLE);                                   // 最小维持连接数
        config.setMaxWaitMillis(MAX_WAIT_MILLIS);                      // 最大等待时间
        config.setTestOnBorrow(TEST_ON_BORROW);                        // 测试通过后返回可用连接
        this.pool = new JedisSentinelPool(MASTER_NAME, sentinels,config); // 构建哨兵连接池
    }
    public Jedis getConnection() {
        Jedis jedis = pool.getResource() ;                             // 通过连接池获取连接对象
        jedis.auth(REDIS_AUTH) ;                                       // 设置认证信息
        return jedis ;                                                 // 返回Jedis对象
    }
    public void close() {
        this.pool.close();                                             // 连接池关闭
    }
}
```

本程序利用 JedisSentinelPool 构建了哨兵连接池，但连接池的配置依然要通过 JedisPoolConfig 指明。在进行哨兵连接池配置时，需将所有的哨兵主机都配置到程序中，这样一旦进行 Master 切换，就可通过哨兵连接池获取 Master 主机信息，进行 Redis 数据库操作。

10.7.5　使用 Jedis 访问 RedisCluster

RedisCluster 可以实现 Redis 集群访问，但与传统中心化的集群配置不同，RedisCluster 在使用时需要将所有的连接地址都配置到程序中，才可以正常获取 Jedis 连接。如果想连接 RedisCluster，必须依靠 JedisCluster 类来处理。此类提供的构造方法如下。

```
public JedisCluster(Set<HostAndPort> jedisClusterNode, int connectionTimeout, int soTimeout,
    int maxAttempts, String password, final GenericObjectPoolConfig poolConfig) {
        super(jedisClusterNode, connectionTimeout, soTimeout, maxAttempts, password, poolConfig);
}
```

构造方法中的参数说明如下。

☑ Set<HostAndPort> jedisClusterNode：所有集群主机的地址。

☑ **int** connectionTimeout：主机连接的超时时间。

☑ **int** soTimeout：两次操作的间隔超时时间。

☑ **int** maxAttempts：重试的连接次数。

☑ String password：认证密码。

☑ **final** GenericObjectPoolConfig poolConfig：连接池配置项。

范例：【mldnspring-redis 项目】连接 RedisCluster 集群。

```
package cn.mldn.mldnspring.util.dbc;
import java.io.IOException;
import java.util.Hashset;
import java.util.set;
import redis.clients.jedis.HostAndPort;
import redis.clients.jedis.JedisCluster;
import redis.clients.jedis.JedisPoolConfig;
public class RedisConnectionUtil {
    public static final int TIMEOUT = 1000;                               // 连接超时时间
    public static final int SO_TIMEOUT = 100;                             // 间隔超时时间
    public static final int MAX_ATTEMPTS = 100;                           // 重试次数
    private static final String REDIS_AUTH = "mldnjava" ;                 // 认证信息
    private static final int MAX_TOTAL = 200 ;                            // 最多允许200个连接
    private static final int MAX_IDLE = 20 ;                              // 没有访问时的最小维持数量
    private static final int MAX_WAIT_MILLIS = 1000 ;                     // 最大等待时间
    private static final boolean TEST_ON_BORROW = true ;                  // 是否要进行连接测试
    private JedisCluster jedisCluster ;                                   // JedisCluster
    public RedisConnectionUtil() {                                        // 构造方法连接数据库
        JedisPoolConfig config = new JedisPoolConfig() ;                  // 进行连接池配置
        config.setMaxTotal(MAX_TOTAL);                                    // 最大连接数
        config.setMaxIdle(MAX_IDLE);                                      // 最小维持连接数
        config.setMaxWaitMillis(MAX_WAIT_MILLIS);                         // 最大等待时间
        config.setTestOnBorrow(TEST_ON_BORROW);                           // 测试通过后返回可用连接
```

```
        // 定义出所有保存RedisCluster集群主机的集合对象
        Set<HostAndPort> allRedisCluster = new Hashset<HostAndPort>();
        allRedisCluster.add(new HostAndPort("redis-cluster-a", 6379));
        allRedisCluster.add(new HostAndPort("redis-cluster-a", 6380));
        allRedisCluster.add(new HostAndPort("redis-cluster-a", 6381));
        allRedisCluster.add(new HostAndPort("redis-cluster-b", 6379));
        allRedisCluster.add(new HostAndPort("redis-cluster-b", 6380));
        allRedisCluster.add(new HostAndPort("redis-cluster-b", 6381));
        allRedisCluster.add(new HostAndPort("redis-cluster-c", 6379));
        allRedisCluster.add(new HostAndPort("redis-cluster-c", 6380));
        allRedisCluster.add(new HostAndPort("redis-cluster-c", 6381));
        this.jedisCluster = new JedisCluster(allRedisCluster, TIMEOUT,
            SO_TIMEOUT, MAX_ATTEMPTS, REDIS_AUTH, config);   // 获取RedisCluster连接
    }
    public JedisCluster getConnection() {                   // 获取JedisCluster
        return this.jedisCluster ;
    }
    public void close() {
        try {
            this.jedisCluster.close();                      // 连接池关闭
        } catch (IOException e) {
            e.printStackTrace();
        }
    }
}
```

使用 RedisCluster 搭建 Redis 时，不再使用 Jedis 对象，而是直接利用 JedisCluster 实现数据处理操作，该类中的方法与 Jedis 相同。

范例：【mldnspring-redis】利用 JedisCluster 操作 Redis 数据。

```
public class TestRedisDataDemo {
    public static JedisCluster jedis = null ;               // 保存Jedis连接
    static {
        RedisConnectionUtil rcu = new RedisConnectionUtil() ;
        jedis = rcu.getConnection() ;                       // 获取Jedis连接对象
    }
    @Test
    public void testStringData() throws Exception {
        jedis.set("mldn", "Java") ;                          // 设置数据
        jedis.setex("mldn-message",3, "helloworld") ;        // 设置数据，3s后失效
        TimeUnit.SECONDS.sleep(4);                           // 延迟4s执行
        System.out.println(jedis.get("mldn"));               // 可以获取数据
        System.out.println(jedis.get("mldn-message"));       // 无法获取数据
```

```
    }
}
```

本程序除了将原始的Jedis对象更换为了JedisCluster对象外，整体处理操作与之前完全相同。

10.8　SpringDataRedis

Redis 是一个高速缓存数据库，所以将一些常用信息保存在 Redis 中，可以满足快速读写的操作需求。Java 处理中，对象可以包含很多数据信息，Redis 虽然支持二进制数据的读写功能，但对于对象数据的直接处理，需要开发者进行序列化与反序列化操作。很明显，这样是非常麻烦的。

使用 Spring 提供的 SpringDataRedis 技术可以简化操作，如图 10-17 所示。利用 SpringData Redis 不仅可以方便地实现对象的序列化与反序列化管理，还可以利用其提供的 RedisTemplate 轻松、方便地实现 Redis 数据处理，同时可以由 Spring 实现连接管理。

图 10-17　SpringDataRedis 操作

1.【mldnspring 项目】修改父 pom.xml 配置文件，追加 Jedis 依赖库配置。

属性配置	<spring-data-redis.version>2.0.4.RELEASE</spring-data-redis.version>
依赖库配置	<dependency> <groupId>org.springframework.data</groupId> <artifactId>spring-data-redis</artifactId> <version>${spring-data-redis.version}</version> </dependency>

2.【mldnspring-redis 项目】在子项目之中配置 Jedis 依赖库。

```
<dependency>
    <groupId>org.springframework.data</groupId>
    <artifactId>spring-data-redis</artifactId>
</dependency>
```

使用 spring-data-redis 依赖库，需要 Jedis 依赖库的支持。下面将利用 SpringDataRedis 实现 Redis 操作。

10.8.1 SpringDataRedis 数据操作

要使用 SpringDataRedis 开发支持，需将 Redis 的全部连接管理都统一交由 Spring 负责。开发者通过 RedisTemplate，即可实现全部的数据处理。

1.【mldnspring-redis 项目】建立 src/main/profiles/dev 源文件夹目录，并且创建 config/redis.properties 配置文件。

redis.host=redis-single	# Redis的连接主机地址
redis.port=6379	# Redis连接端口号
redis.password=mldnjava	# Redis的认证信息，认证信息密码
redis.timeout=2000	# Redis连接的超时时间
redis.pool.maxTotal=100	# 设置最大的可用连接数
redis.pool.maxIdle=20	# 最小维持的可用连接数
redis.pool.maxWaitMillis=2000	# 最大等待时间
redis.pool.testOnBorrow=true	# 是否要返回可用的连接

2.【mldnspring-redis 项目】在 spring/spring-base.xml 配置文件中引入所有的资源文件。

```
<context:property-placeholder location="classpath:config/*.properties"/>
```

3.【mldnspring-redis 项目】创建 src/main/resources/spring/spring-redis.xml 配置文件，配置如下选项。

☑ 使用 JedisPoolConfig 配置 Redis 连接池。

```
<bean id="jedisPoolConfig" class="redis.clients.jedis.JedisPoolConfig">
	<property name="maxTotal"
		value="${redis.pool.maxTotal}" />					<!-- 最大可用连接数 -->
	<property name="maxIdle" value="${redis.pool.maxIdle}" />		<!-- 最小维持连接数 -->
	<property name="maxWaitMillis"
		value="${redis.pool.maxWaitMillis}" />				<!-- 最大等待时间 -->
	<property name="testOnBorrow" value="${redis.pool.testOnBorrow}" />	<!-- 可用连接 -->
</bean>
```

☑ 配置 Redis 连接工厂。

```
<bean id="connectionFactory"
	class="org.springframework.data.redis.connection.jedis.JedisConnectionFactory">
	<property name="poolConfig" ref="jedisPoolConfig" />		<!-- 引用进行连接池的配置项 -->
	<property name="hostName" value="${redis.host}" />		<!-- Redis的连接地址 -->
	<property name="password" value="${redis.password}" />	<!-- 认证密码 -->
	<property name="timeout" value="${redis.timeout}" />		<!-- 连接的超时时间 -->
</bean>
```

☑ SpringDataRedis 主要通过 RedisTemplate 实现数据处理，所以还需要定义 RedisTemplate 配置，同时设置序列化与反序列化处理类。

```
<bean id="redisTemplate" class="org.springframework.data.redis.core.RedisTemplate">
    <property name="connectionFactory" ref="connectionFactory" />    <!-- Redis连接工厂 -->
    <property name="keySerializer">        <!-- 定义序列化key的程序处理类 -->
        <bean
            class="org.springframework.data.redis.serializer.StringRedisSerializer" />
    </property>
    <property name="valueSerializer">      <!-- 处理value数据的操作 -->
        <!--进行value数据保存时，保存对象需要使用JDK提供的序列化处理类 -->
    <bean
    class="org.springframework.data.redis.serializer.JdkSerializationRedisSerializer" />
    </property>
    <property name="hashKeySerializer">
        <bean
            class="org.springframework.data.redis.serializer.StringRedisSerializer" />
    </property>
    <property name="hashValueSerializer">                  <!-- 处理hash数据的保存 -->
        <bean
class="org.springframework.data.redis.serializer.JdkSerializationRedisSerializer" />
    </property>
</bean>
```

4．【mldnspring-redis 项目】建立一个程序类，用于实现序列化处理。

```java
package cn.mldn.mldnspring.vo;
import java.io.Serializable;
import java.util.Date;
@SuppressWarnings("serial")
public class Member implements Serializable {
    private String mid ;
    private String name ;
    private Integer age ;
    private Date birthday ;
    private Double salary ;
    // setter、getter、toString略
}
```

5．【mldnspring-redis 项目】编写测试类，使用 RedisTemplate 实现数据操作。

```java
package cn.mldn.mldnspring.test;
import java.util.Date;
import org.junit.Test;
```

```
import org.junit.runner.RunWith;
import org.springframework.beans.factory.annotation.Autowired;
import org.springframework.data.redis.core.RedisTemplate;
import org.springframework.test.context.ContextConfiguration;
import org.springframework.test.context.junit4.SpringJUnit4ClassRunner;
import cn.mldn.mldnspring.vo.Member;
@ContextConfiguration(locations = {"classpath:spring/spring-*.xml"})
@RunWith(SpringJUnit4ClassRunner.class)                              // 使用Junit进行测试
public class TestMember {
    @Autowired
    private RedisTemplate<String, Object> redisTemplate;            // Redis操作模板
    @Test
    public void testSave() {                                         // 数据保存处理
        Member vo = new Member();
        vo.setMid("mldn-java");
        vo.setBirthday(new Date());
        vo.setName("张三");
        vo.setAge(18);
        vo.setSalary(1.1);
        this.redisTemplate.opsForValue().set("mldn-1", vo);          // 保存对象
    }
    @Test
    public void testLoad() {
        Object obj = this.redisTemplate.opsForValue().get("mldn-1") ; // 进行转型处理
        System.out.println(obj);                                     // 输出对象
    }
}
```

通过程序执行可以发现，使用 RedisTemplate 模板处理类后，所有的操作都可以通过此类提供的方法完成。利用序列化和反序列化操作类，可轻松实现对象的保存与读取。

10.8.2 SpringDataRedis 访问哨兵

除了进行单主机的 Redis 连接外，还可以使用 SpringDataRedis 实现哨兵主机访问，并利用哨兵主机获取相关连接信息，但是这样的整合只针对配置文件，程序类中依然可以利用相同的 RedisTemplate 进行数据操作。

1.【mldnspring-redis 项目】修改 config/redis.properties 配置文件，追加所有的哨兵配置地址。

```
redis.sentinel-1.host=redis-master          # 配置所有哨兵主机地址
redis.sentinel-2.host=redis-slave-a
redis.sentinel-3.host=redis-slave-b
```

redis.sentinel-1.port=26379 redis.sentinel-2.port=26380 redis.sentinel-3.port=26381	# 配置所有哨兵连接端口号
redis.sentinel.master.name=mymaster	# 定义哨兵的master的名称
redis.password=mldnjava	# Redis的认证信息，认证信息密码
redis.timeout=2000	# Redis连接的超时时间
redis.pool.maxTotal=100	# 设置最大的可用连接数
redis.pool.maxIdle=20	# 最小维持的可用连接数
redis.pool.maxWaitMillis=2000	# 最大等待时间
redis.pool.testOnBorrow=true	# 是否要返回可用的连接

2.【mldnspring-redis 项目】修改 spring/spring-redis.xml 配置文件，使用哨兵机制连接 Redis 数据库。

☑　定义所有哨兵主机信息。

```
<!-- 进行所有的哨兵地址的配置项 -->
<bean id="sentinelsConfiguration"
        class="org.springframework.data.redis.connection.RedisSentinelConfiguration">
        <property name="master">                                          <!-- 配置Master的节点名称 -->
                <bean class="org.springframework.data.redis.connection.RedisNode">
                        <!-- 通过资源文件读取出Master的名称进行配置 -->
                        <property name="name" value="${redis.sentinel.master.name}"/>
                </bean>
        </property>
        <property name="sentinels">                                  <!-- 配置所有哨兵的连接地址信息 -->
                <set>
                        <bean class="org.springframework.data.redis.connection.RedisNode">
                                <constructor-arg name="host" value="${redis.sentinel-1.host}"/>
                                <constructor-arg name="port" value="${redis.sentinel-1.port}"/>
                        </bean>
                        <bean class="org.springframework.data.redis.connection.RedisNode">
                                <constructor-arg name="host" value="${redis.sentinel-2.host}"/>
                                <constructor-arg name="port" value="${redis.sentinel-2.port}"/>
                        </bean>
                        <bean class="org.springframework.data.redis.connection.RedisNode">
                                <constructor-arg name="host" value="${redis.sentinel-3.host}"/>
                                <constructor-arg name="port" value="${redis.sentinel-3.port}"/>
                        </bean>
                </set>
        </property>
</bean>
```

☑ 在 Redis 连接工厂配置上，引入哨兵配置 Bean。

```
<!-- 进行ConnectionFactory的配置 -->
<bean id="connectionFactory"
        class="org.springframework.data.redis.connection.jedis.JedisConnectionFactory">
        <constructor-arg name="sentinelConfig" ref="sentinelsConfiguration"/>
        <property name="poolConfig" ref="jedisPoolConfig"/>                <!-- 引用连接池配置项 -->
        <property name="password" value="${redis.password}"/>               <!-- 认证密码 -->
</bean>
```

此时程序将可以根据哨兵来获取数据库连接对象。

10.8.3 SpringDataRedis 访问 RedisCluster

SpringDataRedis 支持 RedisCluster 的数据访问，在使用时需要将所有的 RedisCluster 主机信息全部配置到资源文件中，而后将此配置信息注入到 JedisConnectionFactory 配置类中，就可以实现 RedisCluster 连接管理。

1.【mldnspring-redis 项目】修改 config/redis.properties 配置文件，定义 RedisCluster 集群访问。

```
redis.cluster.max-redirect=2                         # 限制redirect重定向次数
redis.cluster.node-1.host=redis-cluster-a            # 追加RedisCluster中的所有主机信息
redis.cluster.node-2.host=redis-cluster-a
redis.cluster.node-3.host=redis-cluster-a
redis.cluster.node-4.host=redis-cluster-b
redis.cluster.node-5.host=redis-cluster-b
redis.cluster.node-6.host=redis-cluster-b
redis.cluster.node-7.host=redis-cluster-c
redis.cluster.node-8.host=redis-cluster-c
redis.cluster.node-9.host=redis-cluster-c
redis.cluster.node-1.port=6379                       # 配置集群中所有主机的端口号
redis.cluster.node-2.port=6380
redis.cluster.node-3.port=6381
redis.cluster.node-4.port=6379
redis.cluster.node-5.port=6380
redis.cluster.node-6.port=6381
redis.cluster.node-7.port=6379
redis.cluster.node-8.port=6380
redis.cluster.node-9.port=6381
redis.password=mldnjava                              # Redis的认证信息，认证信息密码
redis.timeout=2000                                   # Redis连接的超时时间
redis.pool.maxTotal=100                              # 设置最大的可用连接数
redis.pool.maxIdle=20                                # 最小维持的可用连接数
```

redis.pool.maxWaitMillis=2000	# 最大等待时间
redis.pool.testOnBorrow=true	# 是否要返回可用的连接

2．【mldnspring-redis 项目】修改 spring/spring-redis.xml 配置文件，定义 RedisCluster。

☑　定义 RedisCluster 所有主机信息。

```
<bean id="redisClusterConfiguration"
      class="org.springframework.data.redis.connection.RedisClusterConfiguration">
      <property name="maxRedirects" value="${redis.cluster.max-redirect}"/>
      <property name="clusterNodes">                                    <!-- 配置集群中的所有节点 -->
          <list>
              <bean class="org.springframework.data.redis.connection.RedisNode">
                  <constructor-arg index="0" value="${redis.cluster.node-1.host}"/>
                  <constructor-arg index="1" value="${redis.cluster.node-1.port}"/>
              </bean>
              <bean class="org.springframework.data.redis.connection.RedisNode">
                  <constructor-arg index="0" value="${redis.cluster.node-2.host}"/>
                  <constructor-arg index="1" value="${redis.cluster.node-2.port}"/>
              </bean>
              <!-- 后续操作略，重复配置9台主机即可 -->
          </list>
      </property>
</bean>
```

☑　修改 Jedis 连接工厂配置，注入集群配置项。

```
<bean id="connectionFactory"
class="org.springframework.data.redis.connection.jedis.JedisConnectionFactory">
      <!-- 配置所有的RedisCluster集群主机 -->
      <constructor-arg name="clusterConfig" ref="redisClusterConfiguration"/>
      <property name="poolConfig" ref="jedisPoolConfig"/>              <!-- 引用连接池配置项 -->
      <property name="password" value="${redis.password}"/>            <!-- 认证密码 -->
</bean>
```

此时可以在 RedisCluster 环境下实现 Redis 数据处理操作。

10.9　抢红包案例分析

抢红包是时下很热门的活动，属于高并发访问处理。最好的做法是利用程序将一个完整金额进行指定个数的拆分处理，而后将这些分配好的金额信息保存在 Redis 数据库中。当有多个用户争抢红包时，可采用集合数据弹出模式从里面取出各个红包信息。为了方便记录红包的争抢过程，可以将结果保存在 Redis 数据库中，基本结构如图 10-18 所示。

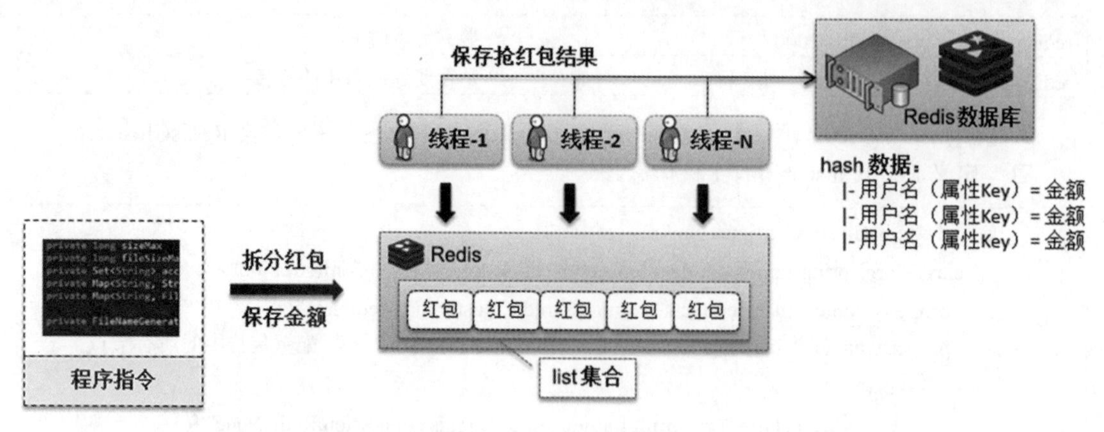

图 10-18　抢红包架构

下面通过具体程序演示抢红包处理过程。为了方便，使用多线程模拟多用户的抢红包场景。

1.【mldnspring-redis 项目】拆分红包时需要考虑数据的四舍五入问题，所以先建立一个数学工具类。

```
package cn.mldn.mldnspring.util;
public class MyMath {
    private MyMath() {}
    /**
     * 数据四舍五入处理
     * @param num 要处理的数字
     * @param scale 保留小数位
     * @return 四舍五入后的数据
     */
    public static double round(double num,int scale) {
        return Math.round(num * Math.pow(10.0, scale)) / Math.pow(10.0, scale) ;
    }
}
```

2.【mldnspring-redis 项目】建立红包拆分工具类，该程序类可以将指定金额拆分为指定个数的红包。

```
package cn.mldn.mldnspring.util;
import java.util.Arraylist;
import java.util.Iterator;
import java.util.list;
import java.util.Random;
public class SplitMoneyUtil {
    private double money;                          // 保存总金额，用于数据验证
    private int amount;                            // 存放数量，数量控制循环的次数
    private int currentAmount;                     // 当前次数
    private double surplusMoney;                   // 剩余资金
    private double currentMoney;                   // 当前处理后的金额
```

```
    // 随机数字使用nextInt方法进行拆分，该方法支持int数据
    private Random rand = new Random();                                    // 准备随机数拆分
    private list<Double> allPackages = new Arraylist<Double>();            // 保存红包信息
    public SplitMoneyUtil(int amount, double money) {                      // 设置红包数据
        this.amount = amount;                                              // 保存红包数量
        this.money = money - (amount / 100.00);                            // 预先处理金额
        this.currentAmount = amount;                                       // 当前处理金额
        this.surplusMoney = money * 100;                                   // 剩余资金等于总资金
        if (this.currentAmount == 1) {                                     // 不拆分，做整体包
            this.allPackages.add(money);                                   // 一个大包
        } else {
            this.handle();                                                 // 拆分处理
        }
    }
    private void handle() {                                                // 处理红包拆分
        int count = (int) this.surplusMoney / this.amount;
        int key = count * 2;
        int rand = this.rand.nextInt(key);                                 // 通过已有数据取出一个内容
        this.surplusMoney -= rand;                                         // 从原始资金中减少部分数据
        this.allPackages.add(rand / 100.00);                               // 保存到最终红包数据
        this.currentMoney += rand;
        if (--this.currentAmount > 1) {                                    // 还没有拆分到指定个数
            this.handle();                                                 // 继续拆分
        } else {
            if (this.currentAmount == 1) {                                 // 余额给最后一个红包
                this.allPackages.add(((this.money * 100) - this.currentMoney) / 100.00);
                return;
            }
        }
    }
    public list<Double> getAllPackages() {                                 // 得到全部红包数据
        list<Double> all = new Arraylist<>();
        Iterator<Double> it = this.allPackages.iterator();
        while (it.hasNext()) {
            double s = it.next();
            all.add(MyMath.round(s + 0.01, 2));                            // 四舍五入处理
        }
        return all;
    }
}
```

3．【mldnspring-redis 项目】定义抢红包业务接口。

```
package cn.mldn.mldnspring.service;
```

```
import java.util.Map;
public interface IRedEnvelopeService {
    /**
     * 实现红包数据保存，所有的红包数据保存在list集合中
     * @param userid 发红包用户名
     * @param amount 红包个数
     * @param money 总金额
     * @return 保存数据key，key定义规则“envelope-用户名-时间戳”
     */
    public String add(String userid,int amount, double money) ;
    /**
     * 抢红包处理，红包到手后要记录在Redis中，使用hash数据类型存储，属性key为用户名
     * hash数据保存的key为“result-红包数据key”
     * @param userid 用户名
     * @param key 集合key
     * @return 返回抢到的红包数据，如果已经抢过了，将会返回-1.0
     */
    public Double grab(String userid,String key) ;
    /**
     * 获取抢红包数据结果
     * @param key hash-key名称
     * @return 所有保存在hash中的数据
     */
    public Map<Object,Object> result(String key) ;
}
```

4.【mldnspring-redis 项目】定义 IRedEnvelopeService 业务接口子类，利用 RedisTemplate 将红包信息保存到 Redis 数据库中，采用 list 数据类型（可以直接保存集合数据），可以使用“envelope-用户名-时间戳”形式作为红包 key 的生成格式，而用户抢红包的结果使用 hash 数据类型保存，key 定义名称为“result-envelope-用户名-时间戳”，里面属性 key 为用户名，Value 为抢红包金额。

```
package cn.mldn.mldnspring.service.impl;
import java.util.list;
import java.util.Map;
import org.springframework.beans.factory.annotation.Autowired;
import org.springframework.data.redis.core.RedisTemplate;
import org.springframework.stereotype.Service;
import cn.mldn.mldnspring.service.IRedEnvelopeService;
import cn.mldn.mldnspring.util.SplitMoneyUtil;
@Service
public class RedEnvelopeServiceImpl implements IRedEnvelopeService {
    @Autowired
```

```
    private RedisTemplate<String, Double> redisTemplate;                    // Redis操作模板
    @Override
    public String add(String userid,int amount, double money) {
        SplitMoneyUtil smu = new SplitMoneyUtil(amount,money) ;       // 设置总金额和拆分个数
        List<Double> result = smu.getAllPackages() ;                         // 获取红包信息
        String key = "envelope-" + userid + "-" + System.currentTimeMillis();
        this.redisTemplate.opsForlist().leftPushAll(key, result) ;           // 保存红包数据
        return key ;
    }
    @Override
    public Double grab(String userid,String key) {
        Double popResult = null ;
        String hashKey = "result-" + key ;                                   // 保存的hash-key
        boolean flag = this.redisTemplate.opsForHash().hasKey(hashKey, userid) ;// 是否抢过
        if (flag == false) {                                                // 不存在，没有抢过红包
            popResult = this.redisTemplate.opsForlist().leftPop(key) ;      // 取出一个数据
            if (popResult != null) {                                        // 保存结果
                this.redisTemplate.opsForHash().put(hashKey, userid, popResult);
            }
        } else {
            popResult = -1.0 ;                                              // 已经抢过了
        }
        return popResult;
    }
    @Override
    public Map<Object, Object> result(String key) {
        return this.redisTemplate.opsForHash().entries(key);                // 获取指定hash数据
    }
}
```

5.【mldnspring-redis 项目】编写测试程序，生成红包数据。

```
package cn.mldn.mldnspring.test;
import java.util.concurrent.CountDownLatch;
import org.junit.Test;
import org.junit.runner.RunWith;
import org.springframework.beans.factory.annotation.Autowired;
import org.springframework.test.context.ContextConfiguration;
import org.springframework.test.context.junit4.SpringJUnit4ClassRunner;
import cn.mldn.mldnspring.service.IRedEnvelopeService;
@ContextConfiguration(locations = {"classpath:spring/spring-*.xml"})
@RunWith(SpringJUnit4ClassRunner.class)                                   // 使用Junit进行测试
public class TestRedEnvelope {
    @Autowired
```

```
    private IRedEnvelopeService redEnvelopeService;                    // Redis操作模板
    @Test
    public void testSave() {                                           // 保存红包信息
        String key = this.redEnvelopeService.add("mldn",5, 66.66) ;
        System.out.println(key);
    }
}
```

程序执行结果	envelope-mldn-1519475635874

本程序执行后，会将红包数据保存在 Redis 数据库中，而后会返回红包数据存放的 key 名称。

6.【mldnspring-redis 项目】在测试程序中创建争抢红包程序，创建 10 个线程对象，模拟抢红包操作。

```
    @Test
    public void testGrab() throws Exception {
        String key = "envelope-mldn-1519475635874" ;
        CountDownLatch latch = new CountDownLatch(10) ;                // 线程阻塞处理
        for (int x = 0; x < 10; x++) {
            new Thread(()->{                                           // 使用线程名作为用户ID
                String userid = Thread.currentThread().getName() ;
                Double result = this.redEnvelopeService.grab(userid, key) ;
                if (result != null) {
                    System.out.println(userid + "抢到红包，金额：" + result);
                } else {
                    System.out.println(userid + "没有抢到红包！");
                }
                latch.countDown();                                     // 减少等待线程量
            },"MLDN用户-" + x) .start();
        }
        latch.await();                                                 // 等到全部抢完后输出信息
        System.out.println("*** 红包已全部抢完，最终结果：");
        this.redEnvelopeService.result("result-" + key).forEach((Object k,Object v)->{
            System.out.println("\t|- " + k + "抢到红包，金额：" + v);
        });
    }
```

程序执行结果	MLDN用户-3没有抢到红包！ MLDN用户-8没有抢到红包！ MLDN用户-2抢到红包，金额：12.34 MLDN用户-4没有抢到红包！ MLDN用户-5没有抢到红包！ MLDN用户-6抢到红包，金额：16.52 MLDN用户-7没有抢到红包！ MLDN用户-0抢到红包，金额：31.63

```
MLDN用户-9抢到红包，金额：2.11
MLDN用户-1抢到红包，金额：4.06
*** 红包已全部抢完，最终结果：
      |- MLDN用户-0抢到红包，金额：31.63
      |- MLDN用户-6抢到红包，金额：2.11
      |- MLDN用户-2抢到红包，金额：12.34
      |- MLDN用户-1抢到红包，金额：4.06
      |- MLDN用户-9抢到红包，金额：16.52
```

本测试程序模拟了 10 个抢红包客户端。如果抢到红包，则返回红包金额；如果没有抢到红包，则返回 null 数据。利用 null 数据，可以判断是否争抢失败。

10.10　本章小结

1．Redis 可以实现分布式高速缓存处理，每秒可承受近 10 万的访问量。

2．Redis 程序包提供的是源代码，需要手动编译后才可以获得程序命令。

3．Redis 支持的数据类型包含 string、hash、list、set、zset 和 GEO。

4．Redis 主从模式可以实现数据的自动备份处理。如果 Master 主机出现问题，利用哨兵机制可重新推选出新的 Master 主机，继续进行主从处理。

5．RedisCluster 是 Redis 官方提供的数据库集群技术，采用去中心化的设计思想，每个节点都可以提供数据服务，同时可以实现 Master 宕机后的重新选举。

6．SpringDataRedis 提供了 Redis 数据库处理功能，利用 RedisTemplate 可以降低开发的复杂度。

第 11 章

JDBC 操作模板

通过本章学习，可以达到以下目标：

1. 理解传统 JDBC 操作问题。
2. 掌握 C3P0 数据源配置。
3. 可以使用 JDBC Template 实现数据库的 CRUD（创建、读取、更新、删除）处理。
4. 掌握 Spring 缓存处理的使用。

Spring 是一个综合性的开发框架，为了便于项目开发，专门提供了 JDBC 操作模板，可以半自动化形式简化 JDBC 处理。本章将为读者讲解 JDBC 模板和数据源组件的使用。

11.1　JDBC 操作模板简介

Spring 设计之初，核心在于简化程序相关代码的开发。但在实际商业项目开发与运行过程中，数据库成了极为重要的数据存储中介，几乎所有项目都是以数据库操作为核心展开的。因此，必须对传统 JDBC 有如下清晰的认识：

☑　JDBC 是 Java 开发中定义的唯一一个与数据库操作相关的标准。只要是与数据库操作相关，一定是 JDBC 在提供默认支持。

☑　JDBC 提供的是一个开发标准，JDBC 处理包括加载数据库驱动程序、连接数据库、操作数据库、关闭数据库等核心操作。

☑　要想安全地执行数据库处理，需要使用 PreparedStatement，这样的做法在项目中必须要使用“?”进行数据内容的设置。

☑　DBC 中如果要执行查询，必须操作 ResultSet 集合，同时利用程序类将访问结果取回。

通过以上分析可以发现，如果所有项目都重复采用传统 JDBC 步骤进行处理，则会出现大量的冗余代码。为了简化这一操作，必须对其进行可重用设计。Spring 提供的 JDBC 可重用设计方案就是 JDBC Template（JDBC 模板）。

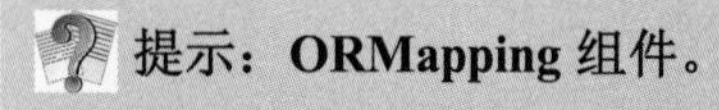
提示：ORMapping 组件。

JDBC 是一个标准，所以注定了其内容会极其琐碎。为了提升项目开发速度，一些开发者提供了大量 ORMapping 组件（对象关系映射），即结合配置文件（或注解）与反射机制实现 JDBC 的可重用定义。例如，JDO、Hibernate、EntityBean、IBatis、MyBatis、JPA 等都属于此类组件。

JDBC Template 是 ORMapping 设计中最小的一种组件。

JDBC Template 是 Spring 提供的一个简单到极致的 JDBC 操作模板组件，利用该组件可有效解决一些重复设计问题。依托 Spring 框架的 IOC 与 AOP 的操作特征，又可以实现连接配置以及事务处理控制。如表 11-1 所示是 JDBC Template 与传统 JDBC 开发的对比。

表 11-1 JDBC Template 与传统 JDBC 对比

	传统 JDBC 开发	JDBC Template
开发步骤	① 进行数据库驱动程序的加载。 ② 取得数据库的连接对象。 ③ 声明要操作的 SQL 语句（需要使用预处理）。 ④ 创建数据库操作对象。 ⑤ 执行 SQL 语句。 ⑥ 处理返回的操作结果（ResultSet）。 ⑦ 关闭结果集对象。 ⑧ 关闭数据库的操作对象（Statement）。 ⑨ 如果执行的是更新则应该进行事务提交或回滚。 ⑩ 关闭数据库连接	① 取得数据库的连接对象。 ② 声明要操作的 SQL 语句（需要使用预处理）。 ③ 执行 SQL 语句。 ④ 处理返回的操作结果（ResultSet）
优点	① 具备固定的操作流程，代码结构简单。 ② JDBC 是一个 Java 的公共服务，属于标准。 ③ 由于没有涉及过于复杂对象操作，所以性能是最高的	① 代码简单，但是又不脱离 JDBC 形式。 ② 由于有 Spring AOP 的支持，用户只关心核心。 ③ 对于出现的程序异常可以采用统一的方式进行处理。 ④ 与 JDBC 的操作步骤或形式几乎雷同
缺点	① 代码的冗余度太高了，每次都需要编写大量的重复操作。 ② 用户需要自己手工进行事务的处理操作。 ③ 所有的操作必须严格按照既定的步骤执行。 ④ 如果出现了执行的异常，则需要用户自己处理	① 与重度包装的 ORMapping 框架不同，不够智能。 ② 处理返回结果的时候不能够自动转化为 VO 类对象，需要由用户自己手工处理结果集

通过对比可以发现，使用 JDBC Template 且基于 Spring 开发管理，虽然程序代码不够智能，但与传统 JDBC 相比，开发者多数情况下只需要关注 SQL 定义以及返回结果的处理即可。

要在项目中使用 JDBC Template 处理，需要修改依赖配置库，同时还需要配置使用的数据库驱动。

1.【mldnspring 项目】修改 pom.xml 配置文件，配置 Spring-jdbc 依赖，以及 MySQL 的数据库驱动程序。

属性配置	<mysql.version>5.1.25</mysql.version>
依赖库配置	<dependency> <groupId>org.springframework</groupId> <artifactId>spring-jdbc</artifactId> <version>${spring.version}</version>

```
        </dependency>
        <dependency>
            <groupId>mysql</groupId>
            <artifactId>mysql-connector-java</artifactId>
            <version>${mysql.version}</version>
        </dependency>
```

2.【mldnspring-jdbc 项目】修改 pom.xml 配置文件引入以上的依赖包。

```
<dependency>
    <groupId>org.springframework</groupId>
    <artifactId>spring-jdbc</artifactId>
</dependency>
<dependency>
    <groupId>mysql</groupId>
    <artifactId>mysql-connector-java</artifactId>
</dependency>
```

项目中引入 spring-jdbc 依赖支持库后，就可以直接使用 JDBC Template 进行项目开发了。

3.【mldnspring-jdbc 项目】由于 JDBC Template 主要是进行数据库操作处理，所以这里将使用如下脚本创建一张新闻数据表，且包含常用的数据类型。

```
DROP DATABASE IF EXISTS mldn ;
CREATE DATABASE mldn CHARACTER SET UTF8 ;
USE mldn ;
CREATE TABLE news(
    nid         BIGINT    AUTO_INCREMENT ,
    title       VARCHAR(50) ,
    pubdate     DATETIME ,
    note        TEXT ,
    price       DOUBLE ,
    readcount   INT ,
    CONSTRAINT pk_nid PRIMARY KEY(nid)
) engine=innodb ;
```

脚本执行后，将利用新闻数据表来讲解 JDBC Template 的相关处理操作。

11.2 配置数据库连接

JDBC 开发标准中，如果要通过程序进行数据库的处理操作，首先要解决的就是数据库连接问题。开发者引入了 spring-jdbc 依赖库后，会提供一个 org.springframework.jdbc.datasource.DriverManager DataSource 程序类，此类为 javax.sql.DataSource 接口的子类，如图 11-1 所示。

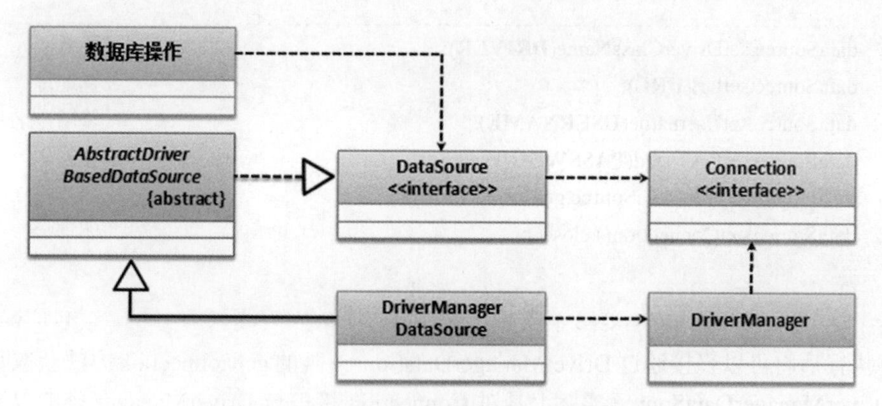

图 11-1　DriverManagerDataSource 类继承结构

利用此类可以直接实现 JDBC 的连接配置，该类中提供的常用操作方法如表 11-2 所示。

表 11-2　DriverManagerDataSource 类方法

编号	方法名称	类型	描述
1	public DriverManagerDataSource()	构造	主要可以用在 Spring 配置文件上处理
2	public DriverManagerDataSource(String url, String username, String password)	构造	设置数据库的连接地址、用户名、密码
3	public void setDriverClassName(String driver Class Name)	普通	设置数据库的驱动程序
4	public Connection getConnection() throws SQLException	普通	获取 JDBC 连接对象
5	public void setUrl(@Nullable String url)	普通	设置连接地址
6	public void setUsername(@Nullable String username)	普通	设置用户名
7	public void setPassword(@Nullable String password)	普通	设置连接密码
8	public void setSchema(@Nullable String schema)	普通	设置数据库的操作模式（模式名 = 用户名）

通过该类定义的方法可以发现，进行数据库连接时，只需填写上驱动程序、访问地址、用户名与密码，就可以通过 getConnection 方法获取数据库连接对象。对于 DriverManagerDataSource 类，可采用如下简单定义形式进行实例化。

范例：【mldnspring-jdbc 项目】实例化 DriverManagerDataSource 类对象。

```
package cn.mldn.mldnspring.dbc;
import org.springframework.jdbc.datasource.DriverManagerDataSource;
public class MySQLDatabaseConnectDemo {
    public static final String DRIVER = "org.gjt.mm.mysql.Driver" ;          // 驱动程序
    public static final String URL = "jdbc:mysql://localhost:3306/mldn" ;    // 连接地址
    public static final String USERNAME = "root" ;                           // 用户名
    public static final String PASSWORD = "mysqladmin" ;                     // 密码
    public static void main(String[] args) throws Exception {
        DriverManagerDataSource dataSource = new DriverManagerDataSource() ;
```

```
        dataSource.setDriverClassName(DRIVER);                    // 设置驱动
        dataSource.setUrl(URL);                                   // 设置地址
        dataSource.setUsername(USERNAME);                         // 设置用户名
        dataSource.setPassword(PASSWORD);                         // 设置密码
        System.out.println(dataSource.getConnection());           // 检测连接
        dataSource.getConnection().close();                       // 关闭连接
    }
}
```

程序执行后将可以直接通过 DriverManagerDataSource 类的 getConnection 方法获取数据库连接，即 DriverManagerDataSource 实际上是对 Connection 接口与 DriverManager 类的包装处理。但是在 Spring 开发过程中，如果直接采用关键字 new 的形式来创建对象是不合理的，最好的做法是将所有的对象管理交由 Spring 容器管理，所以应该在配置文件中进行定义。

1.【mldnspring-jdbc 项目】定义 src/main/resources/spring/spring-jdbc.xml 配置文件，进行数据库连接配置。

```
<bean id="dataSource" class="org.springframework.jdbc.datasource.DriverManagerDataSource">
    <property name="driverClassName" value="org.gjt.mm.mysql.Driver"/>  <!-- 驱动程序 -->
    <property name="url" value="jdbc:mysql://localhost:3306/mldn"/>     <!-- 连接地址 -->
    <property name="username" value="root"/>                            <!-- 用户名 -->
    <property name="password" value="mysqladmin"/>                      <!-- 密码 -->
</bean>
```

2.【mldnspring-jdbc 项目】通过 Spring 配置文件定义完成后，可以直接注入 DataSource 接口对象，编写测试类。

```
package cn.mldn.mldnspring.test;
import javax.sql.DataSource;
import org.junit.Test;
import org.junit.runner.RunWith;
import org.springframework.beans.factory.annotation.Autowired;
import org.springframework.test.context.ContextConfiguration;
import org.springframework.test.context.junit4.SpringJUnit4ClassRunner;
@ContextConfiguration(locations = { "classpath:spring/spring-*.xml" })
@RunWith(SpringJUnit4ClassRunner.class)                              // 设置要使用的测试工具
public class TestDataSource {
    @Autowired
    private DataSource dataSource ;                                  // 注入DataSource对象
    @Test
    public void testConnection() throws Exception {
        System.out.println(this.dataSource.getConnection());
    }
}
```

本程序通过配置文件获取了 DataSource 接口对象，此时即便不清楚具体的子类，也不会影响到数据库连接对象的获取。只有获取了数据库连接之后，才可以实现数据库操作。

提示：可以配置日志组件，观察操作信息。

上述程序执行时，会提示用户没有发现日志组件。开发者可修改 pom.xml 配置文件，添加 log4j-core、slf4j-api、slf4j-log4j12 依赖库，并在 CLASSPATH 路径下定义日志配置文件，即可在程序运行时通过日志信息观察程序执行过程。

11.3　使用 JDBC Template 操作数据库

JDBC Template 属于 JDBC 的轻度包装，使用过程中开发者仅需要考虑 SQL 语句的定义、参数传递与返回值处理。

org.springframework.jdbc.core.JdbcTemplate 类中定义了许多的数据库处理方法，常用的操作方法如表 11-3 所示。

表 11-3　JDBC Template 常用操作方法

编号	方法名称	类型	描述
1	public JdbcTemplate()	构造	定义 JDBC Template 对象，可以在 Spring 配置中使用
2	public JdbcTemplate(DataSource dataSource)	构造	接收 DataSource 对象
3	public void setDataSource(@Nullable DataSource dataSource)	普通	设置要使用的 DataSource
4	public <T> List<T> query(String sql, Object[] args, int[] argTypes, RowMapper<T> rowMapper) throws DataAccessException	普通	执行数据库查询操作
5	public <T> T queryForObject(String sql, Class<T> requiredType, @Nullable Object... args) throws DataAccessException	普通	返回单个查询对象
6	public int update(String sql, @Nullable Object... args) throws DataAccessException	普通	执行更新处理
7	public <T> int[][] batchUpdate(String sql, Collection<T> batchArgs, int batchSize, ParameterizedPreparedStatementSetter<T> pss) throws DataAccessException	普通	数据批量更新
8	public <T> T queryForObject(String sql, @Nullable Object[] args, RowMapper<T> rowMapper) throws DataAccessException	普通	RowMapper 查询转换

JDBC Template 如果想正常执行数据库操作，必须提供 DataSource 接口对象。此操作可以直接通过配置文件进行定义。

范例：【mldnspring-jdbc 项目】修改 spring-jdbc.xml 配置文件，追加 JDBC Template 类配置。

```
<!-- 配置JDBC Template的处理对象，如果想使用该对象，一定要配置数据库连接 -->
<bean id="jdbcTemplate" class="org.springframework.jdbc.core.JdbcTemplate">
    <property name="dataSource" ref="dataSource"/>          <!-- 配置数据源 -->
</bean>
```

配置完成后，JDBC Template 就有了可用的数据源对象。通过依赖注入向使用的程序类中注入 JDBC Template 对象，就可以进行数据库操作了。

范例：【mldnspring-jdbc 项目】编写测试类，注入 JDBC Template 对象，实现数据增加处理。

```
package cn.mldn.mldnspring.test;
import java.util.Date;
import org.junit.Test;
import org.junit.runner.RunWith;
import org.slf4j.Logger;
import org.springframework.beans.factory.annotation.Autowired;
import org.springframework.jdbc.core.JdbcTemplate;
import org.springframework.test.context.ContextConfiguration;
import org.springframework.test.context.junit4.SpringJUnit4ClassRunner;
@ContextConfiguration(locations = { "classpath:spring/spring-*.xml" })
@RunWith(SpringJUnit4ClassRunner.class)                    // 设置要使用的测试工具
public class TestJdbcTemplate {
    private Logger logger = org.slf4j.LoggerFactory.getLogger(TestJdbcTemplate.class) ;
    @Autowired
    private JdbcTemplate jdbcTemplate ;                    // 注入JDBC Template对象
    @Test
    public void testAdd() throws Exception {
        String sql = "INSERT INTO news(title,pubdate,note,price,readcount) VALUES (?,?,?,?,?)" ;
        String title = "MLDN魔乐科技" ;
        Date pubdate = new Date() ;
        String note = "技术教学：www.mldn.cn" ;
        double price = 19800.0 ;
        int readcount = 567000 ;
        int len = this.jdbcTemplate.update(sql, title,
                pubdate, note, price, readcount);          // 数据更新操作
        this.logger.info("更新行数：" + len);              // 日志输出
    }
}
```

程序执行结果	INFO [cn.mldn.mldnspring.test.TestJdbcTemplate] - 更新行数：1

本程序使用 JDBC Template 实现了数据增加操作。首先定义了要增加的 SQL 语句，而后利用 update 方法传入要执行的 SQL 语句与参数内容，即可实现数据追加处理。

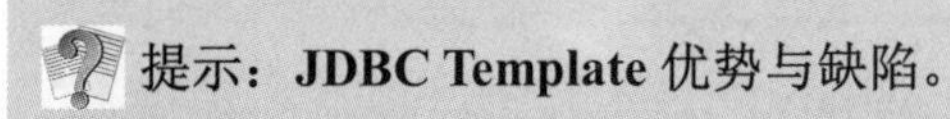

提示：JDBC Template 优势与缺陷。

通过程序可以发现，JDBC Template 的使用非常接近于传统 JDBC。但与传统 JDBC 操作相比，开发者不再需要关注 PreparedStatement 参数的设置问题（全部交由 JDBC Template 封装），所以性能相对较高。唯一的缺陷是：现代开发中常用 VO 或 PO 传递操作数据，需要依次取出对象中的数据，比较麻烦。

在使用 JDBC Template 模板处理时，如有必要，也可以利用 PreparedStatementSetter 接口实现 PreparedStatement 设置数据。

范例：【mldnspring-jdbc 项目】PreparedStatementSetter 操作。

```
@Test
public void testAddPreparedStatementSetter() throws Exception {
	String sql = "INSERT INTO news(title,pubdate,note,price,readcount)" +
				"VALUES (?,?,?,?,?)" ;
	String title = "MLDN魔乐科技" ;
	Date pubdate = new Date() ;
	String note = "技术教学：www.mldn.cn" ;
	double price = 19800.0 ;
	int readcount = 567000 ;
	int len = this.jdbcTemplate.update(sql, new PreparedStatementSetter() {
		@Override
		public void setValues(PreparedStatement ps) throws SQLException {
			ps.setString(1, title);
			ps.setDate(2, new java.sql.Date(pubdate.getTime()));
			ps.setString(3, note);
			ps.setDouble(4, price);
			ps.setInt(5, readcount);
		}
	}) ;
	this.logger.info("更新行数：" + len);			// 日志输出
}
```

一旦使用 PreparedStatementSetter 接口进行数据设置处理，则整体形式会更接近传统 JDBC，同时程序的烦琐程度也会大大提高。所以，如果不是必要，不建议采用此类做法。

本程序所使用的新闻数据表主键采用的是自动增长的处理形式完成的，很多时候开发者往往需要获取增长后的 ID 信息，此时可以使用 org.springframework.jdbc.support.KeyHolder 接口完成。在 JDBC Template 执行 update 方法时，传入 KeyHolder 接口对象即可将增长后的 ID 保存，随后可以通过 getKey 方法获取。KeyHolder 使用结构如图 11-2 所示。

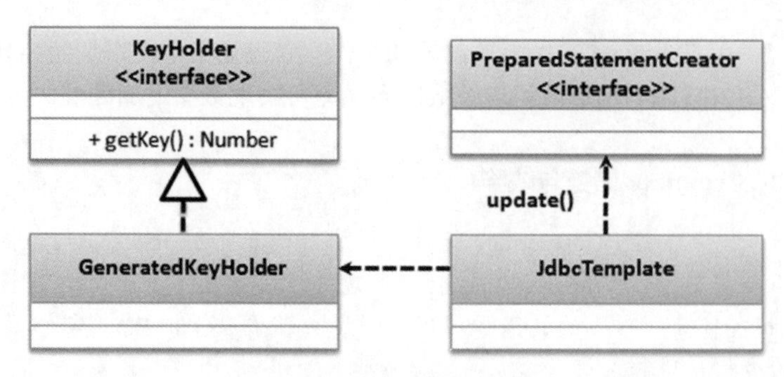

图 11-2　KeyHolder 操作

注意：KeyHolder 需要合适的 JDBC 驱动程序才可以使用。

要想使用 KeyHolder 接口获取增长后的 ID，需要选择合适的数据库驱动程序。否则，会出现程序执行错误或无法获取增长后的 ID。由此可见，此类操作的可适应性与 MyBatis 或 JPA 相比相对较差。

范例：【mldnspring-jdbc 项目】获取增长后的 ID 信息。

```
@Test
public void testAddKeyHolder() throws Exception {
	KeyHolder keyHolder = new GeneratedKeyHolder() ;				// 获得增长后的ID数据
	String sql = "INSERT INTO news(title,pubdate,note,price,readcount) VALUES (?,?,?,?,?)" ;
	String title = "MLDN魔乐科技" ;
	Date pubdate = new Date() ;
	String note = "技术教学：www.mldn.cn" ;
	double price = 19800.0 ;
	int readcount = 567000 ;
	int len = this.jdbcTemplate.update(new PreparedStatementCreator() {
		@Override
		public PreparedStatement createPreparedStatement(Connection con)
				throws SQLException {
			PreparedStatement ps = con.prepareStatement(sql);
			ps.setString(1, title);
			ps.setDate(2, new java.sql.Date(pubdate.getTime()));
			ps.setString(3, note);
			ps.setDouble(4, price);
			ps.setInt(5, readcount);
			return ps;
		}
	}, keyHolder);
	this.logger.info("更新行数： " + len + "、增长后的ID： " + keyHolder.getKey());
}
```

程序执行结果	INFO [cn.mldn.mldnspring.test.TestJdbcTemplate] - 更新行数：1、增长后的ID：3

这里由于需要使用 KeyHolder 接口获取增长后的 ID，所以在进行数据库更新处理中必须使用 PreparedStatementCreator 接口手动创建 PreparedStatement 对象，执行更新后才可以通过 getKey 方法获取当前 ID。

JDBC Template 类中的 update 方法除了可以执行数据增加外，也可以实现数据的修改与删除操作。

范例：【mldnspring-jdbc 项目】使用 JDBC Template 实现数据修改处理。

```
@Test
public void testEdit() throws Exception {
    String sql = "UPDATE news SET title=?,pubdate=?,note=?,price=?,readcount=? WHERE nid=?" ;
    String title = "极限IT程序员" ;
    Date pubdate = new Date() ;
    String note = "线上培训：www.jixianit.com" ;
    double price = 3980.0 ;
    int readcount = 7867000 ;
    long nid = 3 ;
    int len = this.jdbcTemplate.update(sql, title,
            pubdate, note, price, readcount,nid);                // 数据更新操作
    this.logger.info("更新行数：" + len);                        // 日志输出
}
```

程序执行结果	INFO [cn.mldn.mldnspring.test.TestJdbcTemplate] - 更新行数：1

范例：【mldnspring-jdbc 项目】使用 JDBC Template 实现数据删除操作。

```
@Test
public void testDelete() {
    String sql = "DELETE FROM news WHERE nid=?";
    long nid = 2;
    int len = this.jdbcTemplate.update(sql, nid);                // 更新操作
    this.logger.info("更新行数：" + len);                        // 日志输出
}
```

程序执行结果	INFO [cn.mldn.mldnspring.test.TestJdbcTemplate] - 更新行数：1

实际数据库开发过程中，经常需要进行大规模数据更新操作。JDBC Template 中也支持批处理，需要使用 org.springframework.jdbc.core.ParameterizedPreparedStatementSetter 进行内容设置。

范例：【mldnspring-jdbc 项目】实现数据批量增加。

☑　定义一个可以描述新闻数据表结构的 VO 类。

```
package cn.mldn.mldnspring.vo;
import java.io.Serializable;
import java.util.Date;
@SuppressWarnings("serial")
public class News implements Serializable {
    private Long nid ;
    private String title ;
```

```
    private Date pubdate ;
    private Double price ;
    private String note ;
    private Integer readcount ;
    // setter、getter、toString略
}
```

☑ 将要保存的数据保存在 VO 类中，随后通过 ParameterizedPreparedStatementSetter 设置每一条更新语句数据。

```
@Test
public void testBatch() {
    String sql = "INSERT INTO news(title,pubdate,note,price,readcount) VALUES (?,?,?,?,?)";
    List<News> allNews = new ArrayList<News>();
    for (int x = 0; x < 10; x++) {
        News vo = new News();
        vo.setTitle("极限IT程序员" + x);
        vo.setNote("www.jixianit.com");
        vo.setPubdate(new Date());
        vo.setPrice(999.00);
        vo.setReadcount(89765);
        allNews.add(vo);
    }
    int len[][] = this.jdbcTemplate.batchUpdate(sql, allNews, allNews.size(),
            new ParameterizedPreparedStatementSetter<News>() {
                @Override
                public void setValues(PreparedStatement ps, News vo) throws SQLException {
                    ps.setString(1, vo.getTitle());
                    ps.setDate(2, new java.sql.Date(vo.getPubdate().getTime()));
                    ps.setString(3, vo.getNote());
                    ps.setDouble(4, vo.getPrice());
                    ps.setInt(5, vo.getReadcount());
                }
            });
}
```

本程序创建了 10 个 News 类对象，并且通过 JDBC Template 类中的 batchUpdate 方法实现了数据的批量保存。

11.4 数据查询

使用传统 JDBC 进行数据库查询时，返回结果包装在 ResultSet 集合中，利用 ResultSet 中提

供的方法可取出每一行数据，在很多时候还需要将返回结果内容保存到 VO 类中进行返回。JDBC Template 中，也支持同样的处理形式，但需要 org.springframework.jdbc.core.RowMapper 接口支持。此接口的定义如下：

```
@FunctionalInterface
public interface RowMapper<T> {
    @Nullable
    public T mapRow(ResultSet rs, int rowNum) throws SQLException;
}
```

此接口中只定义了一个 mapRow 方法，作用是将 ResultSet 返回的数据转换为指定的 VO 对象。mapRow 方法接收两个参数：ResultSet 结果集对象和数据行号。

范例：【mldnspring-jdbc 项目】根据 id 查询数据。

```
@Test
public void testSingle() {
    String sql = "SELECT nid,title,pubdate,note,price,readcount FROM news WHERE nid=?" ;
    // 数据查询，传入SQL语句与参数（通过对象数组包装），查询还需要通过RowMapper处理
    News vo = this.jdbcTemplate.queryForObject(sql, new Object[] {6} ,
            new RowMapper<News>() {
            @Override
            public News mapRow(ResultSet rs, int rowNum) throws SQLException {
                News vo = new News() ;
                vo.setNid(rs.getLong(1));
                vo.setTitle(rs.getString(2));
                vo.setPubdate(rs.getDate(3));
                vo.setNote(rs.getString(4));
                vo.setPrice(rs.getDouble(5));
                vo.setReadcount(rs.getInt(6));
                return vo;
            }}) ;
    this.logger.info(vo.toString());
}
```

程序执行结果	INFO [cn.mldn.mldnspring.test.TestJdbcTemplate] - News [nid=6, title=MLDN魔乐科技, pubdate=2019-02-18, price=19800.0, note=技术教学：www.mldn.cn, readcount=567000]

在进行新数据查询时，利用 RowMapper 接口将 ResultSet 接口返回的查询结果转为了 VO 类对象，使查询结果与 VO 转换更加方便。

范例：【mldnspring-jdbc 项目】数据分页查询。

```
@Test
public void testSplit() {
    String column = "title";                                    // 查询列
    String keyWord = "极限IT";                                  // 查询关键字
    Long currentPage = 1L;                                      // 当前所在页
```

```
        Integer lineSize = 5;                                    // 每页显示数据行
        String sql = "SELECT nid,title,pubdate,note,price,readcount FROM news WHERE " +
                column + " LIKE ? LIMIT ?,?";
        List<News> allNews = this.jdbcTemplate.query(sql,
                new Object[] { "%" + keyWord + "%", (currentPage - 1) * lineSize, lineSize },
                new RowMapper<News>() {
                    @Override
                    public News mapRow(ResultSet rs, int rowNum) throws SQLException {
                        News vo = new News();
                        vo.setNid(rs.getLong(1));
                        vo.setTitle(rs.getString(2));
                        vo.setPubdate(rs.getDate(3));
                        vo.setNote(rs.getString(4));
                        vo.setPrice(rs.getDouble(5));
                        vo.setReadcount(rs.getInt(6));
                        return vo;
                    }
                });
    }
```

本程序实现了一个数据分页查询处理操作。在定义查询语句时，采用了多个参数，所以在执行查询中必须利用对象数组封装多个参数，由于 RowMapper 是针对某一数据表统一的转换接口，所以本程序直接利用 RowMapper 进行每一行数据的转换，多个结果直接以 List 集合返回。

范例：【mldnspring-jdbc 项目】 数据统计查询。

```
    @Test
    public void testCount() {
        String column = "title";
        String keyWord = "极限IT";
        String sql = "SELECT COUNT(*) FROM news WHERE " + column + " LIKE ?";
        Long count = this.jdbcTemplate.queryForObject(sql, Long.class, "%" + keyWord + "%");
        this.logger.info("数据量统计：" + count);
    }
```

程序执行结果	INFO [cn.mldn.mldnspring.test.TestJdbcTemplate] - 数据量统计：11

本程序实现了 COUNT 统计查询。由于返回的只是一个结果，所以利用 JDBC Template 类中的 queryForObject 方法执行查询，并且指明了返回的类型为 Long。

11.5 Spring 数据缓存

开发过程中，经常会出现多个线程读取同一数据的情况。以 JDBC 访问为例，可能有十个不同的线程读取了同一条新闻数据，那么在默认情况下肯定会进行十次数据读取操作。很明显，

这样的做法存在问题。为了提高性能，可以直接将某些数据进行缓存，这样在进行读取时就可以避免数据的重复加载。

计算机系统运行时，程序指令是交由 CPU 处理的。CPU 首先会通过内存读取需要的数据，内存中没有的数据，则通过本地磁盘或其他网络主机进行读取，如图 11-3 所示。加载本地磁盘时，寻址和网络加载延迟问题会造成数据加载缓慢。所以，将一些可能重用的数据保存在内存中，能有效提升此类反复读取操作的性能。

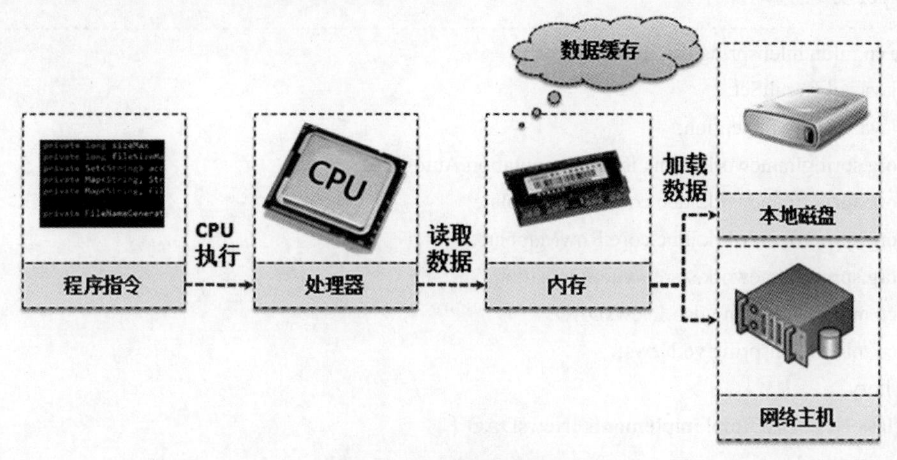

图 11-3 数据缓存机制

Spring 3.x 之后的版本都提供了缓存处理机制。Spring 的缓存机制非常灵活，可以对容器中任意程序类或类中的方法进行缓存，因此这种缓存机制可以在 Java EE 应用的任何层次上进行。

下面将通过一个简单程序介绍下未加入缓存机制时的程序问题。该程序的实现结构如图 11-4 所示，采用业务层与数据层设计，利用 JDBC Template 实现数据查询操作。

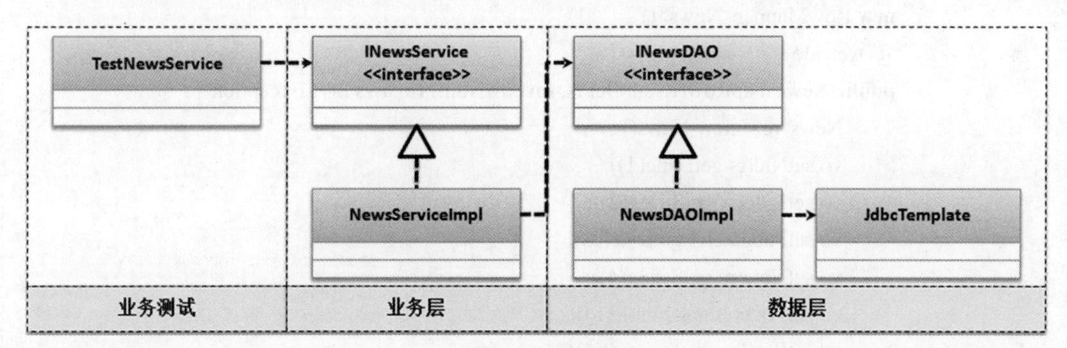

图 11-4 程序逻辑结构

1.【mldnspring-jdbc 项目】建立 INewsDAO 接口，定义数据层操作标准。

```
package cn.mldn.mldnspring.dao;
import cn.mldn.mldnspring.vo.News;
public interface INewsDAO {
    /**
     * 根据ID查询新闻数据
     * @param nid 新闻ID
```

```
     * @return 新闻数据对象
     */
    public News findById(Long nid) ;
}
```

2.【mldnspring-jdbc 项目】建立 NewsDAOImpl 子类，并且注入 JDBC Template 对象，利用 RowMapper 处理查询操作。

```
package cn.mldn.mldnspring.dao.impl;
import java.sql.ResultSet;
import java.sql.SQLException;
import org.springframework.beans.factory.annotation.Autowired;
import org.springframework.jdbc.core.JdbcTemplate;
import org.springframework.jdbc.core.RowMapper;
import org.springframework.stereotype.Repository;
import cn.mldn.mldnspring.dao.INewsDAO;
import cn.mldn.mldnspring.vo.News;
@Repository
public class NewsDAOImpl implements INewsDAO {
    @Autowired
    private JdbcTemplate jdbcTemplate ;
    @Override
    public News findById(Long nid) {
        String sql = "SELECT nid,title,pubdate,note,price,readcount FROM news WHERE nid=?" ;
        News vo = this.jdbcTemplate.queryForObject(sql, new Object[] {nid} ,
            new RowMapper<News>() {
            @Override
            public News mapRow(ResultSet rs, int rowNum) throws SQLException {
                News vo = new News() ;
                vo.setNid(rs.getLong(1));
                vo.setTitle(rs.getString(2));
                vo.setPubdate(rs.getDate(3));
                vo.setNote(rs.getString(4));
                vo.setPrice(rs.getDouble(5));
                vo.setReadcount(rs.getInt(6));
                return vo;
            }}) ;
        return vo ;
    }
}
```

3.【mldnspring-jdbc 项目】建立 INewsService 业务接口。

```
package cn.mldn.mldnspring.service;
import cn.mldn.mldnspring.vo.News;
```

```
public interface INewsService {
    /**
     * 根据ID查询数据
     * @param nid 新闻编号
     * @return 新闻数据对象
     */
    public News get(long nid) ;
}
```

4.【mldnspring-jdbc 项目】建立 NewsServiceImpl 子类，注入 INewsDAO 接口对象实现数据层操作。

```
package cn.mldn.mldnspring.service.impl;
import org.springframework.beans.factory.annotation.Autowired;
import org.springframework.stereotype.Service;
import cn.mldn.mldnspring.dao.INewsDAO;
import cn.mldn.mldnspring.service.INewsService;
import cn.mldn.mldnspring.vo.News;
@Service
public class NewsServiceImpl implements INewsService {
    @Autowired
    private INewsDAO newsDAO ;
    @Override
    public News get(long nid) {
        return this.newsDAO.findById(nid);
    }
}
```

本程序注入了 INewsDAO 接口对象，随后利用 DAO 层提供的 findById 方法实现指定编号的新闻数据查询。

5.【mldnspring-jdbc 项目】建立测试类，测试 INewsService 接口功能，为了方便观察问题，可将 News 类中的 toString 方法取消，即采用 Object 类中默认的 toString 方法，只得到对象编号即可。

```
package cn.mldn.mldnspring.test;
import org.junit.Test;
import org.junit.runner.RunWith;
import org.springframework.beans.factory.annotation.Autowired;
import org.springframework.test.context.ContextConfiguration;
import org.springframework.test.context.junit4.SpringJUnit4ClassRunner;
import cn.mldn.mldnspring.service.INewsService;
@ContextConfiguration(locations = { "classpath:spring/spring-*.xml" })
@RunWith(SpringJUnit4ClassRunner.class)                    // 设置要使用的测试工具
public class TestNewsService {
```

```
        @Autowired
        private INewsService newsService;
        @Test
        public void testFindById() throws Exception {
                System.out.println(this.newsService.get(3L));
                System.out.println(this.newsService.get(3L));
        }
}
```

程序执行结果	cn.mldn.mldnspring.vo.News@34b9f960（第一次查询） cn.mldn.mldnspring.vo.News@791d1f8b（第二次查询，ID相同）

通过程序可以发现，在一个执行线程内使用 INewsService 接口的 get 方法查询数据时，实际上每次都获得了一个新的 News 类对象。所以，在没有数据缓存支持的情况下，将产生重复查询操作。

11.5.1 Spring 缓存实现

Spring 缓存实现过程中，主要定义了两个接口标准：数据缓存（org.springframework.cache.Cache）和缓存管理器（org.springframework.cache.CacheManager）。其中，缓存管理器描述了缓存数据的保存位置。两个接口的关系如图 11-5 所示。

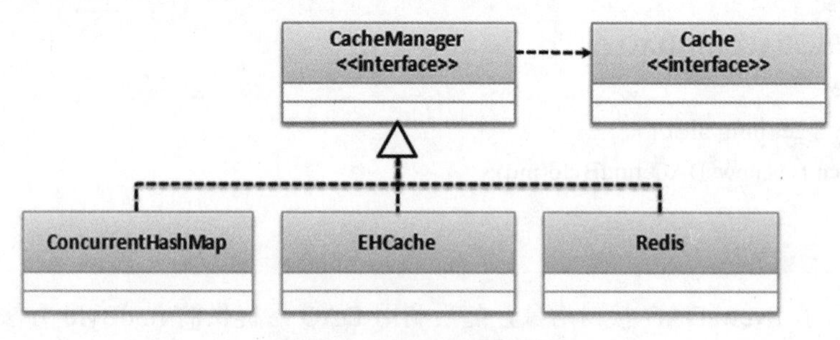

图 11-5 Spring 缓存架构

在项目中引入缓存处理时，除需要配置缓存管理器外，还需要在缓存目标上使用@Cacheable 注解。在此注解中定义了许多配置属性，如表 11-4 所示。

表 11-4 @Cacheable 注解配置属性

编号	属性	描述
1	cacheManager	定义缓存管理器名称
2	cacheNames	配置缓存区名称，该名称是在缓存管理器中配置的缓存区名称
3	key	缓存采用 key-value 形式存储，此处可以设置缓存的 key 名称，可使用 SpEL
4	sync	设置同步缓存，将采用阻塞策略进行缓存更新
5	condition	设置条件缓存
6	unless	设置缓存排除信息

下面将直接在业务层中进行缓存配置，手动实现一个缓存处理。

1.【mldnspring-jdbc 项目】建立 spring/spring-base.xml 配置文件，进行 context 扫描路径配置。

```
<context:component-scan base-package="cn.mldn.mldnspring" />
```

2.【mldnspring-jdbc 项目】建立 spring/spring-cache.xml 配置文件。该配置文件需要引入 cache 命名空间，同时要进行缓存管理器配置（org.springframework.cache.concurrent.ConcurrentMapCacheFactoryBean）。这里选择 ConcurrentHashMap 实现缓存数据存储。

```
<?xml version="1.0" encoding="UTF-8"?>
<beans xmlns="http://www.springframework.org/schema/beans"
    xmlns:xsi="http://www.w3.org/2001/XMLSchema-instance"
    xmlns:cache="http://www.springframework.org/schema/cache"
    xsi:schemaLocation="
        http://www.springframework.org/schema/cache
        http://www.springframework.org/schema/cache/spring-cache-4.3.xsd
        http://www.springframework.org/schema/beans
        http://www.springframework.org/schema/beans/spring-beans-4.3.xsd">
    <cache:annotation-driven cache-manager="cacheManager" />        <!-- 启用缓存注解配置 -->
    <bean id="cacheManager" class="org.springframework.cache.support.SimpleCacheManager">
        <property name="caches">                                    <!-- 定义缓存管理 -->
            <set>                                                   <!-- 设置多个缓存区管理信息 -->
                <bean name="news"
                class="org.springframework.cache.concurrent.ConcurrentMapCacheFactoryBean" />
            </set>
        </property>
    </bean>
</beans>
```

本配置文件中定义了一个 news 缓存处理类对象，在使用缓存注解时必须设置与之相同的名称。

3.【mldnspring-jdbc 项目】修改 INewsService 业务接口，使用@Cacheable 注解定义需要启用缓存的配置。该注解可以定义在接口中，表示所有方法都需要缓存；也可以定义在某个方法上，表示只缓存此方法的数据。

```
public interface INewsService {
    @Cacheable(cacheNames = "news")     // 启用缓存，同时定义使用的缓存管理器中的缓存信息名称
    public News get(long nid) ;
}
```

本程序在 INewsService 接口的 get 方法上使用了@Cacheable 注解，这样一来只有该方法上的操作可以启用缓存。INewsService 接口中如果定义有其他方法，将无法实现数据缓存。缓存配置完成后，可以再次执行测试程序，此时可以发现不管查询多少次，都将只获得一个 News 对象的信息，即只会调用一次数据层操作。这表示该对象已经被成功缓存，而相应的查询性能也将得到提升。

11.5.2 @Cacheable 注解

进行缓存操作时，所有数据都是以 key-value 的形式存储的。默认情况下，Spring 会为每个缓存对象进行命名处理。如果用户有需要，也可以在@Cacheable 注解中设置缓存的 key 生成策略。

范例：【mldnspring-jdbc 项目】使用缓存 key 生成策略，将当前的新闻编号作为缓存 key 名称。

```
@Cacheable(cacheNames = "news", key = "#nid")
public News get(long nid) ;
```

此时程序采用 SpEL 表达式，直接使用 get 方法中的 nid 参数作为缓存的名称。如果此时传递参数为一个对象，也可以采用 key = "#对象参数.属性"形式，使用类中的某一个属性作为缓存名称。

默认情况下，Spring 对标注了@Cacheable 注解的方法不会进行线程同步管理，这样当多个线程同时访问时，就可能造成查询多次的处理问题。为了解决此操作，可以采用同步阻塞的方式进行处理，即当有多个线程使用同一参数同时查询时，Spring 的缓存管理器将采用阻塞的形式只发出一次查询处理，其他线程将进入到阻塞状态。成功取得数据后，所有阻塞的线程将可以直接缓存数据。

范例：【mldnspring-jdbc 项目】使用 sync 实现缓存同步处理。

```
@Cacheable(cacheNames = "news", key = "#nid", sync = true)
public News get(long nid) ;
```

在进行数据缓存处理时，也可以使用 condition 属性定义数据被缓存的触发条件，此属性可以直接利用 SpEL 进行条件判断。

范例：【mldnspring-jdbc 项目】定义条件缓存，新闻 ID 数据大于 50 的才进行数据缓存。

```
@Cacheable(cacheNames = "news", key = "#nid", sync = true, condition="#nid > 50")
public News get(long nid) ;
```

默认情况下 Spring 缓存会针对于配置注解而后进行数据缓存，如果需要设置一些不缓存的数据，可以使用 unless，在此属性中可以利用 SpEL 表达式进行判断。如表 11-5 所示为 Spring 缓存中可以使用的 SpEL 上下文表达式。

表 11-5 SpEL 上下文数据

编号	调用范围目标	位置	范例
1	当前调用的方法名称	root 对象	#root.methodName
2	当前执行的方法	root 对象	#root.method.name
3	当前执行的目标对象	root 对象	#root.target
4	当前执行目标对象所属类	root 对象	#root.targetClass
5	当前调用参数列表	root 对象	#root.args[0] 或 #root.args[1]
6	当前方法调用使用的缓存列表	root 对象	#root.caches[0].name
7	当前调用方法参数，可以是普通参数或者是对象参数	执行上下文	get(long nid)，则为#nid edit(News vo)，则为#vo.nid

续表

编号	调用范围目标	位置	范例
8	方法执行的返回值	执行上下文	#result（#result.属性表示返回对象属性）

范例：【mldnspring-jdbc 项目】定义缓存排除策略。

```
@Cacheable(cacheNames = "news", key = "#nid", condition="#nid > 50",unless="#result==null")
public News get(long nid) ;
```

本程序使用了 unless="#result==null"属性，该配置表示：结果返回为 null 时，不再进行数据缓存。

11.5.3　缓存更新策略

程序中配置了缓存处理策略后，还需要考虑数据更新问题。例如，已经缓存了编号为 3 的新闻数据，当进行新闻数据更新时，缓存中的数据也需要同时进行修改，否则缓存的数据将永远只能是旧数据。Spring 中，缓存数据的更新可以利用@CachePut 注解来实现。该注解相关属性如表 11-6 所示。

表 11-6　@CachePut 相关属性

编号	属性	描述
1	cacheNames	缓存区名称
2	key	要更新的缓存key，可以使用SpEL描述
3	condition	缓存更新条件

下面通过以下的程序步骤为读者演示具体操作。

1．【mldnspring-jdbc 项目】修改 INewsDAO 接口，在数据层中追加数据更新方法。

```
/**
 * 更新新闻数据
 * @param vo 要更新的新闻数据
 * @return 更新成功返回true，否则返回false
 */
public boolean doUpdate(News vo) ;
```

2．【mldnspring-jdbc 项目】在 NewsDAOImpl 子类中利用 JDBC Template 实现数据更新处理。

```
@Override
public boolean doUpdate(News vo) {
    String sql = "UPDATE news SET title=?,pubdate=?,note=?,price=?,readcount=? WHERE nid=?" ;
    int len = this.jdbcTemplate.update(sql, vo.getTitle(), vo.getPubdate(), vo.getNote(),
            vo.getPrice(), vo.getReadcount(), vo.getNid());        // 数据更新操作
    return len > 0 ;
}
```

3．【mldnspring-jdbc 项目】扩充 INewsService 业务接口中提供的方法，由于此时需要进行缓存更新，所以方法上要使用@CachePut 注解。

```
/**
 * 数据更新操作
 * @param vo 要更新的数据
 * @return 更新后的数据，主要用于缓存操作
 */
@CachePut(cacheNames = "news", key = "#vo.nid", unless="#result==null")
public News edit(News vo) ;
```

由于在 INewsService 的 get 方法上使用新闻编号作为缓存 key，所以在修改方法中需要利用 SpEL 使用参数对象中的 nid 作为缓存 key。

4.【mldnspring-jdbc 项目】在 NewsServiceImpl 子类中实现数据修改操作方法。

```
@Override
public News edit(News vo) {
    if (this.newsDAO.doUpdate(vo)) {
        return vo ;
    }
    return null;
}
```

5.【mldnspring-jdbc 项目】编写测试类，实现缓存修改操作。

```
@Test
public void testEdit() throws Exception {
    System.out.println(this.newsService.get(3L));          // 查询指定编号数据，进行缓存
    News vo = new News() ;
    vo.setNid(3L);
    vo.setTitle("MLDN新闻更新");
    vo.setNote("www.mldn.cn");
    vo.setPrice(3980.00);
    vo.setReadcount(97898);
    vo.setPubdate(new Date());
    this.newsService.edit(vo) ;                            // 更新缓存中的数据
    System.out.println(this.newsService.get(3L));          // 获取更新后的内容
}
```

在本程序中，利用 INewsService 业务层中的 edit 方法实现了缓存对象的信息更新，当再次使用 get 方法获取数据时，依然不会通过数据层发出查询指令，并且可以取得新的新闻数据信息。

11.5.4 缓存清除

数据被缓存后，用户再次查询该数据时，不必再通过数据层进行 JDBC 数据加载。但当指定数据需要删除时，相应的缓存数据也应该被系统自动删除。要实现缓存的清除操作，需要使用@CacheEvict 注解。@CacheEvict 注解的相关属性如表 11-7 所示。

表 11-7　@CacheEvict 相关属性

编号	属性	描述
1	cacheNames	缓存区名称
2	key	缓存的 key，可以使用 SpEL 编写
3	condition	缓存删除条件
4	allEntries	是否清除全部缓存数据，如果指定为 true，则方法调用后将立即清空所有缓存
5	beforeInvocation	是否在方法执行前就清空，默认为 false，如果指定为 true，则在方法还没有执行的时候就清空缓存，默认情况下，如果方法执行抛出异常，则不会清空缓存

1．【mldnspring-jdbc 项目】在 INewsDAO 接口中追加数据删除方法。

```
/**
 * 根据编号删除指定新闻数据
 * @param nid 新闻编号
 * @return 删除成功返回true，否则返回false
 */
public boolean doRemove(Long nid) ;
```

2．【mldnspring-jdbc 项目】在 NewsDAOImpl 子类中实现 doRemove 方法，利用 JDBC Template 实现数据删除。

```
@Override
public boolean doRemove(Long nid) {
    String sql = "DELETE FROM news WHERE nid=?";
    int len = this.jdbcTemplate.update(sql, nid);           // 数据删除
    return len > 0 ;
}
```

3．【mldnspring-jdbc 项目】在 INewsService 业务接口中扩充新的业务方法。

```
/**
 * 删除新闻数据
 * @param nid 新闻编号
 * @return 删除成功返回true，否则返回false
 */
@CacheEvict(cacheNames="news", key="#nid")
public boolean delete(long nid) ;
```

在 delete 业务方法中使用了@CacheEvict 注解，并且配置了要删除的缓存 key 名称。这样，当执行此方法后，对应 key 的缓存会被自动删除。

4．【mldnspring-jdbc 项目】在 NewsServiceImpl 子类中实现删除业务。

```
@Override
public boolean doRemove(Long nid) {
```

```
        String sql = "DELETE FROM news WHERE nid=?";
        int len = this.jdbcTemplate.update(sql, nid);                    // 数据删除
        return len > 0 ;
    }
```

5.【mldnspring-jdbc 项目】编写测试类。

```
    @Test
    public void testDelete() throws Exception {
        System.out.println(this.newsService.get(6L));            // 指定数据缓存
        this.newsService.delete(6L) ;                            // 删除指定编号新闻，并且清除缓存
        System.out.println(this.newsService.get(6L));            // 此时将无法获取缓存信息
    }
```

本测试程序中首先进行了指定编号的缓存数据加载，而后删除了指定数据。由于@CacheEvict 注解的作用，数据被删除后，缓存也同样被清除。再次查询指定编号新闻时，将返回 null。如果业务方法上没有使用@CacheEvict 注解，则即使数据删除了，也依然可以被缓存机制继续存储。

11.5.5 @CacheConfig 缓存统一配置

进行业务层方法缓存配置时可以发现，几乎需要为每个@Cacheable 注解定义使用的缓存名称，这样的做法实在是过于重复了。在 Spring 缓存配置中，使用@CacheConfig 可为一个业务层的所有方法进行统一配置。

范例：【mldnspring-jdbc 项目】修改 INewsService 业务接口。

```
package cn.mldn.mldnspring.service;
import org.springframework.cache.annotation.CacheConfig;
import org.springframework.cache.annotation.CacheEvict;
import org.springframework.cache.annotation.CachePut;
import org.springframework.cache.annotation.Cacheable;
import cn.mldn.mldnspring.vo.News;
@CacheConfig(cacheNames="news")                                  // 进行该接口缓存的统一配置
public interface INewsService {
    @Cacheable(key = "#nid", unless="#result==null")
    public News get(long nid) ;
    @CachePut(key = "#vo.nid", unless="#result==null")
    public News edit(News vo) ;
    @CacheEvict(key="#nid")
    public boolean delete(long nid) ;
}
```

本程序在 INewsService 接口上统一使用@CacheConfig 注解配置了要使用的缓存名称，避免了每个业务方法上的重复配置。

11.5.6　多级缓存策略

前面介绍缓存配置时，简单地使用新闻编号作为缓存数据的 key。实际开发中，可能需要通过新闻编号或新闻标题来进行数据缓存 key 的配置，这时需要用@Caching 注解进行配置。@Caching 注解的属性如表 11-8 所示。

表 11-8　@Caching 相关属性

编号	属性	描述
1	cacheable	定义多个缓存配置注解
2	put	定义多个更新缓存配置注解
3	evict	定义多个删除缓存配置注解

范例：【mldnspring-jdbc 项目】在 INewsService 接口上追加多级缓存配置。

```
@CacheConfig(cacheNames="news")                        // 进行该接口缓存的统一配置
public interface INewsService {
    @Caching(put= {
            @CachePut(key = "#vo.nid",unless="#result==null"),
            @CachePut(key = "#vo.title",unless="#result==null")
    })
    public News edit(News vo) ;
    // 其他操作方法略
}
```

本程序在 edit 方法上定义了多个缓存 key，数据更新后可通过新闻编号或新闻标题来获取缓存数据。

这样的缓存配置过于复杂，如果若干个业务方法有相同的缓存配置，使用起来会非常麻烦，可以采用自定义注解的形式来简化业务层调用。

范例：【mldnspring-jdbc 项目】自定义注解。

```
package cn.mldn.mldnspring.cache;
import java.lang.annotation.ElementType;
import java.lang.annotation.Inherited;
import java.lang.annotation.Retention;
import java.lang.annotation.RetentionPolicy;
import java.lang.annotation.Target;
import org.springframework.cache.annotation.CachePut;
import org.springframework.cache.annotation.Caching;
@Caching(put= {
        @CachePut(key = "#vo.nid", unless="#result==null"),
        @CachePut(key = "#vo.title", unless="#result==null")
})
@Target({ElementType.METHOD, ElementType.TYPE})
@Retention(RetentionPolicy.RUNTIME)
```

```
@Inherited
public @interface NewsEditCache {                          // 该注解主要用于新闻数据更新缓存处理
}
```

本程序定义的 NewsEditCache 注解是一个复合注解，包含有缓存相关注解定义。在业务层方法上引入此注解，就可以实现缓存更新注解的使用。

范例：【mldnspring-jdbc 项目】在 INewsService 接口中使用自定义注解。

```
	@NewsEditCache
	public News edit(News vo) ;
```

可以发现，利用自定义注解可以轻松简化业务层的缓存配置操作。如果若干个业务方法缓存更新配置相同，也可以使用此种方式简化重复配置。

11.5.7　整合 EHCache 缓存组件

EHCache 是一个流行的 Java 缓存框架，具有配置简单、结构清晰、访问速度快等优点，广泛应用于 Hibernate 缓存处理中。在 Spring 提供的缓存管理中，允许使用 EHCache 作为缓存管理方案。下面将通过具体开发步骤介绍 SpringCache 与 EHCache 组件整合。

1.【mldnspring 项目】修改父 pom.xml 配置文件，引入 EHCache 缓存依赖库。

属性配置	`<ehcache.version>2.10.4</ehcache.version>`

依赖库配置	`<dependency>`
	`<groupId>net.sf.ehcache</groupId>`
	`<artifactId>ehcache</artifactId>`
	`<version>${ehcache.version}</version>`
	`</dependency>`

2.【mldnspring-jdbc 项目】修改 pom.xml 配置文件，在子模块中引入 ehcache 组件。

```
<dependency>
	<groupId>net.sf.ehcache</groupId>
	<artifactId>ehcache</artifactId>
</dependency>
```

3.【mldnspring-jdbc 项目】建立 src/main/resource/ehcache.xml 缓存配置文件。

```
<?xml version="1.0" encoding="UTF-8"?>
<ehcache>
	<diskStore path="java.io.tmpdir" />                      <!-- 设置临时缓存路径 -->
	<defaultCache                                            ➔ 定义默认缓存区配置
		maxElementsInMemory="10000"                          ➔ 缓存中允许保存的元素个数
		eternal="true"                                       ➔ 是否允许自动失效
		timeToIdleSeconds="120"                              ➔ 缓存失效时间
		timeToLiveSeconds="120"                              ➔ 最大存活时间
		maxElementsOnDisk="10000000"                         ➔ 磁盘最大保存元素个数
```

```
        diskExpiryThreadIntervalSeconds="120"              ➔  磁盘失效时间
        memoryStoreEvictionPolicy="LRU" />                 ➔  定义缓存策略，如FIFO或LRU等
    <cache                                                 ➔  定义缓存区配置
        name="news"                                        ➔  缓存区名称，与缓存注解关联
        maxElementsInMemory="10000"                        ➔  缓存中允许保存的元素个数
        eternal="true"                                     ➔  是否允许自动失效
        overflowToDisk="true"                              ➔  超过保存大小将缓存保存到磁盘
        timeToIdleSeconds="300"                            ➔  缓存失效时间
        timeToLiveSeconds="600" />                         ➔  最大存活时间
</ehcache>
```

4.【mldnspring-jdbc 项目】修改 spring-cache.xml 配置文件，配置 EHCache 缓存管理器。

```
    <cache:annotation-driven cache-manager="cacheManager" />          <!-- 启用缓存注解配置 -->
    <bean id="EHCacheManager"
        class="org.springframework.cache.ehcache.EhCacheManagerFactoryBean">
        <property name="configLocation" value="classpath:ehcache.xml"/>     <!-- 缓存文件路径 -->
    </bean>
    <bean id="cacheManager" class="org.springframework.cache.ehcache.EhCacheCacheManager">
        <property name="cacheManager" ref="EHCacheManager"/> <!-- 引入EHCache缓存管理器 -->
    </bean>
```

本程序中指定了 EHCache 配置文件的路径，同时将缓存管理器更换为 EhCacheCache Manager，这样只需按照之前的方式去使用缓存注解，即可实现 EHCache 缓存管理操作。

11.5.8　整合 Redis 实现缓存管理

在实际项目开发过程中，经常会出现服务器集群情况。为了保证集群中的所有主机实现统一缓存管理，应该使用 Redis 缓存数据库存储。SpringCache 也支持 Redis 缓存配置，这样就可以方便地实现缓存数据共享处理，如图 11-6 所示。

图 11-6　Redis 缓存处理

1.【mldnspring-jdbc 项目】建立 src/main/profiles/dev 源文件目录，并且创建 config/redis.config 配置文件，在此配置文件中定义 Redis 连接信息。

```
redis.host=redis-single                              # Redis的连接主机地址
```

```
redis.port=6379                                # Redis连接端口号
redis.password=mldnjava                        # Redis的认证信息，认证信息密码
redis.timeout=2000                             # Redis连接的超时时间
redis.pool.maxTotal=100                        # 设置最大的可用连接数
redis.pool.maxIdle=20                          # 最小维持的可用连接数
redis.pool.maxWaitMillis=2000                  # 最大等待时间
redis.pool.testOnBorrow=true                   # 是否要返回可用的连接
```

2．【mldnspring-jdbc 项目】修改 spring/spring-base.xml 配置文件，引入 redis.confg 属性配置。

```
<context:property-placeholder location="classpath:config/*.properties"/>
```

3．【mldnspring-jdbc 项目】在 src/main/resource 源文件目录中创建 spring/spring-redis.xml 配置文件。

```
<!-- 定义一个Redis数据库操作模板，可以帮助用户进行Redis处理操作 -->
<bean id="redisTemplate" class="org.springframework.data.redis.core.RedisTemplate">
    <property name="connectionFactory" ref="connectionFactory"/>    <!-- Redis连接工厂 -->
</bean>
<bean id="jedisPoolConfig" class="redis.clients.jedis.JedisPoolConfig">    <!-- Jedis配置 -->
    <property name="maxTotal" value="${redis.pool.maxTotal}"/>    <!-- 最大可用数 -->
    <property name="maxIdle" value="${redis.pool.maxIdle}"/>    <!-- 最小维持数 -->
    <property name="maxWaitMillis" value="${redis.pool.maxWaitMillis}"/><!-- 等待时间 -->
    <property name="testOnBorrow" value="${redis.pool.testOnBorrow}"/>  <!-- 确保可用 -->
</bean>
<!-- 配置ConnectionFactory -->
<bean id="connectionFactory"
    class="org.springframework.data.redis.connection.jedis.JedisConnectionFactory">
    <property name="poolConfig" ref="jedisPoolConfig"/>    <!-- 引用连接池的配置 -->
    <property name="hostName" value="${redis.host}"/>    <!-- 连接地址 -->
    <property name="port" value="${redis.port}"/>    <!-- 连接端口 -->
    <property name="password" value="${redis.password}"/>    <!-- 认证密码 -->
    <property name="timeout" value="${redis.timeout}"/>    <!-- 连接超时时间 -->
</bean>
```

4．【mldnspring-jdbc 项目】要实现 Redis 缓存，需要自定义 Redis 缓存处理类，此类要求实现 Cache 接口。

```
package cn.mldn.mldnspring.util.redis;
import java.util.concurrent.Callable;
import org.springframework.cache.Cache;
import org.springframework.cache.support.SimpleValueWrapper;
import org.springframework.data.redis.core.RedisTemplate;
public class DefaultRedisCache implements Cache {                    // 自定义缓存实现
    private RedisTemplate<Long, Object> redisTemplate;               // 注入Redis操作模板
```

```
private String name ;                                                    // 定义缓存区名称
@Override
public String getName() {                                                // 获取缓存区名称
    return this.name;
}
@Override
public Object getNativeCache() {                                         // 获取原生缓存配置
    return this.redisTemplate;
}
@Override
public ValueWrapper get(Object key) {                                    // 根据key获取数据
    Object object = this.redisTemplate.opsForValue().get(key);
    ValueWrapper obj = (object != null ? new SimpleValueWrapper(object) : null);
    return obj;
}
@Override
public <T> T get(Object key, Class<T> type) {                            // 根据key获取数据并指明类型
    T object = (T) this.redisTemplate.opsForValue().get(key);
    return object ;
}
@Override
public <T> T get(Object key, Callable<T> valueLoader) {
    return null;
}
@Override
public void put(Object key, Object value) {                              // 向缓存保存数据
    this.redisTemplate.opsForValue().set(Long.parseLong(key.toString()), value);
}
@Override
public ValueWrapper putIfAbsent(Object key, Object value) {
    return null;
}
@Override
public void evict(Object key) {                                          // 删除一个缓存数据
    this.redisTemplate.delete(Long.parseLong(key.toString())) ;
}
@Override
public void clear() {                                                    // 清空所有数据
    this.redisTemplate.getConnectionFactory().getConnection().flushDb();
}
public void setRedisTemplate(RedisTemplate<Long, Object> redisTemplate) {
    this.redisTemplate = redisTemplate;
}
```

```
    public void setName(String name) {
        this.name = name;
    }
}
```

5.【mldnspring-jdbc 项目】修改 spring/spring-cache.xml 配置文件，进行缓存定义。

```
<cache:annotation-driven cache-manager="cacheManager" />          <!-- 启用缓存注解配置 -->
<bean id="cacheManager" class="org.springframework.cache.support.SimpleCacheManager">
    <property name="caches">
        <set><!-- 这里可以配置多个redis -->
            <bean class="cn.mldn.mldnspring.util.redis.DefaultRedisCache">
                <property name="redisTemplate" ref="redisTemplate" />
                <property name="name" value="news" />          <!-- 缓存区名称 -->
            </bean>
        </set>
    </property>
</bean>
```

本程序使用了自定义的 Redis 缓存实现类，并且设置了缓存区的名称，这样就可以在缓存注解中利用此名称实现数据缓存处理。

11.6 C3P0 数据库连接池

进行数据库处理操作过程中，如果每次业务处理都需要重复进行数据库的打开与关闭操作，则项目性能一定会受到严重影响。为了解决这一问题，可以在项目中引入数据库连接池配置，利用连接池实现多数据库连接管理与连接对象分配，以提升数据库的操作性能。

C3P0 数据库连接池最早是同 Hibernate 一起发布的，它实现了数据源与 JNDI 绑定处理，并且具有自动回收空闲连接功能。下面将通过具体步骤进行 C3P0 连接池配置。

1.【mldnspring-jdbc 项目】建立 src/main/profiles/dev/config/database.properties 配置文件，保存数据库连接信息。

```
database.driverClass=org.gjt.mm.mysql.Driver          # 定义数据库的驱动程序
databas.jdbcUrl=jdbc:mysql://localhost:3306/mldn      # 定义数据库的连接地址
database.user=root                                    # 定义数据库的用户名
database.password=mysqladmin                          # 定义数据库的访问密码
database.maxPoolSize=1                                # 定义数据库连接池的最大可用连接数
database.initialPoolSize=1                            # 定义数据库连接池的初始化连接数
database.minPoolSize=1                                # 定义数据库连接池的最小保持连接数
database.maxIdleTime=1000                             # 定义数据库连接池获取连接的最大等待时间
```

2.【mldnspring-jdbc 项目】此时 spring-base.xml 配置文件中引入了所有的*.properties 资源文件，所以可以直接修改 spring-jdbc.xml 配置文件，配置 C3P0 数据库连接池，并引入 database.properties 中的配置项。

```
<bean id="dataSource" class="com.mchange.v2.c3p0.ComboPooledDataSource">
    <property name="driverClass" value="${database.driverClass}"/>         <!-- 驱动程序 -->
    <property name="jdbcUrl" value="${database.jdbcUrl}"/>                 <!-- 连接地址 -->
    <property name="user" value="${database.user}"/>                       <!-- 用户名 -->
    <property name="password" value="${database.password}"/>               <!-- 密码 -->
    <property name="maxPoolSize" value="${database.maxPoolSize}"/>         <!-- 最大连接数 -->
    <property name="initialPoolSize"
        value="${database.initialPoolSize}"/>                              <!-- 初始化连接个数 -->
    <property name="minPoolSize" value="${database.minPoolSize}"/>         <!-- 最小连接数 -->
    <property name="maxIdleTime" value="${database.maxIdleTime}"/>         <!-- 等待时间 -->
</bean>
```

由于 ComboPooledDataSource 类属于 DataSource 接口子类，所以可将此数据源对象注入到 JDBC Template 中，随后实现数据库操作。

11.7　本章小结

1．Spring 提供的 JDBC Template 是一种对传统 JDBC 操作的简单包装，利用此类提供的方法可以方便地实现数据库的 CRUD（创建、读取、更新、删除）处理操作。

2．进行查询结果返回时，可利用 RowMapper 接口将 ResultSet 结果集转换为 VO 形式处理。

3．SpringCache 可以利用注解实现与平台无关的缓存配置操作，以提升数据库的查询性能。常用注解包括@Cacheable、@CachePut 和@CacheEvict。

4．SpringCache 可以采用 ConcurrentHashMap、EHCache、Redis 实现缓存数据保存。

5．作为广泛使用的数据库连接池技术，C3P0 可以有效实现数据库连接的管理与回收，提升数据库操作性能。

第 12 章

Spring 事务管理

通过本章学习，可以达到以下目标：

1. 掌握 Spring 中事务管理架构与核心配置接口。
2. 掌握 Spring 事务传播属性与隔离级别控制。
3. 掌握编程式事务控制。
4. 掌握 AOP 声明式事务管理控制。

项目开发中，事务是一个重要的数据库控制手段。利用事务操作，可以保证若干更新操作同时成功或失败，也可以防止多个 Session 并发访问数据库所造成的数据不同步问题。本章将为读者讲解 Spring 提供的事务处理架构及其实现处理。

提示：关于本次讲解的基础程序。

由于事务控制牵扯到业务层与数据层的操作，所以本章讲解的内容将继续沿用第 11 章的新闻数据表程序，并且基于 JdbcTemplate 讲解事务架构。

12.1 传统 JDBC 事务控制概述

事务控制的核心本质指的是对数据库中数据操作的完整性保证，利用事务的 ACID 原则可以保证在一次复杂的更新业务过程中，针对于数据库中的所有更新操作要么一起成功，要么一起失败。传统的关系型数据库最大的特征是支持事务，而这也导致传统关系型数据库的处理性能有限。

提示：ACID 原则。

ACID 指的是事务的 4 种特点：原子性（Atomicity）、一致性（Consistency）、隔离性（Isolation，也称独立性）和持久性（Durabilily）。

☑ **原子性（Atomicity）**：一个事务的所有操作，要么全部完成，要么全部不完成，不可能停滞在中间某个环节。事务在执行过程中发生错误，会回滚（Rollback）到开始前的状态，就像没有执行过一样。

☑ **一致性（Consistency）**：一个事务可以封装状态改变（除非是只读的）。事务必须始终保持系统处于一致的状态，不管在给定时间内并发事务有多少。

☑ **隔离性（Isolation）**：隔离状态执行事务，使其好像是给定时间内系统执行的唯一操作。如果有两个事务运行在相同的时间内，执行着相同的功能，隔离性将确保每个事务都认为只有自己在使用系统。

☑ **持久性（Durability）**：事务完成后，该事务对数据库所进行的更改将持久保存在数据库中，不进行回滚。

JDBC 提供了最原始的数据库操作标准。在 java.sql.Connection 接口中提供了完整的事务控制方法，如表 12-1 所示。

表 12-1 JDBC 提供的事务控制方法

编号	方法	类型	描述
1	public void setAutoCommit(boolean autoCommit) throws SQLException	普通	取消自动的事务提交
2	public void commit() throws SQLException	普通	手动事务提交
3	public void rollback() throws SQLException	普通	手动事务回滚
4	public Savepoint setSavepoint(String name) throws SQLException	普通	设置事务保存点

由于传统 JDBC 在进行数据更新处理时采用的是立即执行模式，所以要先通过 setAutoCommit (false)取消自动提交，而后才可以利用 commit 或 rollback 来实现更新数据的提交或回滚。

传统开发过程中，为了可以合理进行事务处理，往往需要在业务层中进行大量的编码处理，这就导致了开发与扩展的困难。在整个 Spring 里面充分考虑到事务控制的问题，所以将其与 AOP 的特征结合在一起，形成了新的事务处理机制，并且也提供了更加完善的事务处理架构。

12.2 Spring 事务处理架构

Spring 中为了让用户处理事务更加方便，专门设计了一套事务处理架构。这个架构可以有效地与 AOP 进行整合开发，通过 AspectJ 表达式来实现事务控制的切面定义。在 Spring 事务处理中，首先需要清楚地掌握几个重要的处理接口，架构关系如图 12-1 所示。

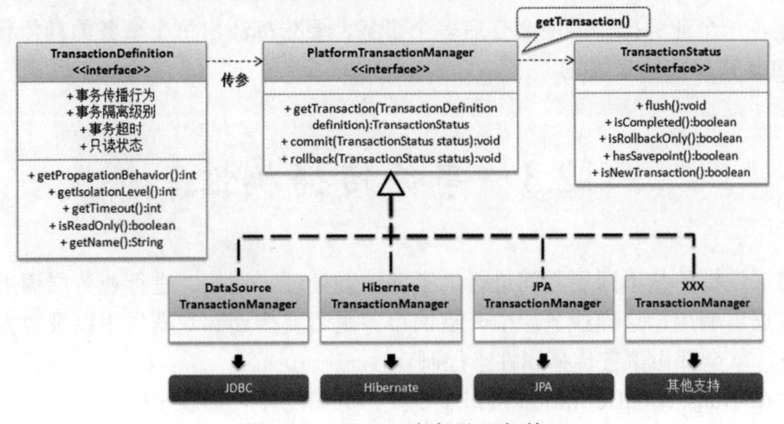

图 12-1 Spring 事务处理架构

1．事务处理的核心标准接口 org.springframework.transaction.PlatformTransactionManager。

此接口定义了事务控制的核心标准，在实际开发过程中可能有若干种数据库操作事务控制都会实现此接口，这样就可以被 Spring 进行统一管理，该接口的具体定义如下。

```
public interface PlatformTransactionManager {
	/**
	 * 根据指定的事务操作标准定义，来获取一个事务的控制状态
	 * @param definition 事务相关处理定义，包括传播属性、隔离级、只读控制
	 * @return 事务的控制状态，所有的提交与回滚都依靠此状态完成
	 * @throws TransactionException 事务处理异常
	 */
	public TransactionStatus getTransaction(@Nullable TransactionDefinition definition)
		throws TransactionException;
	/**
	 * 事务的提交处理操作，真正的Spring开发过程中，事务的提交操作应该由AOP自己负责
	 * @param status 事务的状态
	 * @throws TransactionException 事务处理异常
	 */
	public void commit(TransactionStatus status) throws TransactionException;
	/**
	 * 事务的回滚处理
	 * @param status 根据状态进行回滚
	 * @throws TransactionException 事务处理异常
	 */
	public void rollback(TransactionStatus status) throws TransactionException;
}
```

2．事务的定义 org.springframework.transaction.TransactionDefinition。

本接口中主要定义了事务的传播属性和隔离级别，同时确定了该事务是否只为只读事务以及超时时间配置。

3．事务状态 org.springframework.transaction.TransactionStatus。

Spring 里，一个业务操作可能会开启多个事务。要想标注出每个事务的具体状态，可通过 TransactionStatus 接口实例来描述。

12.3 事务传播属性

事务传播属性指的是不同层之间进行的事务控制处理。例如，进行业务层调用时，往往会执行若干次数据层调用，这样业务层中开启的事务是否要传递到数据层中以及数据层中是否要开启新的事务，就全部由事务传播属性来控制。

传播属性在 TransactionDefinition 接口中定义，共有 7 个，如表 12-2 所示。

表 12-2　TransactionDefinition 定义的事务传播属性

编号	传播属性	描述
1	PROPAGATION_REQUIRED	如果存在事务，则支持当前事务。如果没有事务，则开启新的事务
2	PROPAGATION_SUPPORTS	如果存在事务，则支持当前事务。如果没有事务，则非事务地执行
3	PROPAGATION_MANDATORY	如果存在事务，则支持当前事务。如果没有活动事务，则抛出异常
4	PROPAGATION_REQUIRES_NEW	总是开启新的事务。如果已经存在事务，则将已存在的事务挂起
5	PROPAGATION_NOT_SUPPORTED	总是非事务地执行，并挂起任何存在的事务
6	PROPAGATION_NEVER	总是非事务地执行，如果存在活动事务，则抛出异常
7	PROPAGATION_NESTED	如果存在活动事务，则将其运行在一个嵌套事务中。如果没有活动事务，则按 TransactionDefinition. PROPAGATION_REQUIRED 属性执行

下面针对表 12-2 所定义的传播属性进行分析。为了便于理解，使用 JDBC 伪代码进行操作解释。

1. TransactionDefinition.PROPAGATION_REQUIRED：此属性使用较多。如果业务层中已经存在一个事务，数据层不再开启新事务，继续使用已有事务进行控制；如果业务层中没有事务，则开启新的事务控制。

☑　业务层已经开启了事务。

业务操作调用	JDBC事务处理模式

【S】业务方法A() { // PROPAGATION_REQUIRED
　　...　// 其他处理操作
　　【D】数据方法B() {}
　　...　// 其他处理操作
}
	【D】数据方法B() {
	　　// PROPAGATION_REQUIRED，不再开启新事务
	}

☑　业务层未开启事务，数据层要自己开启事务控制。

业务操作调用	JDBC事务处理模式

【S】业务方法A() { // 没有开启事务
　　...　// 其他处理操作
　　【D】数据方法B() {}
　　...　// 其他处理操作
}
	【D】数据方法B() {　　// PROPAGATION_REQUIRED
	　connection.setAutoCommit(**false**) ; // 1. 关闭自动事务提交
	　try {
	　　pstmt.executeUpdate() ;　　// 2. 进行更新处理操作
	　　connection.commit() ; // 3. 使用commit方法提交事务
	　} **catch** (Exception e) {
	　　connection.rollback() ;　　// 4. 手动回滚
	　} **finally** {
	　　connection.close() ;　　// 5. 关闭连接

	} }

如果现在出现总业务调用了子业务，子业务调用了数据层，那么这个时候也需要考虑传播属性：

```
【A】总业务方法X() {                    // 未开启事务
    【S】子业务方法A() {                // PROPAGATION_REQUIRED
        ...                             // 其他处理操作
        【D】数据方法B() {}             // 继续使用子业务层中的事务处理
        ...                             // 其他处理操作
    }
}
```

2．TransactionDefinition.PROPAGATION_SUPPORTS：如果业务层开启了事务，数据层继续使用该事务；如果业务层没有开启事务，数据层采用非事务状态继续执行。

☑ 业务层未开启事务，数据层采用非事务运行。

```
【S】子业务方法A() {                // PROPAGATION_SUPPORTS，未开启事务
    ...                             // 其他处理操作
    【D】数据方法B() {}             // 采用非事务方式处理
    ...                             // 其他处理操作
}
```

☑ 业务层开启了事务，则继续使用业务层的事务进行处理。

```
【S】子业务方法A() {                // PROPAGATION_SUPPORTS，开启事务
    ...                             // 其他的处理操作
    【D】数据方法B() {}             // 采用事务方式处理
    ...                             // 其他处理操作
}
```

3．TransactionDefinition.PROPAGATION_MANDATORY：如果业务层开启了事务支持，数据层操作正常完成；如果业务层没有事务支持，数据层直接抛出 org.springframework.transaction.IllegalTransactionStateException 异常。

业务操作调用	JDBC事务处理模式

【S】子业务方法A() {   // PROPAGATION_MANDATORY，未开启事务
    ...   // 其他的处理操作
    【D】数据方法B() {}// 必须采用事务方式处理
    ...   // 其他处理操作
}
	【D】数据方法B() { // PROPAGATION_MANDATORY
	    // 未开启，抛出异常（非法事务状态异常）；已开启，向下执行事务控制
	    connection.setAutoCommit(**false**) ; // 1. 关闭自动事务提交
	    try {
	        pstmt.executeUpdate() ; // 2. 数据库更新
	        connection.commit() ; // 3. 提交事务

	} **catch** (Exception e) {
	　　connection.rollback() ;　　// 4. 手工回滚
	} **finally** {
	　　connection.close() ;　　// 5. 关闭连接
	}
	}

4．TransactionDefinition.PROPAGATION_REQUIRES_NEW：永远开启新的事务，并将原有事务挂起。

业务操作调用：	JDBC事务处理模式：

【S】子业务方法A() {　　// PROPAGATION_REQUIRES_NEW，开启事务
　...　// 其他处理操作
　【D】数据方法B() {}// 有无事务，都重新开启
　...　// 其他处理操作
}
	【D】数据方法B() { // PROPAGATION_REQUIRES_NEW
	　// 将业务层事务处理暂时挂起，开启新的事务
	　connection.setAutoCommit(**false**) ; // 1. 关闭自动事务提交
	　try {
	　　pstmt.executeUpdate() ;　　// 2. 数据更新
	　　connection.commit() ;　　// 3. 提交事务
	　} **catch** (Exception e) {
	　　connection.rollback() ;　　// 4. 手工回滚
	　} **finally** {
	　　connection.close() ;　　// 5. 关闭连接
	　}
	}

5．TransactionDefinition.PROPAGATION_NOT_SUPPORTED：以非事务方式运行。如果存在事务，则将已有事务挂起后执行。

6．TransactionDefinition.PROPAGATION_NEVER：永远以非事务的方式运行。如果存在事务，则抛出异常。

7．TransactionDefinition.PROPAGATION_NESTED：如果有事务存在，则在内部开启一个新的事务，作为嵌套事务存在；如果没有事务存在，则与 TransactionDefinition.PROPAGATION_REQUIRED 执行相同。

```
【D】数据方法B() {                                    // PROPAGATION_ NESTED
    // 将业务层的事务处理暂时挂起，开启新的事务
    connection.setAutoCommit(false) ;
    // T1：开启第一个T1事务，负责外部处理操作
    try {
        pstmt.executeUpdate() ;                       // T1更新
        // T1：设置一个事务处理的保存点SavePoint
        try {
            // T2：开启第二个T2事务，负责内部嵌套处理操作
            connection.setAutoCommit(false) ;
```

```
            pstmt.executeUpdate() ;                                     // T2更新
            connection.commit() ;                                       // T2提交
        } catch (Exception e) {
            connection.rollback() ;                                     // T2回滚
        }
        connection.commit() ;                                           // T1提交
    } catch (Exception e) {
        connection.rollback() ;                                         // T1回滚
    } finally {
        connection.close() ;                                            // 关闭连接
    }
}
```

在实际开发中，复杂的业务处理操作往往会出现总业务层与子业务层事务控制，此时可利用嵌套事务的传播属性进行控制。

12.4 事务隔离级别

进行数据库事务处理操作时，需要考虑并发状态下数据读取的正确性。该操作属于事务的隔离级别，隔离的主要的目的是为了防止脏读、不可重复读、幻读的问题。

1. 脏读

脏读，是指一个事务正在访问数据并对数据进行了修改，当这种修改还没来得及提交到数据库中时，另一个事务也访问了这个数据，然后使用了这个数据。例如，在一个人事系统中，A 用户修改某一雇员工资时，B 用户直接按照原始工资进行工资处理，这时 B 用户将使用错误的雇员工资进行数据处理。

2. 不可重复读

不可重复读，是指一个事务需要多次读取同一数据，该事务还没结束时，另外一个事务也访问了该数据并进行了修改，那么，第一个事务两次读到的数据就有可能是不一样的。例如，一位编辑人员两次读取同一文档，但在两次读取之间，作者重写了该文档。当编辑人员第二次读取文档时，文档已更改，原始读取不可重复读。如果限制必须在作者全部完成编写后编辑人员才可以读取该文档，则可以避免此类问题。

同一个数据两次的查询读取结果不同。同样的条件两次读取的数据不同，重点在于修改。看一下下面的示例。

第一个 Session 读取雇员 SMITH 的信息，返回 800。

```
SELECT sal FROM emp WHERE ename='SMITH' ;
```

第二个 Session 更新 SMITH 的工资，但这个时候第一个 Session 还没有关闭连接。

```
UPDATE emp SET sal=9000 WHERE ename='SMITH' ;
```

第一个 Session 重复读取了一次 SMITH 的工资，返回 9000。

```
SELECT sal FROM emp WHERE ename='SMITH' ;
```

3．幻读

幻读是事务非独立执行时发生的一种现象。例如，第一个事务对一个表中的数据进行了修改，这种修改涉及表中的全部数据行。同时，第二个事务也对表中的数据进行了修改，这种修改是向表中插入一行新数据。那么，操作第一个事务的用户就会发现，表中竟然还有未修改的数据行，就像发生了幻觉一样。

再举个例子。一位编辑人员更改作者提交的文档，但当生产部门将其更改内容合并到该文档的主复本时，发现作者已将未编辑的新材料添加到该文档中。如果在编辑人员和生产部门完成对原始文档的处理之前，任何人都不能将新材料添加到文档中，则可以避免该问题。

两次读取数据结果不同。同样的数据两次的记录不同，主要在于增加和删除处理上。

例如，要读取工资大于 2000 元的员工，第一个 Session 读取数据时返回了 10 行数据。第二个 Session（未关闭第一个 Session）增加了一行新数据，工资为 2500 元。第一个 Session 重复读取时就变为了 11 行数据。

为了解决如上问题，TransactionDefinition 接口中定义了 5 个不同的事务隔离级别，如表 12-3 所示。

表 12-3　事务隔离级别

编号	事务隔离级别	描述
1	ISOLATION_DEFAULT	默认的隔离级别，使用数据库默认的事务隔离级别
2	ISOLATION_READ_UNCOMMITTED	最低隔离级别，允许其他事务访问当前事务未提交的数据。会产生脏读、不可重复读、幻读
3	ISOLATION_READ_COMMITTED	当前事务修改的数据提交后才能被下一个事务读取。下一个事务不能读取当前事务未提交的数据。可防止脏读，可能出现不可重复读、幻读
4	ISOLATION_REPEATABLE_READ	可防止脏读、不可重复读，可能出现幻读
5	ISOLATION_SERIALIZABLE	最可靠的事务隔离级别，事务处理为顺序执行。可防止脏读、不可重复读、幻读

从实际开发角度来看，一般都会基于数据库自身的隔离级别进行处理，所以 ISOLATION_DEFAULT 作为应用首选。

12.5　编程式事务控制

清楚了 Spring 提供的事务处理架构以及各个核心意义后，下面通过具体的程序编码来实现事务的控制处理操作。

1.【mldnspring 项目】修改父 pom.xml 配置文件，引入相关依赖库。

```
<dependency>
        <groupId>org.springframework</groupId>
```

```
        <artifactId>spring-tx</artifactId>
        <version>${spring.version}</version>
    </dependency>
```

2．【mldnspring-jdbc 项目】在子模块中导入依赖库。

```
    <dependency>
        <groupId>org.springframework</groupId>
        <artifactId>spring-tx</artifactId>
    </dependency>
```

3．【mldnspring-jdbc 项目】建立 src/main/resources/spring/spring-transaction.xml 配置文件，定义事务管理器。

```
    <!-- 定义事务管理的配置，必须配置PlatformTransactionManager接口子类-->
    <bean id="transactionManager"
        class="org.springframework.jdbc.datasource.DataSourceTransactionManager">
        <property name="dataSource" ref="dataSource"/>          <!-- 定义要管理的DataSource -->
    </bean>
```

这里使用的 DataSourceTransactionManager 事务管理器是 PlatformTransactionManager 接口的子类。

4.【mldnspring-jdbc 项目】在业务层实现类 NewsServiceImpl 中注入 PlatformTransactionManager 对象，并实现数据读取与数据更新的事务控制。

☑ 在类中注入 PlatformTransactionManager 接口对象。

```
    @Autowired    // 注入的只是一个接口实例化对象，还需要配置隔离级别、传播属性
    private PlatformTransactionManager transactionManager;
```

☑ 在删除方法中开启更新事务。

```
    @Override
    public boolean delete(long nid) {
        boolean flag = false;
        // TransactionDefinition通过此类设置传播属性、隔离级别、是否为只读、超时访问
        DefaultTransactionDefinition transactionDefinition =
            new DefaultTransactionDefinition();
        // 设置传播属性，必须有一个事务存在
        transactionDefinition.setPropagationBehavior(
            TransactionDefinition.PROPAGATION_REQUIRED);
        // 使用默认隔离级别进行控制
        transactionDefinition.setIsolationLevel(TransactionDefinition.ISOLATION_DEFAULT);
        TransactionStatus transactionStatus =
            this.transactionManager.getTransaction(transactionDefinition); // 获取事务状态
        try {
            flag = this.newsDAO.doRemove(nid);
```

```
            this.transactionManager.commit(transactionStatus);          // 提交事务
        } catch (Exception e) {
            this.transactionManager.rollback(transactionStatus);        // 回滚到指定提交点
        }
        return flag;
    }
```

☑　在查询方法中设置只读事务。

```
    @Override
    public News get(long nid) {
        News vo = null;
        // TransactionDefinition通过此类设置传播属性、隔离级别、是否为只读、超时访问
        DefaultTransactionDefinition transactionDefinition =
            new DefaultTransactionDefinition();
        // 设置传播属性，必须有一个事务启动
        transactionDefinition.setPropagationBehavior(
            TransactionDefinition.PROPAGATION_REQUIRED);
        // 隔离级别一定要使用默认的隔离级别进行控制
        transactionDefinition.setIsolationLevel(TransactionDefinition.ISOLATION_DEFAULT);
        transactionDefinition.setReadOnly(true);                        // 只读操作
        TransactionStatus transactionStatus =
            this.transactionManager.getTransaction(transactionDefinition); // 获取事务状态
        try {
            vo = this.newsDAO.findById(nid);
            this.transactionManager.commit(transactionStatus);          // 提交事务
        } catch (Exception e) {
            this.transactionManager.rollback(transactionStatus);        // 回滚到指定的提交点
        }
        return vo;
    }
```

本程序采用手动方式实现了事务的控制处理，设置的传播属性统一为 **PROPAGATION_REQUIRED**，这样将会一直提供事务支持，以保证数据操作的正确性。如果某一个业务方法只是进行查询处理，则可以使用 setReadOnly 方法将该业务方法的事务设置为只读，此时将无法进行数据库更新处理。

12.6　@Transactional 事务控制注解

虽然 Spring 支持手动方式进行编程式事务控制，但此类控制过于烦琐，在所有业务方法中进行定义会造成大量的代码重复操作。为了简化事务的控制操作，在 Spring 中提供了@Transactional 注解，在业务方法上直接使用此注解即可实现事务控制。@Transactional 注解支持的属性如表 12-4 所示。

表 12-4　@Transactional 注解属性

编号	属性	描述
1	propagation	设置事务的传播属性
2	isolation	设置事务隔离级别
3	timeout	设置事务超时时间，如果为-1，则表示永不超时
4	readOnly	设置为只读事务，设置为 true 表示只读，设置为 false 表示读写
5	rollbackFor	设置回滚的异常类型，可设置多个类型。 单一异常：@Transactional(rollbackFor=RuntimeException.class) 多个异常： @Transactional(rollbackFor={RuntimeException.class, Exception.class})
6	rollbackFor ClassName	设置回滚异常类名称。 单一异常：@Transactional(rollbackForClassName="RuntimeException") 多个异常： @Transactional(rollbackForClassName={"RuntimeException","Exception"})
7	noRollbackFor	设置不回滚的异常类型。 单一异常：@Transactional(noRollbackFor=RuntimeException.class) 多个异常： @Transactional(noRollbackFor={RuntimeException.class, Exception.class})
8	noRollbackFor ClassName	设置不回滚异常类名称。 单一异常：@Transactional(noRollbackForClassName="RuntimeException") 多个异常： @Transactional(noRollbackForClassName={"RuntimeException","Exception"})

要开启@Transactional 注解支持，需要启动事务的注解配置操作，下面通过具体步骤进行配置说明。

1.【mldnspring-jdbc 项目】修改 spring/spring-transaction.xml 配置文件，追加 tx 命名空间，如图 12-2 所示。

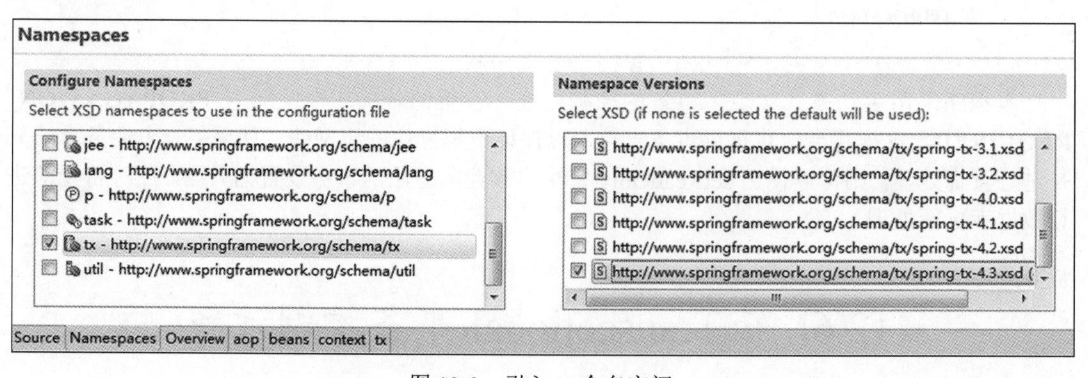

图 12-2　引入 tx 命名空间

2.【mldnspring-jdbc 项目】修改 spring/spring-transaction.xml 配置文件，启用事务注解配置。启用时，必须明确设置要使用的事务管理器配置。

```
<tx:annotation-driven transaction-manager="transactionManager" />
```

3.【mldnspring-jdbc 项目】修改 INewsService 接口，追加事务控制。

```
public interface INewsService {
    @Cacheable(key = "#nid", unless="#result==null")
    @Transactional(propagation = Propagation.REQUIRED, readOnly = true)
    public News get(long nid) ;
    @NewsEditCache
    @Transactional(propagation=Propagation.REQUIRED)
    public News edit(News vo) ;
    @CacheEvict(key="#nid")
    @Transactional(propagation = Propagation.REQUIRED)
    public boolean delete(long nid) ;
}
```

此时程序将可以直接利用注解简化烦琐的 PlatformTransactionManager 编程式事务控制。

12.7　声明式事务控制

在进行事务控制的处理过程中，不管采用的是编程式事务控制，还是基于@Transactional 注解进行控制，都需要在业务方法上进行一系列定义。因此，在一个大型项目中对事务进行控制就会显得非常烦琐。为了进一步简化事务管理操作，可以利用 Spring AOP 技术的切面表达式与方法名称约定，利用配置文件对所有业务方法进行统一的事务控制，这样将可以极大地简化业务层定义，即业务层只用关注功能的定义。

1.【mldnspring-jdbc 项目】简化 INewsService 接口只定义核心方法，不使用任何事务相关定义。

```
@CacheConfig(cacheNames="news")                                    // 统一配置接口缓存
public interface INewsService {
    @Cacheable(key = "#nid", unless="#result==null")
    public News get(long nid) ;
    @NewsEditCache
    public News edit(News vo) ;
    @CacheEvict(key="#nid")
    public boolean delete(long nid) ;
}
```

2.【mldnspring-jdbc 项目】修改 spring/spring-transaction.xml 配置文件，利用 AOP 进行事务控制。

```
<tx:annotation-driven transaction-manager="transactionManager" />
<!-- 事务控制专门提供了切面程序类 -->
<tx:advice id="txAdvice" transaction-manager="transactionManager">
    <tx:attributes>  <!-- 配置事务控制属性，以业务层方法为主，方法名按照统一标准定义 -->
```

```
            <tx:method name="add*" propagation="REQUIRED"/>          <!-- 设置事务传播属性 -->
            <tx:method name="delete*" propagation="REQUIRED"/>       <!-- 设置事务传播属性 -->
            <tx:method name="edit*" propagation="REQUIRED"/>         <!-- 设置事务传播属性 -->
            <tx:method name="get*" propagation="REQUIRED" read-only="true"/>
        </tx:attributes>
    </tx:advice>
    <aop:config>
        <!-- 定义事务要控制的切面表达式，即在此切面范围内定义的业务方法事务控制生效 -->
        <aop:pointcut expression="execution(public * cn.mldn..service..*.*(..))"
            id="myPointcut"/>
        <aop:advisor advice-ref="txAdvice" pointcut-ref="myPointcut"/>
    </aop:config>
```

本配置文件中，针对不同种类的业务方法进行了统一事务配置。后期进行功能扩充时，只需要定义出方法的前缀，即可对业务方法自动实现业务管理。

12.8 本章小结

1．Spring 事务控制是对传统 JDBC 的包装，利用 PlatformTransactionManager 作为事务处理公共标准，而后针对不同的数据层操作均有事务支持。

2．Spring 事务处理中需要考虑事务的传播属性与隔离级别，利用 AOP 可轻松实现业务方法的事务控制。

第 13 章

SpringDataJPA

通过本章学习，可以达到以下目标：

1. 掌握 JPA 的主要作用与核心结构。
2. 掌握 JPA 常用注解。
3. 可以使用 JPA 实现数据的 CRUD 处理。
4. 掌握 JPA 缓存处理。
5. 掌握 JPA 锁处理机制。
6. 掌握 JPA 数据关联技术。
7. 掌握 Spring 与 JPA 整合定义。
8. 掌握 SpringDataJPA 开发框架的使用。

JPA（Java Persistence API，Java 持久化 API）是 Java 提供的数据层标准，也是一个较为流行的 ORMapping 开发框架。由于 Spring 技术的发展，SpringDataJPA 为 JPA 提供了更加方便的处理支持，可以大量减少数据层的定义。本章将为读者详细讲解 JPA 开发框架的使用方法以及 SpringDataJPA 技术。

13.1 JPA 简介

Java 持久化 API（Java Persistence API，JPA）是早期 SUN 公司（现为 Oracle 公司提供）在 Java EE 5 规范中提出的 Java 持久化接口。JPA 吸取了 Java 持久化技术的优点，利用 Java 对象可以方便地实现持久层开发。

JPA 由 EJB 3.0 软件专家组开发，作为 JSR-220 实现的一部分，但它又不限于 EJB 3.0，JPA 拥有如下特点。

☑ **标准化：**JPA 是 JCP 组织发布的 Java EE 标准之一。

☑ **对容器级特性的支持：**支持大数据集、事务、并发等容器级事务。

☑ **简单易用，集成方便：**实体类创建简单，注解统一。

☑ **可媲美 JDBC 的查询能力：**支持常用 SQL 标准语法。

☑ **支持面向对象的高级特性：**支持继承、多态和类之间的复杂关系。

JPA 是一个开源标准，所以并不局限于某一个具体的 ORMapping 开发框架，各企业厂商，如 Oracle、Redhat、Eclipse 等都提供了 JPA 的相关实现，产品包括 Hibernate、Eclipselink、Toplink、Spring Data JPA 等，如图 13-1 所示。本节将使用 Hibernate 框架进行讲解。

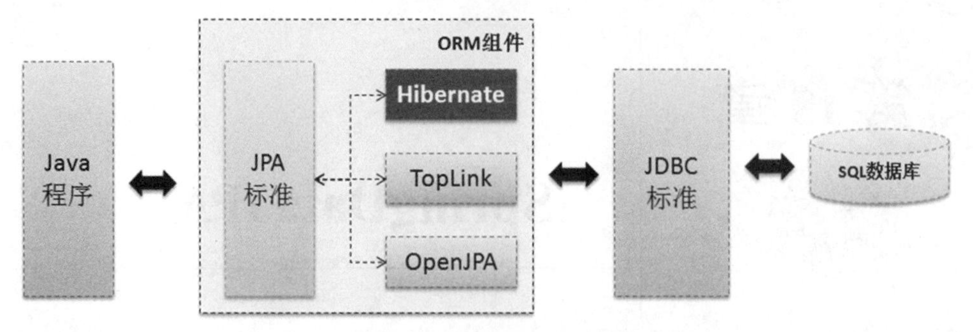

图 13-1 JPA 标准实现

提示：JPA 的发展历史。

☑ 早期版本的 EJB，定义持久层结合使用 javax.ejb.EntityBean 接口作为业务逻辑层。
☑ 引入 EJB 3.0 的持久层分离，并指定为 JPA 1.0。这个 API 规范随着 Java EE 5 对 2006 年 5 月 11 日使用 JSR-220 规范发布。
☑ JPA 2.0 规范发布于 2009 年 12 月 10 日并成为 Java Community Process JSR-317 的一部分。
☑ JPA 2.1 使用 JSR-338 的 Java EE 7 的规范发布于 2013 年 4 月 22 日。

JPA 作为 Java EE 开源实现标准，其所有的接口与类都定义在 javax.persistence 开发包中，核心类库如表 13-1 所示。

表 13-1 JPA 核心类库

编号	单元	描述
1	EntityManagerFactory	EntityManager 的工厂类，用于创建并管理多个 EntityManager 实例
2	EntityManager	接口，用于管理持久化操作的对象，工作原理类似工厂的查询实例
3	Entity	实体，是持久性对象存储在数据库中的记录
4	EntityTransaction	与 EntityManager 是一对一的关系。EntityManager 的操作由 EntityTransaction 类维护
5	Persistence	包含静态方法，以获取 EntityManagerFactory 实例
6	Query	实现 JPA 数据查询

表 13-1 所给出的类与接口之间的关联如图 13-2 所示。

提示：JPA 标准与原始 Hibernate 类库的对应关系。

☑ **Configuration（javax.persistence.Persistence）**：主要负责读取配置文件。
☑ **SessionFactory（javax.persistence.EntityManagerFactory）**：数据库连接管理，表示打开 Session。
☑ **Session（javax.persistence.EntityManager）**：表示每个用户的具体操作。不同用户拥有不同的 Session，每个 Session 控制自己的操作事务（javax.persistence.Entity Transaction）、实体数据处理以及 Query 查询处理。

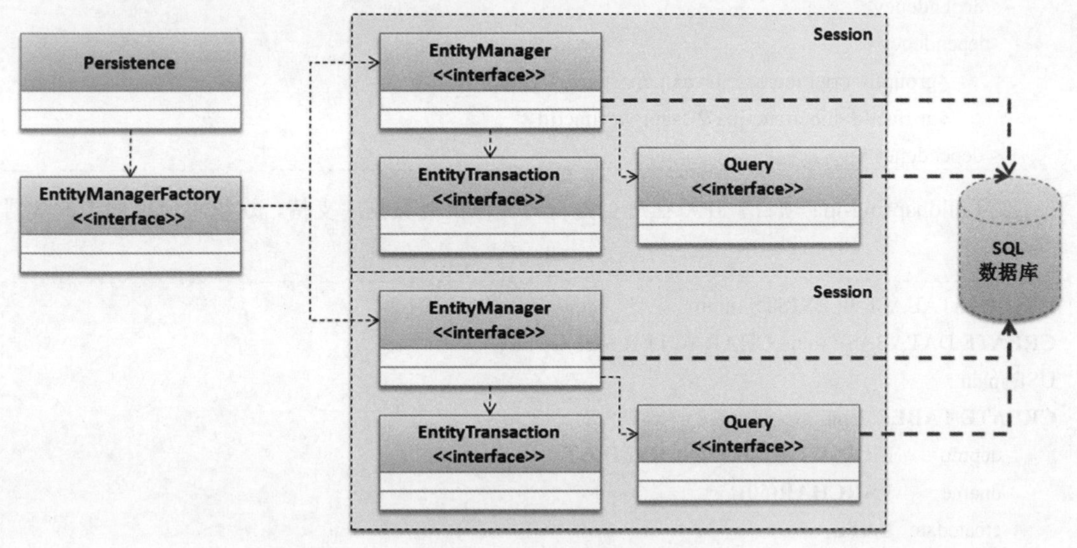

图 13-2　JPA 开发架构

13.2　JPA 编程起步

JPA 是一个技术标准。要实现该技术标准，需要有对应的技术实现产品。下面将使用 Hibernate 作为 JPA 实现类展开讲解。如果要在项目中使用 JPA，则一定要修改依赖库配置。

1.【mldnspring 项目】修改父 pom.xml 配置文件，追加 JPA 相关依赖库。

属性配置	<hibernate.version>5.2.13.Final</hibernate.version> <hibernate-jpa-2.1-api.version>1.0.2.Final</hibernate-jpa-2.1-api.version>
依赖库配置	<dependency> <groupId>org.hibernate</groupId> <artifactId>hibernate-c3p0</artifactId> <version>${hibernate.version}</version> </dependency> <dependency> <groupId>org.hibernate.javax.persistence</groupId> <artifactId>hibernate-jpa-2.1-api</artifactId> <version>${hibernate-jpa-2.1-api.version}</version> </dependency>

由于 JPA 属于数据层操作，所以依然需要配置 MySQL 驱动程序。

2.【mldnspring-jpa 项目】修改子 pom.xml 配置文件，追加依赖库配置。

```
<dependency>
    <groupId>org.hibernate</groupId>
    <artifactId>hibernate-c3p0</artifactId>
```

```
    </dependency>
    <dependency>
        <groupId>org.hibernate.javax.persistence</groupId>
        <artifactId>hibernate-jpa-2.1-api</artifactId>
    </dependency>
```

3.【mldnspring-jpa 项目】JPA 在进行处理时无法离开数据库支持，这里将使用如下的数据库脚本。

```
DROP DATABASE IF EXISTS mldn ;
CREATE DATABASE mldn CHARACTER SET UTF8 ;
USE mldn ;
CREATE TABLE dept(
    deptno      BIGINT   AUTO_INCREMENT ,
    dname       VARCHAR(50) ,
    createdate  DATE,
    num         INT ,
    avgsal      DOUBLE ,
    CONSTRAINT pk_deptno PRIMARY KEY(deptno)
) ;
```

13.2.1 JPA 基础实现

要实现 JPA 程序开发，需要配置数据库的连接以及持久类，且应在 persistence.xml 配置文件中进行定义。下面通过具体步骤来实现一个基础的 JPA 程序。

1.【mldnspring-jpa 项目】在 src/main/resources 下创建 META-INF 目录，在该目录中创建 persistence.xml 配置文件，如图 13-3 所示。

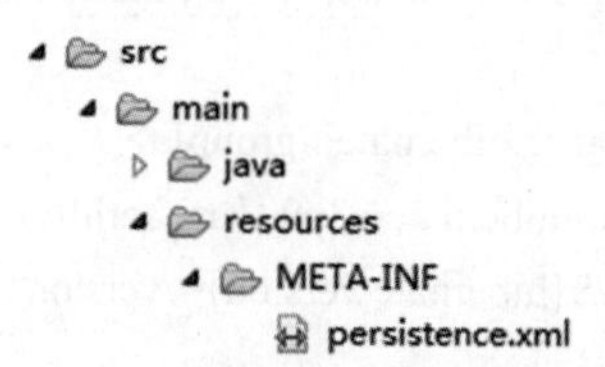

图 13-3 persistence.xml 配置文件

2.【mldnspring-jpa 项目】编辑 persistence.xml 文件内容。

```
<?xml version="1.0" encoding="UTF-8"?>
<persistence version="2.1"
    xmlns="http://xmlns.jcp.org/xml/ns/persistence"
    xmlns:xsi="http://www.w3.org/2001/XMLSchema-instance"
    xsi:schemaLocation="http://xmlns.jcp.org/xml/ns/persistence
    http://xmlns.jcp.org/xml/ns/persistence/persistence_2_1.xsd">
    <persistence-unit name="MLDNJPA">                          <!-- 定义持久化单元 -->
        <properties>
```

```
            <property name="javax.persistence.jdbc.driver"
                value="com.mysql.jdbc.Driver" />                    <!-- 驱动程序 -->
            <property name="javax.persistence.jdbc.url"
                value="jdbc:mysql://localhost:3306/mldn" />         <!-- 连接地址 -->
            <property name="javax.persistence.jdbc.user" value="root" /> <!-- 用户名 -->
            <property name="javax.persistence.jdbc.password"
                value="mysqladmin" />                               <!-- 密码 -->
            <property name="hibernate.show_sql" value="true" />     <!-- 显示执行SQL -->
            <property name="hibernate.format_sql" value="true" />   <!-- 格式化SQL语句 -->
        </properties>
    </persistence-unit>
</persistence>
```

本配置中，最为重要的配置项为持久化单元配置<persistence-unit name="MLDNJPA">，在随后的程序中需要通过 MLDNJPA 来创建 EntityManagerFactory 对象。

3．【mldnspring-jpa 项目】定义持久化程序类 Dept。

```
package cn.mldn.mldnspring.po;
import java.io.Serializable;
import java.util.Date;
import javax.persistence.Entity;
import javax.persistence.GeneratedValue;
import javax.persistence.GenerationType;
import javax.persistence.Id;
import javax.persistence.Temporal;
import javax.persistence.TemporalType;
@SuppressWarnings("serial")
@Entity                                                     // 持久化类
public class Dept implements Serializable {
    @Id                                                     // 主键列
    @GeneratedValue(strategy = GenerationType.IDENTITY)     // 主键生成方式
    private Long deptno;                                // 字段映射（属性名称=字段名称）
    private double avgsal;
    @Temporal(TemporalType.DATE)                            // 类型描述
    private Date createdate;
    private String dname;
    private int num;
    // setter、getter、toString略
}
```

Dept 属于实体类，所以在类定义时使用了@Entity 注解。主键列需要使用@Id 注解，其他类中的属性将自动根据名称与 dept 表进行匹配。

4．【mldnspring-jpa 项目】编写测试程序，实现 dept 表（部门表）数据追加。

```java
package cn.mldn.mldnspring.test;
import java.util.Date;
import javax.persistence.EntityManager;
import javax.persistence.EntityManagerFactory;
import javax.persistence.Persistence;
import org.junit.Test;
import cn.mldn.mldnspring.po.Dept;
public class TestDeptPersistence {
    @Test
    public void testAdd() {
        // 1. 定义EntityManagerFactory工厂类对象
        EntityManagerFactory entityManagerFactory =
                Persistence.createEntityManagerFactory("MLDNJPA");
        // 2. 通过工厂类产生EntityManager（Session）
        EntityManager entityManager = entityManagerFactory.createEntityManager(); // 数据库操作
        // 3. 更新事务支持，通过EntityManager打开一个事务处理
        entityManager.getTransaction().begin();                  // 开启事务
        // 4. 将要保存的数据设置到PO类对象中
        Dept dept = new Dept();                                  // 定义持久化对象
        dept.setDname("MLDN教学管理部");                          // 设置属性内容
        dept.setAvgsal(8968.88);                                 // 设置属性内容
        dept.setCreatedate(new Date());                          // 设置属性内容
        dept.setNum(8);                                          // 设置属性内容
        // 5. 执行数据的持久化处理（保存）
        entityManager.persist(dept);                             // 数据持久化
        // 6. 进行事务提交控制
        entityManager.getTransaction().commit();
        // 7. 关闭数据库连接
        entityManager.close();                                   // 关闭Session操作
        entityManagerFactory.close();                            // 关闭工厂连接
    }
}
```

程序执行结果

```
Hibernate:
    insert
    into
        Dept
        (avgsal, createdate, dname, num)
    values
        (?, ?, ?, ?)
```

程序执行后，将根据持久化单元名称 MLDNJPA 自动加载相关配置项，而后利用 EntityManager 类中的 persist 方法实现持久化对象的保存。由于配置有格式化显示执行 SQL 的选项，所以将返回执行的 SQL 语句。

13.2.2　定义 JPA 连接工厂类

在 JPA 代码执行过程中，EntityManagerFactory 与 EntityManager 接口是整体程序的运行核心。为了简化这两个接口对象的实例化操作，可以单独创建一个 JPAEntityFactory 工具类，在该类中利用 ThreadLocal 进行多线程操作管理，如图 13-4 所示。

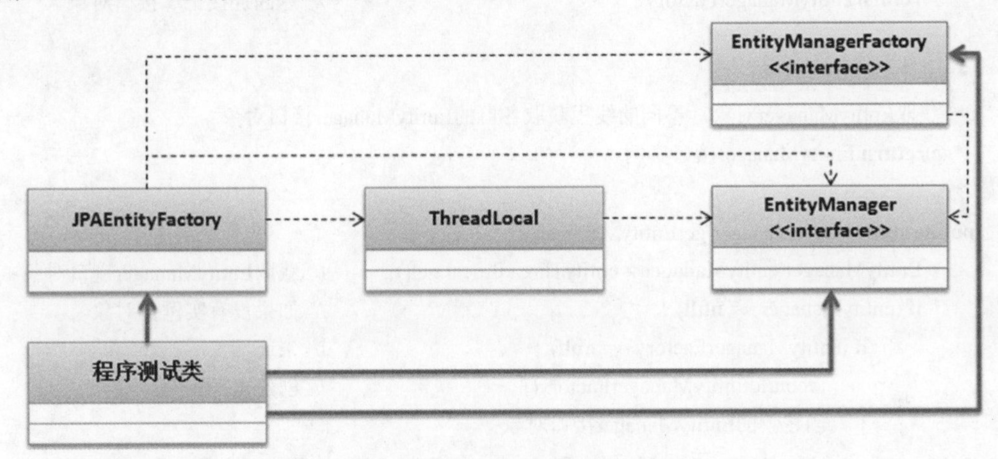

图 13-4　JPAEntityFactory 工厂类

1.【mldnspring-jpa 项目】编写 JPAEntityFactory 工具类。

```
package cn.mldn.mldnspring.util;
import javax.persistence.EntityManager;
import javax.persistence.EntityManagerFactory;
import javax.persistence.Persistence;
/**
 * 定义一个用于操作JPA的工厂程序类，主要功能是负责
 * EntityManger与EntityManagerFactory接口的对象管理
 * @author 李兴华
 */
public class JPAEntityFactory {
    private static final String PERSISTENCE_UNIT = "MLDNJPA" ; // 持久化单元名称
    private static EntityManagerFactory entityManagerFactory ;    // 定义连接工厂类
    private static ThreadLocal<EntityManager> entityThreadLocal =
            new ThreadLocal<EntityManager>() ;          // 保存EntityManager接口对象
    static {                                            // 静态代码块获取EntityManagerFactory实例
        rebuildEntityManagerFactory() ;                 // 实例化EntityManagerFactory接口对象
    }
    private JPAEntityFactory() {}                       // 将构造方法私有化，通过静态方法获取相应对象
    /**
     * 获取EntityManagerFactory接口对象
     * @return EntityManagerFactory接口实例
```

```
    */
    public static EntityManagerFactory getEntityManagerFactory() {
        if (entityManagerFactory == null) {                          // 未连接工厂
            rebuildEntityManagerFactory() ;                          // 创建工厂实例
        }
        return entityManagerFactory ;                                // 返回连接工厂接口对象
    }
    /**
     * 获取EntityManager对象，不同的线程获取各自的EntityManager接口对象
     * @return EntityManager接口实例
     */
    public static EntityManager getEntityManager() {
        EntityManager entityManager = entityThreadLocal.get() ;      // 获取EntityManager接口对象
        if (entityManager == null) {                                 // 如果没有实例化对象
            if (entityManagerFactory == null) {                      // 未连接工厂
                rebuildEntityManagerFactory() ;                      // 创建工厂实例
            } // 创建新的EntityManager接口对象
            entityManager = entityManagerFactory.createEntityManager() ;
            entityThreadLocal.set(entityManager);                    // 保存对象信息
        }
        return entityManager ;
    }
    /**
     * 关闭EntityManager连接
     */
    public static void close() {
        EntityManager entityManager = entityThreadLocal.get() ;
        if (entityManager != null) {                                 // 已保存EntityManager
            entityManager.close();
            entityThreadLocal.remove();                              // 从ThreadLocal中删除对象
        }
    }
    /**
     * 该方法的主要功能是获取EntityManagerFactory接口对象
     */
    private static void rebuildEntityManagerFactory() {
        entityManagerFactory = Persistence.createEntityManagerFactory(PERSISTENCE_UNIT) ;
    }
}
```

2.【mldnspring-jpa 项目】可直接通过 JPAEntityFactory 类获取 EntityManagerFactory 与 EntityManager 接口对象。

```
public class TestDeptPersistence {
    @Test
    public void testAdd() {
        JPAEntityFactory.getEntityManager().getTransaction().begin();        // 开启事务
        Dept dept = new Dept();                                              // 定义持久化对象
        dept.setDname("MLDN教学管理部");                                      // 设置属性内容
        dept.setAvgsal(8968.88);                                             // 设置属性内容
        dept.setCreatedate(new Date());                                      // 设置属性内容
        dept.setNum(8);                                                      // 设置属性内容
        JPAEntityFactory.getEntityManager().persist(dept);                   // 数据持久化
        JPAEntityFactory.getEntityManager().getTransaction().commit();       // 事务提交
        JPAEntityFactory.close();                                            // 关闭连接
    }
}
```

使用 JPAEntityFactory 类封装后，程序调用处可以简化许多重复代码的编写，同时由于 ThreadLocal 类的使用，也将避免多线程访问冲突。

13.2.3　DDL 自动更新

在使用 JPA 进行程序开发的过程中，往往正常的方式是先创建数据表，而后再创建持久化 PO 类，但是在 JPA 中也可以修改 persistent.xml 配置文件，实现动态数据表创建或更新。

DDL 自动更新的配置属性为 hibernate.hbm2ddl.auto，其设置内容有如下几种。

☑ **create**：每次加载时都删除上一次生成的表，然后根据用户定义的持久化类重新生成新的数据表。由于每次执行都会进行表的重新创建，所以执行后将导致原始数据丢失。

☑ **create-drop**：每次加载时，根据实体类生成数据表。EntityManagerFactory 实例一关闭，对应的数据表将自动删除。

☑ **update**：最常用的 DDL 操作属性。第一次加载时根据持久化类自动创建数据表（需先创建好数据库），再次加载 JPA 程序时将根据持久化类的结构自动更新表结构，同时保留原始数据记录。需要注意的是，部署到服务器后，表结构不会马上被建立起来，要等应用第一次运行起来后建立。

☑ **validate**：每次加载 Hibernate 时，验证创建数据库表结构，只会和数据库中的表进行比较，不会创建新表，但是会插入新值。

范例：【mldnspring-jpa 项目】修改 persistent.xml 配置文件，实现数据表动态更新。

```
<property name="hibernate.hbm2ddl.auto" value="update"/>
```

本程序中配置了 DDL 更新处理，当数据表不存在时执行增加操作，会出现如下提示信息。

```
【创建数据表】Hibernate: create table Dept (deptno bigint not null auto_increment, avgsal double precision
not null, createdate date, dname varchar(255), num integer not null, primary key (deptno)) engine=MyISAM
【执行增加】Hibernate: insert into Dept (avgsal, createdate, dname, num) values (?, ?, ?, ?)
```

可以发现，由于程序执行时数据表不存在，因此会先根据持久化 Dept 类创建数据表，随后才会执行数据增加处理。如果数据表已经存在，则不会重新创建。Dept 持久化类更新时，可以动态更新数据表结构。

13.2.4 JPA 常用注解

在 JPA 标准中，持久化类是最重要的组成部分。作为数据表控制的核心结构，持久化类需要通过注解进行配置。在 JPA 中主要使用如下 5 种注解。

☑ @Entity 注解

标注一个实体类，如果没有进行任何其他配置，则表名称与类名称一致。

范例：【mldnspring-jpa 项目】使用@Entity 注解。

```
@Entity
public class Dept implements Serializable {}
```

本程序在 Dept 类上定义了@Entity 注解，以后访问时将以 dept 作为表名称。

☑ @Table 注解

如果现在表名称与持久化类名称不同，则利用此注解进行表名称定义。

范例：【mldnspring-jpa 项目】使用@Table 注解。

```
@Entity
@Table(name="mldn_dept")
public class Dept implements Serializable {}
```

本程序配置了@Table 注解，在操作时使用的表名称为 mldn_dept。

☑ @Column 注解

明确表示每一个列，定义实体类时即便不使用此注解，也可以实现列的映射定义。当用户执行 DDL 自动更新处理时，没有加入@Column 注解的列将会自动按照该类型的最大长度进行创建。

范例：【mldnspring-jpa 项目】使用@Column 注解。

```
    @Column(length=50,nullable=true)
    private String dname;
```

dname 字段使用了@Column 注解，所以采用 DDL 更新处理时，所创建数据表的 dname 字段可有效限制字段的长度。

☑ @Temporal 注解

java.util.Date 类中包含日期与时间，所以对于数据表字段上需要一个描述日期与日期时间类型的定义，这样就可以通过@Temporal 注解中的 TemporalType 枚举来定义列类型保存结构，如 DATE、TIME、TIMESTAMP。

范例：【mldnspring-jpa 项目】使用@Temporal 注解。

```
    @Temporal(TemporalType.DATE)
    private Date createdate;
```

createdate 列只保存日期数据，不保存日期时间数据。

范例：【mldnspring-jpa 项目】使用@Temporal 注解。

```
@Temporal(TemporalType.TIMESTAMP)
private Date createdate;
```

由于配置的是 TemporalType.**TIMESTAMP**，所以 createdate 列将保存日期时间数据。

☑　@Transient 注解

默认情况下，只要是实体 Bean 中编写的属性，都会成为 SQL 生成后的字段描述。有时类里需要写一些临时变量，并不需要进行持久化处理，这时可以使用@Transient 注解标注。

范例：【mldnspring-jpa 项目】使用@Transient 注解。

```
@Transient
private String loc ;
```

因为 loc 字段使用了@Transient 注解定义，所以在进行 Dept 持久化处理时不会保存该字段的任何数据。

13.2.5　JPA 主键生成策略

在实体类中进行主键列属性定义时，可以使用@Id 注解完成。如果该属性上未添加任何注解，则表示此主键由开发者自行控制。例如，用户表中用户名作为主键，则用户名要通过用户定义。如果要采用自动方式实现主键定义，则必须使用@GeneratedValue 注解进行配置。

在@GeneratedValue 注解中，提供了 strategy 属性，该属性用于配置主键的生成策略。JPA 支持的主键生成策略由 javax.persistence.GenerationType 枚举类定义，有如下 4 种类型。

☑　**IDENTITY**：采用数据库 ID 自增长的方式来自增主键字段，Oracle 不支持这种方式。

☑　**AUTO**：JPA 自动选择合适的策略，是默认选项。

☑　**SEQUENCE**：通过序列产生主键，通过@SequenceGenerator 注解指定序列名，MySQL 不支持这种方式。

☑　**TABLE**：通过表产生主键，框架借由表模拟序列产生主键，使用该策略可以使应用更易于数据库移植。

范例：【mldnspring-jpa 项目】在 dept 表中采用自动增长方式定义主键。

```
@Id                                                    // 主键列
@GeneratedValue(strategy = GenerationType.IDENTITY)    // 主键生成方式
private Long deptno;                                   // 字段映射（属性名称=字段名称）
```

由于数据表在 MySQL 数据库中保存，所以可以直接使用 GenerationType.**IDENTITY**)，利用 MySQL 提供的 AUTO_INCREMENT 字段来生成主键。

除了使用传统数据库支持的方式外，也可以利用一张数据表来统一保存所有数据表的主键生成数据，如图 13-5 所示。此时，可以利用@TableGenerator 与@GeneratedValue 两个注解来实现。

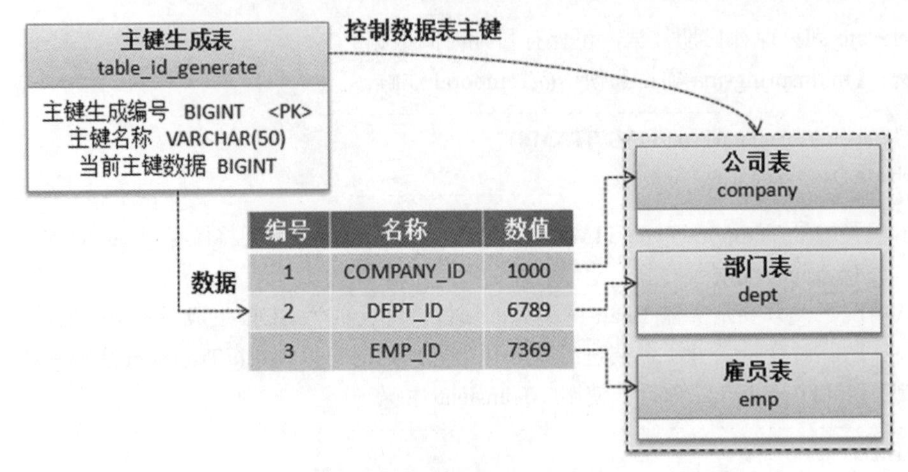

图 13-5　通过数据表控制主键

1.【mldnspring-jpa 项目】编写数据库脚本。

```
DROP DATABASE IF EXISTS mldn ;
CREATE DATABASE mldn CHARACTER SET UTF8 ;
USE mldn ;
CREATE TABLE dept(
    deptno          BIGINT ,
    dname           VARCHAR(50) ,
    createdate  DATE,
    num             INT ,
    avgsal          DOUBLE ,
    CONSTRAINT pk_deptno PRIMARY KEY(deptno)
) engine = innodb ;
CREATE TABLE table_id_generate(
    digid       BIGINT    AUTO_INCREMENT ,
    id_key          VARCHAR(50) ,
    id_value    BIGINT(50) ,
    CONSTRAINT pk_digid PRIMARY KEY(digid)
) engine = innodb ;
INSERT INTO table_id_generate(id_key,id_value) VALUES ('COMPANY_ID',1000) ;
INSERT INTO table_id_generate(id_key,id_value) VALUES ('DEPT_ID',6789) ;
INSERT INTO table_id_generate(id_key,id_value) VALUES ('EMP_ID',7369) ;
```

此时 dept 表中的主键不再采用 AUTO_INCREMENT 自动增长的处理形式，而是利用 table_id_generate 数据表中保存的 deptno 列的内容进行处理。

2.【mldnspring-jpa 项目】修改 Dept 实体类中的主键生成策略。

```
    @Id                                                 // 主键列
    @TableGenerator(
            name="DEPT_GENERATOR" ,                     // 定义主键生成器的名称
```

```
            table="table_id_generate" ,                          // 负责生成主键的数据表名称
            pkColumnName="id_key" ,                              // 要获取的行信息
            pkColumnValue="DEPT_ID" ,                            // 获取指定一行的信息
            valueColumnName="id_value" ,                         // 主键操作的内容字段
            allocationSize=1 )                                   // 每次增长步长
    @GeneratedValue(
            strategy = GenerationType.TABLE,
            generator = "DEPT_GENERATOR")                        // 根据名称引用配置的主键生成器
    private Long deptno;                                         // 字段映射（属性名称=字段名称）
```

本程序修改了 deptno 的主键生成策略，这样当进行数据增加时，会在控制台返回如下提示信息。

```
【操作分析】数据库操作1：查询主键生成数据表中指定列的数据。
Hibernate: select tbl.id_value from table_id_generate tbl where tbl.id_key=? for update
【操作分析】数据库操作2：更新数据表中指定列中的数据，根据原有数值与步长修改数据。
Hibernate: update table_id_generate set id_value=?  where id_value=? and id_key=?
【操作分析】数据库操作3：将获取到的主键数据设置到dept表的deptno字段中。
Hibernate: insert into Dept (avgsal, createdate, dname, num, deptno) values (?, ?, ?, ?, ?)
```

此时 dept 表中的主键字段（deptno）内容会根据 table_id_generate 表与步长进行控制，全部处理操作都会由 JPA 负责实现。

提示：不建议采用数据表控制主键列数据。

虽然 JPA 提供了数据表主键控制处理，但从性能角度来讲，不建议如此操作。因为增加数据的同时涉及 table_id_generate 表的两次 SQL 操作，这样的做法将严重影响数据库的操作性能，在高并发访问下不可能采用。

最好的做法是基于 Redis 数据库创建一个主键生成程序，每次进行数据增加时通过此程序获取要操作的主键信息。这样的管理既高效又适合分布式开发。

13.3 JPA 数据操作

JPA 中所有数据操作都需要通过 EntityManager 接口的方法进行处理。该接口中提供的方法全部是基于持久化类的，常用方法如表 13-2 所示。

表 13-2 EntityManager 接口提供的数据操作方法

编号	方法	类型	描述
1	public void persist(Object entity)	普通	数据持久化（保存）
2	public T merge(T entity)	普通	数据更新处理，如果数据不存在，则持久化处理

续表

编号	方法	类型	描述
3	public void remove(Object entity)	普通	数据删除
4	public <T> T find(Class<T> entityClass, Object primaryKey)	普通	根据 ID 进行查询，如果没有指定数据，返回 null
5	public <T> T getReference(Class<T> entityClass, Object primaryKey)	普通	根据 ID 进行查询
6	public Query createQuery(String sqlString)	普通	创建 Query 接口对象
7	public CriteriaBuilder getCriteriaBuilder()	普通	获取 CriteriaBuilder 接口对象
8	public Query createNativeQuery(String sqlString, Class resultClass)	普通	创建原生 SQL 查询

下面将针对 EntityManager 实现数据操作。

13.3.1 EntityManager 数据操作

EntityManager 接口提供的方法是基于持久化类的。下面将在已有数据表上实现数据的增加、修改、删除与 ID 查询操作。

范例：【mldnspring-jpa 项目】根据 ID 查询数据。

```
@Test
public void testFind() throws Exception {
    // 查询部门编号为3（deptno=3）的部门数据信息，并设置接收目标对象类型
    Dept dept = JPAEntityFactory.getEntityManager().find(Dept.class, 3L);
    System.out.println(dept);
}
```

程序执行结果

【操作分析】数据操作：根据ID进行部门信息查询。

```
Hibernate: select dept0_.deptno as deptno1_0_0_, dept0_.avgsal as avgsal2_0_0_,
dept0_.createdate as createda3_0_0_, dept0_.dname as dname4_0_0_, dept0_.num as num5_0_0_
from Dept dept0_ where dept0_.deptno=?
```

【操作分析】数据查询后将查询结果以PO的形式返回，随后调用toString获取对象信息。

```
Dept [deptno=3, avgsal=7868.88, createdate=2019-11-11, dname=MLDN教学研发中心, num=8]
```

EntityManager 接口里提供的 find 方法可以根据 ID 内容自动找到主键列（@Id 注解标注属性）进行查询。除了使用 find 方法外，也可以使用 getReference 方法查询。

范例：【mldnspring-jpa 项目】根据 ID 查询数据。

```
@Test
public void testReference() throws Exception {
    // 查询部门编号为3（deptno=3）的部门数据信息，并设置接收目标对象类型
    Dept dept = JPAEntityFactory.getEntityManager().getReference(Dept.class, 3L);
    System.out.println(dept);
}
```

此时程序将根据部门编号进行数据查询，执行效果与 find 方法完全相同。

提示：find 与 getReference 查询的区别。

通过以上范例可以发现，ID 数据存在时，两者的查询效果是完全一样的。如果要查询的 ID 数据不存在，getReference 方法将抛出 javax.persistence.EntityNotFoundException 异常，find 方法将返回 null。本书建议使用 find 方法进行查询。

EntityManager 接口支持 ID 更新处理操作方法 merge 方法，利用此方法可根据当前持久化类中的 ID 主键列进行更新。

范例：【mldnspring-jpa 项目】数据更新。

```
@Test
public void testEdit() throws Exception {
    JPAEntityFactory.getEntityManager().getTransaction().begin();        // 开启事务
    Dept dept = new Dept();                                               // 定义持久化对象
    dept.setDeptno(1L);                                                   // 设置主键
    dept.setDname("MLDN教学研发中心");                                    // 设置属性内容
    dept.setAvgsal(7868.88);                                              // 设置属性内容
    dept.setCreatedate(new SimpleDateFormat("yyyy-MM-dd")
            .parse("2006-11-11"));                                        // 设置属性内容
    dept.setNum(8);                                                       // 设置属性内容
    // 此时会返回一个当前的持久化对象
    Dept mergeDept = JPAEntityFactory.getEntityManager().merge(dept);        // 数据更新
    JPAEntityFactory.getEntityManager().getTransaction().commit();           // 事务提交
    JPAEntityFactory.close();                                                // 关闭连接
}
```

程序执行结果	**【操作分析】数据操作1：**由于对象是新创建的，所以首先判断此ID是否为实体数据。 Hibernate: select dept0_.deptno as deptno1_0_0_, dept0_.avgsal as avgsal2_0_0_, dept0_.createdate as createda3_0_0_, dept0_.dname as dname4_0_0_, dept0_.num as num5_0_0_ from Dept dept0_ where dept0_.deptno=? **【操作分析】数据操作2：**进行数据更新操作，将根据部门编号进行单行数据修改。 Hibernate: update Dept set avgsal=?, createdate=?, dname=?, num=? where deptno=?

本程序使用 merge 方法实现了数据更新处理。更新时需要接收一个持久化对象，由于该对象并不是查询出来的，而是新的 Dept 对象，所以会根据当前数据的 ID（deptno）查询此实体数据是否存在。如果存在，则执行更新处理。

提示：数据不存在，则执行增加。

merge 方法默认采用的是根据 ID 进行数据更新，如果在执行更新的时候没有设置 ID 数据（dept.setDeptno(1L);），则表示执行的是 INSERT 处理。在后续讲解 SpringDataJPA 技术时，可利用 merge 方法来代替 persist 方法。

之所以需在更新数据前进行数据查询操作，主要是由持久化对象状态决定的。如果此时先通过 find 方法进行查询，将不会出现 SELECT 查询语句，而是直接进行更新。

EntityManager 接口中提供的 remove 方法在执行时也是根据 ID 列进行删除的。使用此方法时需要先获得一个指定 ID 的持久化对象，而后才可以正常执行更新。

范例：【mldnspring-jpa 项目】删除数据。

```
@Test
public void testRemove() {
    JPAEntityFactory.getEntityManager().getTransaction().begin();            // 开启事务
    // 根据ID查询指定数据行的持久化对象
    Dept dept = JPAEntityFactory.getEntityManager().find(Dept.class, 5L) ;
    JPAEntityFactory.getEntityManager().remove(dept);                          // 数据删除
    JPAEntityFactory.getEntityManager().getTransaction().commit();             // 事务提交
    JPAEntityFactory.close();                                                  // 关闭连接
}
```

程序执行结果	**【操作分析】数据操作一：根据ID获取一个持久化对象。** Hibernate: select dept0_.deptno as deptno1_0_0_, dept0_.avgsal as avgsal2_0_0_, dept0_.createdate as createda3_0_0_, dept0_.dname as dname4_0_0_, dept0_.num as num5_0_0_ from Dept dept0_ where dept0_.deptno=? **【操作分析】数据操作二：进行指定编号的数据删除。** Hibernate: delete from Dept where deptno=?

通过以上程序可以发现，EntityManager 接口提供的修改与删除操作方法需要提供持久化对象才可以正常完成，这种做法的性能相对较低。要想获得更加丰富的数据查询与更新处理，可通过 Query 接口结合 JPQL 进行处理。

13.3.2 JPQL 语句

要想更好地实现数据的查询与更新处理操作，最好使用 JPQL（Java Persistence Query Language，Java 持久化查询语言）完成。

JPQL 提供的查询语法主要包括 3 类：查询用的 SELECT 语法、更新用的 UPDATE 语法以及删除用的 DELETE 语法。基本结构如下所示。

```
SELECT 子句 FROM 字句 [WHERE 子句] [GROUP BY 子句] [HAVING 子句] [ORDER BY 子句]
```

JPQL 还支持 BETWEEN...AND、LIKE、IN、NULL、EMPTY、EXISTS 等运算符的使用。而要想执行 JPQL 语句需要通过 Query 接口完成，该接口定义的常用方法如表 13-3 所示。

表 13-3 Query 接口常用方法

编号	方法	类型	描述
1	public List getResultList()	普通	查询数据，返回全部实体类对象
2	public default Stream getResultStream()	普通	查询数据，返回 Stream 对象
3	public Object getSingleResult()	普通	查询数据，返回单个查询结果
4	public int executeUpdate()	普通	执行数据库更新（只支持 UPDATE、DELETE）
5	public Query setParameter(int position, Object value)	普通	设置占位符参数

续表

编号	方法	类型	描述
6	public Query setParameter(String name, Object value)	普通	设置指定名称的参数
7	public Query setFirstResult(int startPosition)	普通	设置开始数据行
8	public Query setMaxResults(int maxResult)	普通	设置取出的数据行数

范例：【mldnspring-jpa 项目】查询全部部门数据。

```
@Test
public void testSelectAll() throws Exception {
	String jpql = "SELECT d FROM Dept AS d" ;			// 查询全部数据，以实体类返回
	// 查询全部数据，并设置返回的实体类型。如果不设置，将返回Object类型
	Query query = JPAEntityFactory.getEntityManager().createQuery(jpql, Dept.class) ;
	List<Dept> allDepts = query.getResultList() ;			// 查询全部数据
	allDepts.forEach((dept) -> {					// 迭代输出集合
		System.out.println(dept);				// 输出每一个PO对象
	});
	JPAEntityFactory.close();					// 关闭连接
}
```

本程序使用 JPQL 定义了查询语句。在 JPQL 中需要定义实体类对象（通过实体类映射数据表），JPQL 查询需要通过 Query 接口实现。为了避免向下转型问题，创建 Query 接口对象时设置好返回数据类型，这样当使用 getResultList 方法查询时就可以直接返回指定类型的 List 集合对象。

使用 JPQL 查询时，也可以使用 WHERE 子句设置查询条件。在条件定义时可以利用实体类的别名来访问类中的属性。如果使用 Query 接口中提供的 getSingleResult 方法，返回的数据类型将为 Object，这样依然要产生向下转型问题，所以建议使用 Query 的子接口 TypedQuery。

范例：【mldnspring-jpa 项目】根据 ID 查询部门信息。

```
@Test
public void testSelectSingle() throws Exception {
	String jpql = "SELECT d FROM Dept AS d WHERE d.deptno=?" ;	// 条件查询
	TypedQuery<Dept> query = JPAEntityFactory.getEntityManager()
		.createQuery(jpql, Dept.class) ;			// 创建TypedQuery接口对象
	query.setParameter(0, 1L) ;					// 设置第一个参数
	Dept dept = query.getSingleResult() ;				// 获取一个对象
	System.out.println(dept);					// 输出对象
	JPAEntityFactory.close();					// 关闭连接
}
```

程序执行结果	**【操作分析】**数据操作：**Query**接口将**JPQL**定义的语句转为关系型数据库的**SQL**语句。 Hibernate: select dept0_.deptno as deptno1_0_, dept0_.avgsal as avgsal2_0_, dept0_.createdate as createda3_0_, dept0_.dname as dname4_0_, dept0_.num as num5_0_ from Dept dept0_ where dept0_.deptno=?

```
【操作分析】获取一个对象信息。
Dept [deptno=1, avgsal=7868.88, createdate=2006-11-11, dname=MLDN教学研发中心, num=8]
```

本程序使用了 TypedQuery 接口，这样在使用 getSingleResult 方法时就可以根据定义的泛型类型转为指定类对象，避免了向下转型操作。

提示：使用参数名称定义查询。

在 JPQL 查询时可以使用“?”作为参数占位符，这种做法是由传统 JDBC 沿用下来的处理方式，而在 JPQL 语句中也可以使用参数名称来进行参数匹配。

范例：【mldnspring-jpa 项目】使用参数名称定义查询参数。

```
@Test
public void testSelectSingleParameterName() throws Exception {
    String jpql = "SELECT d FROM Dept AS d WHERE d.deptno=:pdeptno" ;
    TypedQuery<Dept> query = JPAEntityFactory.getEntityManager()
        .createQuery(jpql, Dept.class) ;                // 创建TypedQuery接口对象
    query.setParameter("pdeptno", 1L) ;                 // 设置指定名称的参数
    Dept dept = query.getSingleResult() ;               // 获取一个对象
    System.out.println(dept);                           // 输出对象
    JPAEntityFactory.close();                           // 关闭连接
}
```

本程序在定义 JPQL 时使用了 d.deptno=:pdeptno，其中 pdeptno 作为要设置的参数名称，这样在使用 setParameter 方法设置参数内容时可以通过参数名称设置。

JPA 提供的是一个公共的数据层开发标准，所以在 Query 接口中提供了数据分页操作方法。

范例：【mldnspring-jpa 项目】实现数据分页处理。

```
@Test
public void testSelectSplit() throws Exception {
    int currentPage = 2 ;                                           // 当前所在页
    int lineSize = 10 ;                                             // 每页显示数据行数
    String keyWord = "MLDN" ;                                       // 查询关键字
    String jpql = "SELECT d FROM Dept AS d WHERE d.dname LIKE ?";// 数据查询
    TypedQuery<Dept> query = JPAEntityFactory.getEntityManager()
            .createQuery(jpql, Dept.class) ;                        // 创建TypedQuery接口对象
    query.setParameter(0, "%" + keyWord + "%");                     // 设置查询关键字
    query.setFirstResult((currentPage - 1) * lineSize) ;            // 开始行
    query.setMaxResults(lineSize) ;                                 // 取出记录个数
    List<Dept> allDepts = query.getResultList() ;                   // 查询全部数据
    allDepts.forEach((dept) -> {                                    // 迭代输出集合
        System.out.println(dept);                                   // 输出每一个PO对象
    });
    JPAEntityFactory.close();                                       // 关闭连接
}
```

程序执行结果	Hibernate: select dept0_.deptno as deptno1_0_, dept0_.avgsal as avgsal2_0_, dept0_.createdate as createda3_0_, dept0_.dname as dname4_0_, dept0_.num as num5_0_ from Dept dept0_ where dept0_.dname like ? limit ?, ? 返回部门数据信息略

通过查询执行的 SQL 语句可以发现，设置的分页参数会自动转换为 MySQL 数据库中的 LIMIT 语句。因此编写分页查询时，不要在 JPQL 中定义任一指定数据库的分页语法。

范例：【mldnspring-jpa 项目】数据统计查询。

```
@Test
public void testSelectCount() {
    String keyWord = "MLDN" ;                                          // 查询关键字
    String jpql = "SELECT COUNT(d) FROM Dept AS d WHERE d.dname LIKE ?" ;   // 统计查询
    TypedQuery<Long> query = JPAEntityFactory.getEntityManager()
            .createQuery(jpql,Long.class) ;
    query.setParameter(0, "%" + keyWord + "%");                        // 设置查询关键字
    Long count = query.getSingleResult() ;                             // 获取一个查询结果
    System.out.println("数据行数：" + count);                           // 输出记录个数
    JPAEntityFactory.close();                                          // 关闭连接
}
```

程序执行结果	Hibernate: select count(dept0_.deptno) as col_0_0_ from Dept dept0_ where dept0_.dname like ? 数据行数：33

进行数据统计时，返回的一定是一个记录数据，所以这里使用 getSingleResult 方法获取结果数据。

在实际的开发过程中，虽然建议 JPQL 查询结果以实体类对象的形式返回，但有时候开发者可能只需要里面的几个字段，此时依然可以实现部分字段查询，查询的返回结果将以对象数组（Object[]）的形式返回。

范例：【mldnspring-jpa 项目】查询部分字段。

```
@Test
public void testSelectPart() throws Exception {
    String jpql = "SELECT d.deptno,d.dname FROM Dept AS d" ;           // 查询部分字段
    TypedQuery<Object[]> query = JPAEntityFactory.getEntityManager()
        .createQuery(jpql, Object[].class) ;                           // 创建TypedQuery接口对象
    List<Object[]> allDepts = query.getResultList() ;                  // 取得全部数据信息
    allDepts.forEach((data)->{
        System.out.println(Arrays.toString(data));                     // 将对象数组变为字符串
    });
    JPAEntityFactory.close();                                          // 关闭连接
}
```

程序执行结果	Hibernate: select dept0_.deptno as col_0_0_, dept0_.dname as col_1_0_ from Dept dept0_

本程序只查询了部门编号（d.deptno）与部门名称（d.dname），由于取得的不是持久化类对象，所以将使用对象数组保存返回结果（保存顺序为数据字段的查询顺序）。

使用 JPQL 语句除了可进行数据查询操作外，还可以进行数据更新与删除处理操作。这样两种处理支持有效解决了 EntityManager 接口中提供的 merge 与 remove 方法需要实体类对象的操作缺陷。

范例：【mldnspring-jpa 项目】使用 JPQL 修改数据。

```
@Test
public void testEdit() throws Exception {
    String jpql = "UPDATE Dept AS d SET d.dname=:dname,d.num=:num WHERE
d.deptno=:deptno" ;
    Query query = JPAEntityFactory.getEntityManager().createQuery(jpql) ;  // 创建Query对象
    query.setParameter("dname", "MLDNJAVA") ;                              // 设置参数内容
    query.setParameter("num", 50) ;                                        // 设置参数内容
    query.setParameter("deptno", 3L) ;                                     // 设置参数内容
    JPAEntityFactory.getEntityManager().getTransaction().begin();          // 开启事务
    int len = query.executeUpdate() ;                                      // 执行后返回影响的数据行数
    JPAEntityFactory.getEntityManager().getTransaction().commit();         // 提交事务
    JPAEntityFactory.close();                                              // 关闭连接
    System.out.println("更新行数：" + len);
}
```

程序执行结果	Hibernate: update Dept set dname=?, num=? where deptno=? 更新行数：1

本程序定义了一个 UPDATE 更新语句，随后可以设置相应的内容进行指定数据的更新。与 EntityManager 中的 merge 方法相比，此类更新可以像普通 SQL 那样，只更新部分字段的数据。

范例：【mldnspring-jpa 项目】数据删除。

```
@Test
public void testDelete() throws Exception {
    String jpql = "DELETE FROM Dept AS d WHERE d.deptno=:pdeptno" ;
    Query query = JPAEntityFactory.getEntityManager().createQuery(jpql) ;  // 创建Query对象
    query.setParameter("pdeptno", 3L) ;                                    // 设置参数内容
    JPAEntityFactory.getEntityManager().getTransaction().begin();          // 开启事务
    int len = query.executeUpdate() ;                                      // 执行后返回影响的数据行数
    JPAEntityFactory.getEntityManager().getTransaction().commit();         // 提交事务
    JPAEntityFactory.close();                                              // 关闭连接
    System.out.println("更新行数：" + len);
}
```

程序执行结果	Hibernate: delete from Dept where deptno=? 更新行数：1

本程序执行了根据 ID 删除数据的功能实现，避免了 EntityManager 接口中 remove 方法先查询后删除持久化对象的做法，性能是最好的。

13.3.3　Criteria 查询

JPQL 使用类 SQL 语法结构实现数据查询的定义。在 JPA 开发框架中，考虑到语句安全问题，还提供了 Criteria 查询操作。Criteria 采用对象处理形式实现数据查询定义，即可以使用方法来配置查询语句。

在整体定义中，主要涉及如下几个接口。

☑　CriteriaBuilder：Criteria 查询构建器，利用此接口可以创建 Predicate 接口、CriteriaQuery 接口对象。

☑　Predicate：Expression 子接口，可以进行查询条件设置，并且允许设置多个条件。

☑　CriteriaQuery：定义 Criteria 查询处理，同时可根据 Predicate 设置的限定条件进行查询。

☑　Root 接口：可以简单理解为一个 FROM 子句与持久化属性定位。

以上操作接口与 EntityManager 接口之间的对应结构关系如图 13-6 所示。

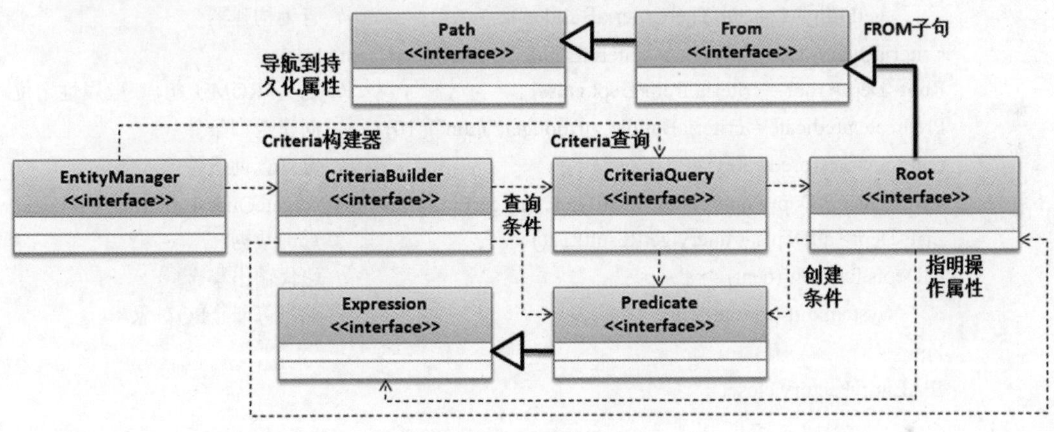

图 13-6　Criteria 查询

范例：【mldnspring-jpa 项目】使用 Criteria 实现查询全部操作。

```
@Test
public void testSelectAll() {
    CriteriaBuilder criteriaBuilder = JPAEntityFactory
        .getEntityManager().getCriteriaBuilder() ;                  // 查询构建器
    CriteriaQuery<Dept> criteria = criteriaBuilder.createQuery(Dept.class) ; // 创建查询
    criteria.from(Dept.class) ;                                    // 定义FROM子句
    TypedQuery<Dept> query = JPAEntityFactory.getEntityManager().createQuery(criteria) ;
    List<Dept> allDepts = query.getResultList() ;                  // 查询数据
    allDepts.forEach((dept) -> {                                   // 迭代输出集合
        System.out.println(dept);                                  // 输出每一个PO对象
    });
    JPAEntityFactory.close();
}
```

程序执行结果	Hibernate: select dept0_.deptno as deptno1_0_, dept0_.avgsal as avgsal2_0_, dept0_.createdate as createda3_0_, dept0_.dname as dname4_0_, dept0_.num as num5_0_

```
from Dept dept0_
部门输出数据略
```

本程序简单使用了 Criteria 查询处理，可以发现，在整体程序中不再明确编写任何 JPQL 语句，所有的语句都是通过类中的方法创建的。但不管如何处理，最终执行查询时依然需要通过 Query 接口完成。

要想在查询中设置查询条件，必须通过 Predicate 接口对象来表示。要想创建 Predicate 接口对象，必须依靠 CriteriaBuilder 接口完成，且在创建时通过 CriteriaBuilder 接口的方法设置查询条件。

范例：【mldnspring-jpa 项目】设置查询条件（num>10）。

```
@Test
public void testSelectWhere() {
    CriteriaBuilder criteriaBuilder = JPAEntityFactory
        .getEntityManager().getCriteriaBuilder() ;              // 查询构建器
    CriteriaQuery<Dept> criteria = criteriaBuilder.createQuery(Dept.class) ;
    Root<Dept> root = criteria.from(Dept.class) ;              // 定义FROM子句，获取属性定位
    Predicate predicate = criteriaBuilder.gt(root.get("num"), 10) ;  // 创建查询条件
    criteria.where(predicate) ;                                // 设置查询条件
    TypedQuery<Dept> query = JPAEntityFactory.getEntityManager().createQuery(criteria) ;
    List<Dept> allDepts = query.getResultList() ;              // 查询数据
    allDepts.forEach((dept) -> {                               // 迭代输出集合
        System.out.println(dept);                              // 输出每一个PO对象
    });
    JPAEntityFactory.close();
}
```

程序执行结果	Hibernate: select dept0_.deptno as deptno1_0_, dept0_.avgsal as avgsal2_0_, dept0_.createdate as createda3_0_, dept0_.dname as dname4_0_, dept0_.num as num5_0_ from Dept dept0_ where dept0_.num>10 部门输出数据略

本程序实现了 WHERE 条件查询。进行条件创建时，通过 CriteriaBuilder 接口的方法设置了一个大于条件，同时将此条件信息保存在 Predicate 接口对象中。这样，在创建 CriteriaQuery 查询时，可直接利用 where 方法实现条件配置。

范例：【mldnspring-jpa 项目】设置多个查询条件。

```
@Test
public void testSelectWhereMulti() {
    CriteriaBuilder criteriaBuilder = JPAEntityFactory
        .getEntityManager().getCriteriaBuilder() ;              // 查询构建器
    CriteriaQuery<Dept> criteria = criteriaBuilder.createQuery(Dept.class) ;
    Root<Dept> root = criteria.from(Dept.class) ;              // 定义FROM子句，获取属性定位
    List<Predicate> predicatesList = new ArrayList<Predicate>(); // 保存查询条件
    // 设置两个查询条件，使用OR连接：deptno为1L（deptno=1）与
```

```
        // avgsal使用between..and（avgsal between 999.00 and 9999.00）
        predicatesList.add(
                criteriaBuilder.or(criteriaBuilder.equal(root.get("deptno"), 1L),
                criteriaBuilder.between(root.get("avgsal"), 999.00, 9999.00)));
        predicatesList.add(criteriaBuilder.like(root.get("dname"), "%MLDN%")) ; // 模糊查询
        predicatesList.add(criteriaBuilder.gt(root.get("num"), 10)) ;          // num>10
        criteria.where(predicatesList.toArray(new Predicate[] {})) ;          // 设置查询条件
        TypedQuery<Dept> query = JPAEntityFactory.getEntityManager().createQuery(criteria) ;
        List<Dept> allDepts = query.getResultList() ;                        // 查询数据
        allDepts.forEach((dept) -> {                                          // 迭代输出集合
            System.out.println(dept);                                        // 输出每一个PO对象
        });
        JPAEntityFactory.close();
    }
```

程序执行结果	Hibernate: select dept0_.deptno as deptno1_0_, dept0_.avgsal as avgsal2_0_, dept0_.createdate as createda3_0_, dept0_.dname as dname4_0_, dept0_.num as num5_0_ from Dept dept0_ where (dept0_.deptno=1 or dept0_.avgsal between 999.0 and 9999.0) and (dept0_.dname like ?) and dept0_.num>10 部门输出数据略

本程序由于需要设置多个查询条件，所以创建了一个 List 集合，该集合中保存了所有的 Predicate 条件对象，利用对象的组织关系可以实现复杂条件定义。

提示：不推荐使用 Criteria 查询。

虽然 Criteria 查询比 JPQL 查询更安全，也更符合面向对象的设计思想，但通过上述一系列程序，相信读者已经发现，Criteria 开发的复杂度较大，同时这类查询也不便于后期代码维护。所以，如果不是必需，不推荐使用 Criteria 查询。需要明确的是，Criteria 有一个非常大的优点：处理 IN 查询非常方便，可避免 JPQL 利用循环处理查询语句的麻烦。

范例：【mldnspring-jpa 项目】使用 IN 查询（查询编号为 1、3、5 的部门信息）。

```
    @Test
    public void testSelectIN() {
        Set<Long> deptnos = new HashSet<Long>() ;                    // 保存部门编号集合
        deptnos.addAll(Arrays.asList(1L,3L,5L)) ;                    // 设置要查询的数据范围
        CriteriaBuilder criteriaBuilder = JPAEntityFactory
                .getEntityManager().getCriteriaBuilder() ;           // 查询构建器
        CriteriaQuery<Dept> criteria = criteriaBuilder.createQuery(Dept.class) ;
        Root<Dept> root = criteria.from(Dept.class) ;               // 定义FROM子句，获取属性定位
```

```
        Predicate predicate = root.get("deptno").in(deptnos) ;              // IN查询处理
        criteria.where(predicate) ;                                         // 设置条件
        criteria.orderBy(criteriaBuilder.desc(root.get("deptno")));        // 结果排序
        TypedQuery<Dept> query = JPAEntityFactory.getEntityManager().createQuery(criteria) ;
        List<Dept> allDepts = query.getResultList() ;                       // 查询数据
        allDepts.forEach((dept) -> {                                        // 迭代输出集合
            System.out.println(dept);                                       // 输出每一个PO对象
        });
        JPAEntityFactory.close();
    }
```

程序执行结果	Hibernate: select dept0_.deptno as deptno1_0_, dept0_.avgsal as avgsal2_0_, dept0_.createdate as createda3_0_, dept0_.dname as dname4_0_, dept0_.num as num5_0_ from Dept dept0_ where dept0_.deptno in (1 , 3 , 5) order by dept0_.deptno desc 部门输出数据略

本程序首先将需要查询的数据保存在了 Set 集合中，这样在使用 Root 创建 Predicate 时就可以利用 in 方法将所需要的数据设置在查询中。利用此种方式，可以简化 JPQL 处理相同操作时的循环处理逻辑。

13.3.4 SQL 原生查询

使用 JPQL 进行数据查询时，除了可查询出所需要的数据，还可能会引发数据级联问题。JPQL 是以标准 SQL 语法结构为主的，在使用数据库中一些特殊的查询处理时还需要通过原生 SQL 实现查询。

提示：不用过多考虑数据库的可移植性。

JPA 是一个技术标准，Hibernate 是一个广泛使用的、可移植性高的 ORMapping 开发框架。随着技术发展，开发者对数据库可移植性的要求在减弱，更关心的是程序的处理性能。所以，即使程序采用原生 SQL 查询，也不会产生设计结构问题。

范例：【mldnspring-jpa 项目】使用原生 SQL 查询全部。

```
    @Test
    public void testSelectAll() {
        String sql = "SELECT * FROM dept" ;                                 // 原始SQL语句
        // 创建原生SQL查询，该操作只能够使用Query接口进行接收
        Query query = JPAEntityFactory.getEntityManager().createNativeQuery(sql,Dept.class) ;
        List<Dept> allDepts = query.getResultList() ;                       // 查询数据
        allDepts.forEach((dept) -> {                                        // 迭代输出集合
            System.out.println(dept);                                       // 输出每一个PO对象
        });
        JPAEntityFactory.close();
    }
```

程序执行结果	Hibernate: SELECT * FROM dept 部门输出数据略

本程序使用了原生 SQL 查询，对于查询结果，利用 Query 接口将其封装为持久化对象的形式返回。

对于数据库中的分页处理操作方法，即使没有使用 JPQL，实际上也可以继续采用。

范例：【mldnspring-jpa 项目】数据分页。

```
@Test
public void testSelectSplit() {
    int currentPage = 1 ;                                              // 当前所在页
    int lineSize = 10 ;                                                // 每页显示数据行数
    String keyWord = "MLDN" ;                                          // 查询关键字
    String sql = "SELECT * FROM dept WHERE dname LIKE :pkw" ;   // 原始SQL语句
    Query query = JPAEntityFactory.getEntityManager().createNativeQuery(sql,Dept.class) ;
    query.setParameter("pkw", "%" + keyWord + "%");                    // 设置查询关键字
    query.setFirstResult((currentPage - 1) * lineSize) ;               // 开始行
    query.setMaxResults(lineSize) ;                                    // 取出记录个数
    List<Dept> allDepts = query.getResultList() ;                      // 查询数据
    allDepts.forEach((dept) -> {                                       // 迭代输出集合
        System.out.println(dept);                                      // 输出每一个PO对象
    });
    JPAEntityFactory.close();
}
```

程序执行结果	Hibernate: SELECT * FROM dept WHERE dname LIKE ? limit ? 部门输出数据略

本程序在定义 SQL 语句时没有指明分页处理，但利用 Query 接口中提供的分页支持，可以根据数据库自动设置分页命令。

范例：【mldnspring-jpa 项目】使用原生 SQL 查询部分数据。

```
@Test
public void testSelectPart() {
    String sql = "SELECT deptno,dname,num FROM dept" ;                 // 原始SQL语句
    Query query = JPAEntityFactory.getEntityManager().createNativeQuery(sql) ;
    List<Object[]> allDepts = query.getResultList();
    allDepts.forEach((data) -> {                                       // 迭代输出集合
        System.out.println(Arrays.toString(data));                     // 输出数据
    });
    JPAEntityFactory.close();
}
```

程序执行结果	Hibernate: SELECT deptno,dname,num FROM dept 部门输出数据略

本程序实现了数据表中部分字段的数据查询处理，返回的结果为对象数组，根据字段的顺序可采用索引形式取出所需要的数据内容。

13.4　JPA 数据缓存

为了尽可能提升操作性能，ORMapping 开发框架提供了缓存解决方案。JPA 中数据缓存分为两类：一级缓存和二级缓存。

13.4.1　一级缓存

一级缓存指的是在 EntityManager 接口操作中提供的缓存机制，是 JPA 默认的缓存支持，并且会一直存在。利用一级缓存，可以有效地解决相同数据的重复查询问题，提升单个线程的数据处理能力。

范例：【mldnspring-jpa 项目】查询两次编号相同的部门信息。

```
@Test
public void testFind() throws Exception {
    Dept deptA = JPAEntityFactory.getEntityManager().find(Dept.class, 3L) ;
    System.out.println("第一次使用find()查询：" + deptA);
    System.out.println("-------------------- 华丽的分割线 --------------------");
    Dept deptB = JPAEntityFactory.getEntityManager().find(Dept.class, 3L) ;
    System.out.println("第二次使用find()查询：" + deptB);
}
```

程序执行结果	Hibernate: select dept0_.deptno as deptno1_0_0_, dept0_.avgsal as avgsal2_0_0_, dept0_.createdate as createda3_0_0_, dept0_.dname as dname4_0_0_, dept0_.num as num5_0_0_ from Dept dept0_ where dept0_.deptno=? **【操作分析】第一次调用EntityManager接口中find方法执行查询，发出数据库查询指令。** 第一次使用find()查询：Dept [deptno=3, avgsal=8968.88, createdate=2018-02-27, dname=MLDN教学管理部, num=8] -------------------- 华丽的分割线 -------------------- **【操作分析】由于一级缓存存在，所以查询相同编号记录时将不会重复发出查询指令。** 第二次使用find()查询：Dept [deptno=3, avgsal=8968.88, createdate=2018-02-27, dname=MLDN教学管理部, num=8]

通过以上查询操作可以发现，在使用 EntityManager 接口的 find 方法根据主键 ID 进行数据查询时，会向数据库发出查询指令。第二次再发出同样查询时，由于一级缓存作用，将不会向数据库发出查询命令，而是直接引用已有的持久化对象。所以，一级缓存是永远都会存在的。

范例：【mldnspring-jpa 项目】修改第一次查询结果。

```
@Test
public void testFind() throws Exception {
```

```
        Dept deptA = JPAEntityFactory.getEntityManager().find(Dept.class, 3L) ;
        System.out.println("第一次使用find()查询：" + deptA);
        deptA.setDname("修改第一次查询结果的部门名称");                  // 修改数据
        deptA.setNum(-1);                                               // 修改数据
        System.out.println("-------------------- 华丽的分割线 --------------------");
        Dept deptB = JPAEntityFactory.getEntityManager().find(Dept.class, 3L) ;
        System.out.println("第二次使用find()查询：" + deptB);
    }
```

程序执行结果	Hibernate: select dept0_.deptno as deptno1_0_0_, dept0_.avgsal as avgsal2_0_0_, dept0_.createdate as createda3_0_0_, dept0_.dname as dname4_0_0_, dept0_.num as num5_0_0_ from Dept dept0_ where dept0_.deptno=? 第一次使用find()查询：Dept [deptno=3, avgsal=8968.88, createdate=2018-02-27, dname=MLDN教学管理部, num=8] -------------------- 华丽的分割线 -------------------- 第二次使用find()查询：Dept [deptno=3, avgsal=8968.88, createdate=2018-02-27, dname=修改第一次查询结果的部门名称, num=-1]

本程序中对第一次查询返回的持久化对象进行了数据修改，由于一级缓存的作用，第二次查询获取的对象信息是修改后的数据。要想避免一级缓存对第二次查询操作产生影响，可以使用 EntityManager 接口提供的 refresh 方法（public void refresh(Object entity)）进行刷新处理。

范例：【mldnspring-jpa 项目】刷新一级缓存。

```
    @Test
    public void testFind() throws Exception {
        Dept deptA = JPAEntityFactory.getEntityManager().find(Dept.class, 3L) ;
        System.out.println("第一次使用find()查询：" + deptA);
        deptA.setDname("修改第一次查询结果的部门名称");                  // 修改数据
        deptA.setNum(-1);                                               // 修改数据
        System.out.println("-------------------- 华丽的分割线 --------------------");
        JPAEntityFactory.getEntityManager().refresh(deptA);             // 刷新缓存数据，重新查询
        Dept deptB = JPAEntityFactory.getEntityManager().find(Dept.class, 3L) ;
        System.out.println("第二次使用find()查询：" + deptB);
    }
```

程序执行结果	**【操作分析】第一次数据查询，此时查询返回的数据将被自动保存在一级缓存中。** Hibernate: select dept0_.deptno as deptno1_0_0_, dept0_.avgsal as avgsal2_0_0_, dept0_.createdate as createda3_0_0_, dept0_.dname as dname4_0_0_, dept0_.num as num5_0_0_ from Dept dept0_ where dept0_.deptno=? 第一次使用find()查询：Dept [deptno=3, avgsal=8968.88, createdate=2018-02-27, dname=MLDN教学管理部, num=8] -------------------- 华丽的分割线 -------------------- **【操作分析】刷新缓存后将发出第二次查询。** Hibernate: select dept0_.deptno as deptno1_0_0_, dept0_.avgsal as avgsal2_0_0_, dept0_.createdate as createda3_0_0_, dept0_.dname as dname4_0_0_, dept0_.num as

```
num5_0_0_ from Dept dept0_ where dept0_.deptno=?
第二次使用find()查询：Dept [deptno=3, avgsal=8968.88, createdate=2018-02-27, dname=MLDN教学管理部, num=8]
```

可以发现，使用 refreh 方法后需要刷新缓存中的数据，第二次使用 find 方法查询时会发出新的查询指令。

13.4.2 JPA 对象状态

一级缓存不仅在数据查询操作中有着重要作用，在 JPA 状态的维护处理上也发挥着作用。JPA 中一共定义了 4 种对象状态，状态的切换可以通过 EntityManager 接口提供的方法完成，具体操作如图 13-7 所示。

☑ **瞬时态对象（New）**：一个新的持久化类对象，此时该对象并未实现持久化存储（也可能尚未有 ID），和 Persistence Context（持久化上下文）建立关联的对象。

☑ **持久态对象（Managed）**：数据库中存在相关 ID 数据，同时保存在一级缓存中，由于该对象已经与 Persistence Context 建立了关联，所以对该对象所做的修改将可以影响到数据库中的相关数据（需要进行事务提交）。

☑ **游离态对象（Datached）**：数据库中存在相关 ID 数据，但是有可能当前的 EntityManager 已经关闭，所以该对象没有和 Persistence Context 存在关联，此时对持久化对象的修改不会影响到数据库中相关数据。

☑ **删除态对象（Removed）**：该持久化对象和 Persistence Context 有关联，但是其对应的数据库数据已经被删除。

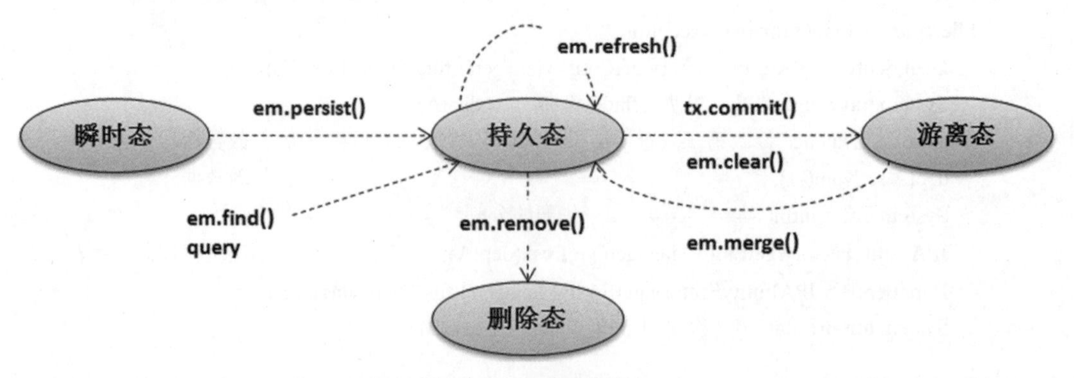

图 13-7　JPA 对象状态

范例：【mldnspring-jpa 项目】观察对象持久态操作。

```
@Test
public void testFind() throws Exception {
    // 通过find方法查询出来的数据会自动保存在一级缓存中，同时该对象状态也属于持久态
    Dept dept = JPAEntityFactory.getEntityManager().find(Dept.class, 3L) ;
    JPAEntityFactory.getEntityManager().getTransaction().begin();        // 开启事务
    dept.setDname("MLDN教学研发中心");                                    // 修改持久态对象的属性
    dept.setNum(30);                                                      // 修改持久态对象的属性
    JPAEntityFactory.getEntityManager().getTransaction().commit();       // 提交事务
```

```
        JPAEntityFactory.close();                                  // 关闭连接
    }
```

程序执行结果

Hibernate: select dept0_.deptno as deptno1_0_0_, dept0_.avgsal as avgsal2_0_0_, dept0_.createdate as createda3_0_0_, dept0_.dname as dname4_0_0_, dept0_.num as num5_0_0_ from Dept dept0_ where dept0_.deptno=?

【操作分析】持久态下的更新，只要修改了持久化对象属性，将自动更新相应数据。

Hibernate: update Dept set avgsal=?, createdate=?, dname=?, num=? where deptno=?

本程序通过 EntityManager 接口的 find 方法得到了一个 Dept 类的持久态对象。如果在持久态状态下修改对象属性，则会自动实现对应数据的变更操作。

范例：【mldnspring-jpa 项目】瞬时态对象转为持久态对象。

```
@Test
public void testAdd() {
    Dept dept = new Dept() ;                                          // 创建持久态对象
    dept.setDname("极限IT教学部");                                     // 设置属性
    dept.setAvgsal(8968.88);                                          // 设置属性
    dept.setCreatedate(new Date());                                   // 设置属性
    dept.setNum(888);                                                 // 设置属性
    JPAEntityFactory.getEntityManager().getTransaction().begin();     // 开启事务
    // 数据持久化，对象将由瞬时态变为持久态（EntityManager关闭前），并自动获取ID
    JPAEntityFactory.getEntityManager().persist(dept);
    JPAEntityFactory.getEntityManager().getTransaction().commit();    // 提交事务
    System.out.println("*** 新增数据的ID是：" + dept.getDeptno());   // 自动获得增长后ID
    // 将保存的对象信息根据ID查询出来，由于一级缓存作用，不会向数据库发出查询指令
    Dept selectDept = JPAEntityFactory.getEntityManager()
            .find(Dept.class, dept.getDeptno()) ;
    System.out.println(selectDept);
    JPAEntityFactory.close();                                         // 关闭连接
}
```

程序执行结果

【操作分析】对象由瞬时态变为持久态后将自动保存在一级缓存中。

Hibernate: insert into Dept (avgsal, createdate, dname, num) values (?, ?, ?, ?)

【操作分析】JPA会自动获取增长后的ID主键数据。

*** 新增数据的ID是：6

【操作分析】由于数据持久化时会自动加入一级缓存中，所以此时将不会发出数据库查询操作。

Dept [deptno=6, avgsal=8968.88, createdate=Tue Feb 27 07:42:12 CST 2018, dname=极限IT教学部, num=888]

本程序实现了一个瞬时态到持久态的转换操作（persist()持久化），由于一级缓存的作用，在 EntityManager 连接过程中会将持久化对象自动保存在一级缓存中，这样当再次根据 ID 进行数据查询时，会直接通过一级缓存获取，而不会再执行数据库查询操作。

每个数据增加后都会被一级缓存保存，因此，当大量数据批量增加的时候，缓存会占用大量的内存空间。此时，最好的做法是进行缓存数据的刷新与清空处理。

范例：【mldnspring-jpa 项目】缓存刷新。

```
@Test
public void testBatch() {
    JPAEntityFactory.getEntityManager().getTransaction().begin();       // 开启事务
    for (int x = 0 ; x < 1000 ; x ++) {
        Dept dept = new Dept() ;                                        // 瞬时态对象
        dept.setDname("极限IT教学部 - " + x);                            // 设置数据
        dept.setAvgsal(8968.88 + x);                                    // 设置数据
        dept.setCreatedate(new Date());                                 // 设置数据
        dept.setNum(888 + x);                                           // 设置数据
        JPAEntityFactory.getEntityManager().persist(dept);              // 数据持久化自动获取ID
        if (x % 10 == 0) {                                              // 每10条记录清除一次缓存
            JPAEntityFactory.getEntityManager().flush();                // 强制刷新
            JPAEntityFactory.getEntityManager().clear();                // 清空缓存
        }
    }
    JPAEntityFactory.getEntityManager().getTransaction().commit();      // 提交事务
    JPAEntityFactory.close();                                           // 关闭连接
}
```

本程序考虑到一级缓存的影响，所以每持久化 10 条数据后都会采用 flush 方法强制性刷新缓存，同时利用 clear 方法清空缓存空间，这样就可以在不受缓存限制的状态下实现数据的批量持久化处理。

13.4.3 二级缓存

一级缓存只针对一个 Session 有效。因此，如果某个数据允许多个 EntityManager 共享，则必须使用二级缓存。二级缓存是针对 EntityManagerFactory 接口级别的配置，需要在项目中利用 EHCache 组件实现。下面给出具体的操作步骤。

1.【mldnspring 项目】修改父 pom.xml 配置文件，引入相关缓存组件。

属性配置	`<ehcache.version>2.10.4</ehcache.version>`

依赖库配置	`<dependency>`
	`<groupId>org.hibernate</groupId>`
	`<artifactId>hibernate-ehcache</artifactId>`
	`<version>${hibernate.version}</version>`
	`</dependency>`
	`<dependency>`
	`<groupId>net.sf.ehcache</groupId>`
	`<artifactId>ehcache</artifactId>`
	`<version>${ehcache.version}</version>`
	`</dependency>`

2.【mldnspring-jpa 项目】修改子 pom.xml 配置文件，追加 ehcache 相关依赖库。

```
<dependency>
    <groupId>org.hibernate</groupId>
    <artifactId>hibernate-ehcache</artifactId>
</dependency>
<dependency>
    <groupId>net.sf.ehcache</groupId>
    <artifactId>ehcache</artifactId>
</dependency>
```

3.【mldnspring-jpa 项目】ehcache 组件需要在 CLASSPATH 下定义 ehcache.xml 配置文件，这里将配置文件保存在 src/main/resources 源文件目录中。ehcache.xml 定义的内容如下。

```
<?xml version="1.0" encoding="UTF-8"?>
<ehcache>
    <diskStore path="java.io.tmpdir" />                 <!-- 设置临时缓存路径 -->
    <defaultCache                                        ➔ 定义默认缓存区配置
        maxElementsInMemory="10000"                      ➔ 缓存中允许保存的元素个数
        eternal="true"                                   ➔ 是否允许自动失效
        timeToIdleSeconds="120"                          ➔ 缓存失效时间
        timeToLiveSeconds="120"                          ➔ 最大存活时间
        maxElementsOnDisk="10000000"                     ➔ 磁盘最多保存的元素个数
        diskExpiryThreadIntervalSeconds="120"            ➔ 磁盘失效时间
        memoryStoreEvictionPolicy="LRU" />               ➔ 定义缓存策略，如FIFO或LRU等
</ehcache>
```

4.【mldnspring-jpa 项目】修改 persistence.xml 配置文件，增加启用二级缓存配置属性。

```
<property name="hibernate.cache.use_second_level_cache" value="true" />   <!-- 启用二级缓存 -->
<property name="hibernate.cache.region.factory_class"
    value="org.hibernate.cache.ehcache.EhCacheRegionFactory" />          <!-- 二级缓存处理类 -->
```

5.【mldnspring-jpa 项目】修改 persistence.xml 配置文件，定义二级缓存模式。

```
<shared-cache-mode>ENABLE_SELECTIVE</shared-cache-mode>
```

<shared-cache-mode/>元素可以使用的配置项如下。

☑ ALL：所有的实体类都被缓存。

☑ NONE：所有的实体类都不被缓存。

☑ ENABLE_SELECTIVE：标识 @Cacheable(true) 注解的实体类将被缓存。

☑ DISABLE_SELECTIVE：缓存除标识 @Cacheable(false) 以外的所有实体类。

☑ UNSPECIFIED：默认值，JPA 产品默认值将被使用。

6.【mldnspring-jpa 项目】修改 Dept 实体类定义，追加二级缓存注解。

```
@Cacheable(true)
@Entity
```

```
public class Dept implements Serializable {}
```

7.【mldnspring-jpa 项目】编写测试程序，创建两个 EntityManager 接口对象，以观察二级缓存处理。

```
@Test
public void testFind() throws Exception {
    EntityManager entityManagerA = JPAEntityFactory
        .getEntityManagerFactory().createEntityManager() ;          // 创建第一个EntityManager
    System.out.println(entityManagerA.find(Dept.class, 3L));        // 第一个EntityManager查询
    entityManagerA.close();                                         // 关闭连接
    System.err.println("-------------------- 华丽丽的分割线 --------------------");
    EntityManager entityManagerB = JPAEntityFactory
        .getEntityManagerFactory().createEntityManager() ;          // 创建第二个EntityManager
    System.out.println(entityManagerB.find(Dept.class, 3L));        // 第二个EntityManager查询
    entityManagerB.close();                                         // 关闭连接
}
```

程序执行结果	Hibernate: select dept0_.deptno as deptno1_0_0_, dept0_.avgsal as avgsal2_0_0_, dept0_.createdate as createda3_0_0_, dept0_.dname as dname4_0_0_, dept0_.num as num5_0_0_ from Dept dept0_ where dept0_.deptno=? Dept [deptno=3, avgsal=8968.88, createdate=2018-02-27, dname=MLDN教学管理部, num=8] -------------------- 华丽丽的分割线 -------------------- Dept [deptno=3, avgsal=8968.88, createdate=2018-02-27, dname=MLDN教学管理部, num=8]

通过查询结果可以发现，由于二级缓存的使用，即便是两个不同 EntityManager 实现的数据查询也可以进行数据信息共享，这样可以提升不同线程对同一条数据查询的操作性能。

13.4.4 查询缓存

使用 EntityManager 接口的 find 方法，将默认实现数据的查询缓存。实际开发中，数据的查询需求是非常复杂的，使用 JPQL 进行数据查询是最方便的。想要实现 Query 查询缓存，必须进行额外配置。

1.【mldnspring-jpa 项目】修改 persistence.xml 配置文件，配置查询缓存。

```
<property name="hibernate.cache.use_query_cache" value="true"/>        <!-- 启用查询缓存 -->
```

2.【mldnspring-jpa 项目】进行查询缓存的配置，还需要使用 QueryHints 操作。

```
@Test
public void testFind() throws Exception {
    String jpql = "SELECT d FROM Dept AS d WHERE d.deptno=:pdeptno" ;
    EntityManager entityManagerA = JPAEntityFactory
        .getEntityManagerFactory().createEntityManager() ;          // 创建第一个EntityManager
    TypedQuery<Dept> queryA = entityManagerA.createQuery(jpql, Dept.class)
        .setHint(QueryHints.HINT_CACHEABLE, true) ;
```

```
queryA.setParameter("pdeptno", 3L) ;
System.out.println(queryA.getSingleResult());                          // 第一个EntityManager查询
entityManagerA.close();                                               // 关闭连接
System.err.println("-------------------- 华丽丽的分割线 --------------------");
EntityManager entityManagerB = JPAEntityFactory
    .getEntityManagerFactory().createEntityManager() ;                 // 创建第二个EntityManager
TypedQuery<Dept> queryB = entityManagerB.createQuery(jpql, Dept.class)
    .setHint(QueryHints.HINT_CACHEABLE, true) ;
queryB.setParameter("pdeptno", 3L) ;
System.out.println(queryB.getSingleResult());                          // 第二个EntityManager查询
entityManagerB.close();                                               // 关闭连接
}
```

程序执行结果	Hibernate: select dept0_.deptno as deptno1_0_, dept0_.avgsal as avgsal2_0_, dept0_.createdate as createda3_0_, dept0_.dname as dname4_0_, dept0_.num as num5_0_ from Dept dept0_ where dept0_.deptno=? Dept [deptno=3, avgsal=8968.88, createdate=2018-02-27, dname=MLDN教学管理部, num=8] -------------------- 华丽丽的分割线 -------------------- Dept [deptno=3, avgsal=8968.88, createdate=2018-02-27, dname=MLDN教学管理部, num=8]

本程序由于使用了查询缓存，所以即使有两个不同的 Query 对象，进行同一数据查询时也只会查询一次。

13.5　JPA 锁机制

数据库并发访问的时候，为了保证操作数据的完整性，往往会对并发数据的更新做出限制。例如，允许一个 Session 进行更新处理，其他 Session 必须等此 Session 更新完成后才可以进行更新处理，这样的机制就称为数据库锁机制。JPA 中也支持锁机制处理，且主要支持两类锁。

☑ **悲观锁（Pessimistic）**：假设数据访问一直存在并发更新。悲观锁一直都存在，依靠的是数据库的锁机制。

☑ **乐观锁（Optimistic）**：假设不会进行数据并发访问（不会同时出现数据更新处理）。乐观锁主要依靠算法来实现。设置一个版本号，通过版本号来判断当前的 Session 能否进行更新。

要实现 JPA 锁机制，可以通过 EntityManager 接口来完成。EntityManager 接口定义的数据锁操作如表 13-4 所示。

表 13-4　EntityManager 接口定义的数据锁操作

编号	方法	类型	描述
1	public <T> T find(Class<T> entityClass, Object primaryKey, LockModeType lockMode)	普通	使用锁模式查询数据
2	public void lock(Object entity, LockModeType lockMode)	普通	为持久化对象设置锁

锁处理主要是通过 javax.persistence.LockModeType 枚举类定义锁模式。下面将通过具体的程序代码进行说明。

13.5.1 悲观锁

悲观锁认为，用户的并发访问会一直发生，因此在整个处理中会采用锁机制对事务内的操作数据进行锁定，这样其他的事务就无法进行该数据的更新操作。

LockModeType 枚举类对悲观锁提供了 4 种处理模式。

☑ **PESSIMISTIC_READ**：只要事务读实体，实体管理器就会锁定实体，直到事务完成（提交或回滚）后才解锁处理。这种锁模式不会阻碍其他事务读取数据。

☑ **PESSIMISTIC_WRITE**：只要事务更新实体，实体管理器就会锁定实体。这种锁模式强制尝试修改实体数据的事务串行化，当多个并发更新事务出现，导致更新失败几率较高时，使用这种锁模式。

☑ **PESSIMISTIC_FORCE_INCREMENT**：事务读实体时，实体管理器就锁定实体。事务结束后，增加实体的版本属性，即使实体未做修改。

☑ **NONE**：不使用锁。

范例：【mldnspring-jpa 项目】使用悲观锁。

```
@Test
public void testFind() throws Exception {
    JPAEntityFactory.getEntityManager().getTransaction().begin();          // 开启事务
    JPAEntityFactory.getEntityManager().find(Dept.class, 3L,
        LockModeType.PESSIMISTIC_WRITE) ;                                  // 悲观锁形式查询
    JPAEntityFactory.getEntityManager().getTransaction().rollback();       // 可以回滚或提交
}
```

程序执行结果	Hibernate: select dept0_.deptno as deptno1_0_0_, dept0_.avgsal as avgsal2_0_0_, dept0_.createdate as createda3_0_0_, dept0_.dname as dname4_0_0_, dept0_.num as num5_0_0_ from Dept dept0_ where dept0_.deptno=? for update

查询语句中，使用数据库锁机制 FOR UPDATE 语句进行了数据锁定，这样当此事务提交或回滚前，该数据都无法被其他 Session 修改。

13.5.2 乐观锁

JPA 早期提供的锁机制是乐观锁。乐观锁通常假设不存在多个事务修改同一条数据的情况。乐观锁需要在数据表上增加一个表示数据版本的编号，进行数据更新时通过此编号判断是否有其他 Session 进行过数据更新。判断过程如下：

☑ 第一个 Session 与第二个 Session 同时读取一条数据，该数据的版本号为 100。

☑ 第一个 Session 更新数据时（此时第二个 Session 未关闭），自动将版本号由 100 修改为 101。

☑ 第二个 Session 读取时版本编号为 100，保存时数据版本编号为 101，由于版本编号不同，无法进行数据更新处理。

对于乐观锁，LockModeType 枚举类有如下两种锁处理模式。

☑ **OPTIMISTIC**：和 READ 锁模式相同，JPA 2.0 仍然支持 READ 锁模式，但明确指出在新应用程序中推荐使用 OPTIMISTIC。

☑ **OPTIMISTIC_FORCE_INCREMENT**：和 WRITE 锁模式相同，JPA 2.0 仍然支持 WRITE 锁模式，在新应用程序中明确推荐使用 OPTIMISTIC_FORCE_INCREMENT。

1.【mldnspring-jpa 项目】定义数据库脚本，为表中追加版本编号。

```
DROP DATABASE IF EXISTS mldn ;
CREATE DATABASE mldn CHARACTER SET UTF8 ;
USE mldn ;
CREATE TABLE dept(
    deptno          BIGINT    AUTO_INCREMENT ,
    dname           VARCHAR(50) ,
    createdate  DATE,
    num             INT ,
    vseq            BIGINT DEFAULT 0,
    avgsal          DOUBLE ,
    CONSTRAINT pk_deptno PRIMARY KEY(deptno)
) engine = innodb ;
```

在数据库脚本中为 dept 表增加一个 vseq 的乐观锁版本编号属性，该编号的内容由 JPA 自动维护。

2.【mldnspring-jpa 项目】在 Dept 持久类上追加 vseq 属性定义，同时指明该属性为乐观锁版本控制属性。

```
@Version
private Long vseq ;
```

3.【mldnspring-jpa 项目】编写程序，采用乐观锁处理。

```
@Test
public void testFind() throws Exception {
    JPAEntityFactory.getEntityManager().getTransaction().begin();          // 开启事务
    Dept dept = JPAEntityFactory.getEntityManager().find(Dept.class, 1L,
            LockModeType.OPTIMISTIC_FORCE_INCREMENT) ;                     // 乐观锁
    dept.setNum(9999);                                                     // 修改数据
    JPAEntityFactory.getEntityManager().getTransaction().commit();         // 可以回滚或提交
}
```

程序执行结果	**【操作分析】根据指定ID查询出数据。** Hibernate: select dept0_.deptno as deptno1_0_0_, dept0_.avgsal as avgsal2_0_0_, dept0_.createdate as createda3_0_0_, dept0_.dname as dname4_0_0_, dept0_.num as num5_0_0_, dept0_.vseq as vseq6_0_0_ from Dept dept0_ where dept0_.deptno=? **【操作分析】进行数据更新时会判断版本号。** Hibernate: update Dept set avgsal=?, createdate=?, dname=?, num=?, vseq=? where deptno=? and vseq=?

	【操作分析】更新版本号数据。 Hibernate: update Dept set vseq=? where deptno=? and vseq=?

如果此时有多个 Session 进行数据更新处理，当版本号不匹配时，将出现 org.hibernate.StaleStateException 异常。

13.6 JPA 数据关联

在 JPA 开发中除了可以实现单表实体映射外，还可以实现数据关联技术，例如，一对一关联、一对多关联、多对多关联，利用这些关联技术可以方便地实现级联数据的获取，简化重复代码编写。

13.6.1 一对一数据关联

一对一数据关联是数据表数据的垂直拆分，即为了提高数据库操作性能，可以将一张信息内容很多的数据表拆分为若干张数据表。例如，本次将创建一张公司信息表与公司信息详情表，如图 13-8 所示。

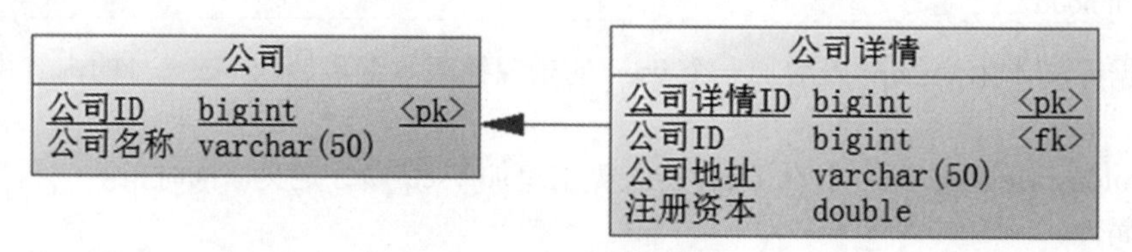

图 13-8 一对一关联模型

1.【mldnspring-jpa 项目】编写数据库创建脚本。

```
DROP DATABASE IF EXISTS mldn ;
CREATE DATABASE mldn CHARACTER SET UTF8 ;
USE mldn ;
CREATE TABLE company(
    cid             BIGINT    AUTO_INCREMENT ,
    cname           VARCHAR(50) ,
    CONSTRAINT pk_cid PRIMARY KEY(cid)
) engine = innodb ;
CREATE TABLE details(
    did             BIGINT    AUTO_INCREMENT ,
    address         VARCHAR(50) ,
    capital         DOUBLE,
    cid             BIGINT ,
    CONSTRAINT pk_did PRIMARY KEY(did) ,
    CONSTRAINT fk_cid FOREIGN KEY(cid) REFERENCES company(cid) ON DELETE
CASCADE
) engine = innodb ;
```

2.【mldnspring-jpa 项目】定义 Company 持久化类。

```
@SuppressWarnings("serial")
@Entity
public class Company implements Serializable {
    @Id
    @GeneratedValue(strategy=GenerationType.IDENTITY)
    private Long cid ;
    private String cname ;
    @OneToOne(mappedBy="company",cascade=CascadeType.ALL)        // 一对一数据关联，级联更新
    private Details details ;                                      // 公司详情信息
    // setter、getter、toString略
}
```

3.【mldnspring-jpa 项目】 定义 Details 持久化类。

```
@Entity
@SuppressWarnings("serial")
public class Details implements Serializable {
    @Id
    @GeneratedValue(strategy=GenerationType.IDENTITY)
    private Long did ;
    private String address ;
    private Double capital ;
    @OneToOne                                                      // 一对一关联
    @JoinColumn(name="cid",
        referencedColumnName="cid",unique=true)                   // 设置关联数据列
    private Company company ;                                      // 公司详情属于一个公司
    // setter、getter、toString略
}
```

通过这两个持久化类的关系不难看出，各自的实体类都保存了对方的引用，这样就可以通过类的关系体现类对象的一对一关联。Company 类定义中，为了可以在持久化 Company 对象时同时保存 Details 数据，使用了 cascade 级联配置。所有的级联关系都在 javax.persistence.CascadeType 枚举类中定义，使用 ALL 表示所有操作都进行级联。也可以选择持久化级联（CascadeType.PERSIST）、更新级联（CascadeType.MERGE）或删除级联（CascadeType.REMOVE）等配置项。

4.【mldnspring-jpa 项目】编写测试类，增加新数据。

```
    @Test
    public void testAdd() throws Exception {
        Company company = new Company() ;                          // 创建持久化类对象
        company.setCname("魔乐科技软件学院");                      // 设置属性
        Details details = new Details() ;                          // 创建持久化类对象
```

```
        details.setAddress("北京市天安门");                             // 设置属性
        details.setCapital(500000.00);                                 // 设置属性
        company.setDetails(details);                                   // 设置一对一关联
        details.setCompany(company);                                   // 设置一对一关联
        JPAEntityFactory.getEntityManager().getTransaction().begin();  // 开始事务
        JPAEntityFactory.getEntityManager().persist(company);          // 持久化数据
        JPAEntityFactory.getEntityManager().getTransaction().commit(); // 提交事务
        JPAEntityFactory.close();                                      // 关闭连接
    }
```

程序执行结果	**【操作分析】向company数据表中保存数据，此时的cid为自动增长。** Hibernate: insert into Company (cname) values (?) **【操作分析】向details表中保存数据，同时保存company.cid内容。** Hibernate: insert into Details (address, capital, cid) values (?, ?, ?)

本程序中配置了一对一关联关系，所以当通过持久化类设置好 Company 与 Details 对象引用后，就可以自动实现数据持久化关系匹配。

5.【mldnspring-jpa 项目】Company 数据信息保存在两张数据表中，下面执行数据查询处理。

```
    @Test
    public void testFind() throws Exception {
        Company company = JPAEntityFactory.getEntityManager().find(Company.class, 1L) ;
        JPAEntityFactory.close();
    }
```

程序执行结果	Hibernate: select company0_.cid as cid1_0_0_, company0_.cname as cname2_0_0_, details1_.did as did1_1_1_, details1_.address as address2_1_1_, details1_.capital as capital3_1_1_, details1_.cid as cid4_1_1_ from Company company0_ left outer join Details details1_ on company0_.cid=details1_.cid where company0_.cid=?

通过查询可以发现，虽然查询的是 Company 实体对象，但由于 Company 与 Details 中的数据是一个整体，所以默认会使用多表连接同时查询 Details 对应的数据。这种多表连接在数据量大时会产生庞大的笛卡儿积，造成访问性能的下降。为了解决这个问题，最好的做法是使用两次查询。

6.【mldnspring-jpa 项目】修改持久化类定义，避免多表关联查询。

修改Company类	`@OneToOne(mappedBy="company",cascade=CascadeType.ALL,`
	`        fetch=FetchType.LAZY)        // 一对一数据关联，级联更新`
	`private Details details ;        // 公司详情信息`

修改Details类	`@OneToOne(fetch=FetchType.LAZY)        // 一对一关联`
	`@JoinColumn(name="cid",`
	`    referencedColumnName="cid",unique=true)  // 设置关联数据列`
	`private Company company ;        // 公司详情属于一个公司`

进行关联配置时，可以定义数据的抓取策略，该策略通过 javax.persistence.FetchType 枚举类设置，可以定义为同时抓取（FetchType.**EAGER**）或延迟抓取（FetchType.**LAZY**）。这里配置为

延迟抓取，这样再次执行查询程序时可以得到如下的执行信息。

```
【操作分析】第一次只查询company数据表。
Hibernate: select company0_.cid as cid1_0_0_, company0_.cname as cname2_0_0_ from Company
company0_ where company0_.cid=?
【操作分析】第二次只查询details数据表，并且采用company.cid作为限定查询条件。
Hibernate: select details0_.did as did1_1_0_, details0_.address as address2_1_0_, details0_.capital as
capital3_1_0_, details0_.cid as cid4_1_0_ from Details details0_
where details0_.cid=?
```

在实际开发中，经过两次查询处理是标准做法。在进行一对一关联中，此配置尤为重要。

13.6.2　一对多数据关联

项目开发中，一对多是最常见的数据关联形式。例如，一个公司有多个部门，就属于一种典型的一对多关联，如图 13-9 所示。

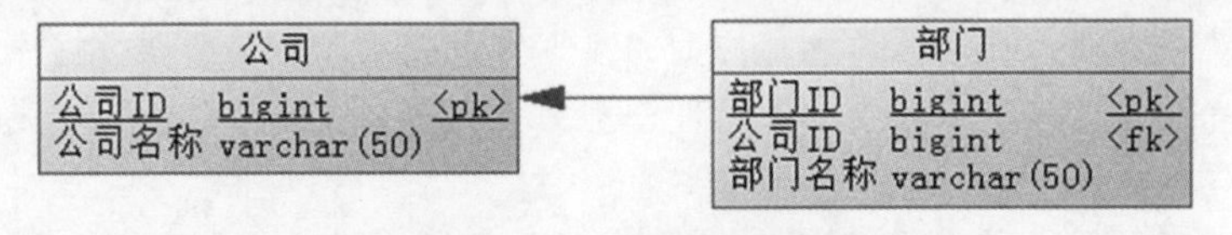

图 13-9　一对多数据关联

1.【mldnspring-jpa 项目】定义一对多关联数据库创建脚本。

```sql
DROP DATABASE IF EXISTS mldn ;
CREATE DATABASE mldn CHARACTER SET UTF8 ;
USE mldn ;
CREATE TABLE company(
    cid            BIGINT    AUTO_INCREMENT ,
    cname          VARCHAR(50) ,
    CONSTRAINT pk_cid PRIMARY KEY(cid)
) engine = innodb ;
CREATE TABLE dept(
    deptno         BIGINT    AUTO_INCREMENT ,
    dname          VARCHAR(50) ,
    cid            BIGINT ,
    CONSTRAINT pk_deptno PRIMARY KEY(deptno) ,
    CONSTRAINT fk_cid FOREIGN KEY(cid) REFERENCES company(cid) ON DELETE
CASCADE
) engine = innodb ;
```

2.【mldnspring-jpa 项目】定义 Company 实体类。

```java
@SuppressWarnings("serial")
@Entity
public class Company implements Serializable {
```

```
    @Id
    @GeneratedValue(strategy=GenerationType.IDENTITY)
    private Long cid ;
    private String cname ;
    @OneToMany(mappedBy="company", cascade=CascadeType.PERSIST)        //一对多关联
    private List<Dept> depts;                                          // 公司包含部门信息
    // setter、getter、toString略
}
```

本程序中使用 List 集合描述多个部门间的对应关系，同时定义了 cascade=CascadeType.**PERSIST** 持久化时的数据级联处理。这样在保存 Company 对象数据时，如果 depts 集合有数据，则会自动保存所有部门信息。

提示：实际开发处理。

实际开发中，很少会在一对多关联中配置级联 CascadeType.**PERSIST**，因为多数情况下是先存在“一”方数据，而后再添加对应的“多”方数据。以一条新闻数据保存为例，一条新闻会属于一个分类，一个分类有多条新闻，那么肯定是分类这个“一”方先存在才可以创建“多”方新闻数据。

本程序配置的级联主要是为读者加深理解 JPA 特点准备的，即可以自动实现子表关联增加。

3.【mldnspring-jpa 项目】定义 Dept 实体类。

```
@Entity
@SuppressWarnings("serial")
public class Dept implements Serializable {
    @Id
    @GeneratedValue(strategy=GenerationType.IDENTITY)
    private Long deptno;
    private String dname;
    @ManyToOne                                             // 多对一关联
    @JoinColumn(name="cid")                                // 设置关联字段
    private Company company;
    // setter、getter、toString略
}
```

4.【mldnspring-jpa 项目】编写测试程序，增加公司与部门数据。

```
    @Test
    public void testAdd() throws Exception {
        Company company = new Company();                   // 实例化“一”方对象
        company.setCname("魔乐科技软件学院");                // 设置数据
        List<Dept> allDepts = new ArrayList<Dept>();       // 保存所有的部门信息
```

```
        for (int x = 0; x < 3; x++) {
            Dept dept = new Dept();                                              // 实例化“多”方对象
            dept.setDname("教学研发" + x + "部");
            dept.setCompany(company);                                            // 需要获得cid信息
            allDepts.add(dept);                                                  // 保存集合信息
        }
        company.setDepts(allDepts);                                              // 一对多关联
        JPAEntityFactory.getEntityManager().getTransaction().begin();            // 开始事务
        JPAEntityFactory.getEntityManager().persist(company);                    // 持久化数据
        JPAEntityFactory.getEntityManager().getTransaction().commit();           // 提交事务
        JPAEntityFactory.close();                                                // 关闭连接
    }
```

程序执行结果	【操作分析】保存**company**数据信息。 Hibernate: insert into Company (cname) values (?) 【操作分析】保存**3**个**dept**数据信息，由于存在引用关系自动配置**company.cid**。 Hibernate: insert into Dept (cid, dname) values (?, ?) Hibernate: insert into Dept (cid, dname) values (?, ?) Hibernate: insert into Dept (cid, dname) values (?, ?)

本程序首先通过 Company 与 Dept 定义了彼此之间的引用关系，随后由于存在级联配置，所以当保存 Company 对象时会自动保存所有的部门信息。

提示：先保存“多”方。

在 JPA 开发中会自动帮助开发者维护关联关系，如果说现在创建了一个新的部门与一个新的公司，并且保存的是 dept 类对象，则在 dept 与 company 数据增加完成后自动更新 dept 表（关联字段 cid）以维护关联关系，但是这样的做法明显不符合正常逻辑，有兴趣的读者可以自行实验。

5.【mldnspring-jpa 项目】查询公司信息。

```
    @Test
    public void testFind() throws Exception {
        Company company = JPAEntityFactory.getEntityManager().find(Company.class, 1L) ;
        JPAEntityFactory.close();                                                // 关闭数据库连接
    }
```

程序执行结果	Hibernate: select company0_.cid as cid1_0_0_, company0_.cname as cname2_0_0_ from Company company0_ where company0_.cid=?

本程序实现了公司信息查询，但是根据查询语句来看，只是查询了 company 一张数据表的信息，并没有查询 dept 表信息，这是因为在 JPA 中考虑到一对多关联问题，所以默认启动了数据延迟加载，如果此时需要加载部门信息，则可以调用 Company 类中的集合方法（自动加载“多”方数据）。

6.【mldnspring-jpa 项目】查询公司信息与对应的全部部门信息。

```
@Test
public void testFind2() throws Exception {
    Company company = JPAEntityFactory.getEntityManager().find(Company.class, 2L) ;
    company.getDepts().size() ;                                    // 获取全部部门数据
    JPAEntityFactory.close();                                      // 关闭数据库连接
}
```

程序执行结果	**【操作分析】查询company数据表。** Hibernate: select company0_.cid as cid1_0_0_, company0_.cname as cname2_0_0_ from Company company0_ where company0_.cid=? **【操作分析】查询company对应的dept数据信息，company.getDepts().size()发出。** Hibernate: select depts0_.cid as cid3_1_0_, depts0_.deptno as deptno1_1_0_, depts0_.deptno as deptno1_1_1_, depts0_.cid as cid3_1_1_, depts0_.dname as dname2_1_1_ from Dept depts0_ where depts0_.cid=?

本程序在 EntityManager 关闭之前利用 company.getDepts().size()获取了“多方”数据，但是如果说此操作发生在 EntityManager 关闭后，则程序会抛出 org.hibernate.LazyInitialization Exception 异常。

7.【mldnspring-jpa 项目】查询指定部门数据。

```
@Test
public void testFindDept() throws Exception {
    Dept dept = JPAEntityFactory.getEntityManager().find(Dept.class, 3L) ;
    JPAEntityFactory.close();
}
```

程序执行结果	Hibernate: select dept0_.deptno as deptno1_1_0_, dept0_.cid as cid3_1_0_, dept0_.dname as dname2_1_0_, company1_.cid as cid1_0_1_, company1_.cname as cname2_0_1_ from Dept dept0_ left outer join Company company1_ on dept0_.cid=company1_.cid where dept0_.deptno=?

本程序查询了指定部门数据，但是通过查询结果可以发现，默认情况下为了保证 Dept 与 Company 的整体性，所以查询采用多表关联一次性查询出指定编号的部门与对应的公司信息，这样做的性能并不高，所以需要修改 Dept 抓取策略。

8.【mldnspring-jpa 项目】修改 Dept 持久化类配置抓取策略。

```
@ManyToOne(fetch=FetchType.LAZY)                                   // 多一对关联
@JoinColumn(name="cid")                                            // 设置关联字段
private Company company;
```

本程序设置了 Company 数据的抓取策略，再次执行时会发现只产生部门数据，不会查询公司数据，而在 EntityManager 未关闭前可以利用 Dept 类中的 company 属性实现公司数据加载。

9.【mldnspring-jpa 项目】加载部门与公司数据。

```
@Test
public void testFindDept() throws Exception {
    Dept dept = JPAEntityFactory.getEntityManager().find(Dept.class, 3L) ;   // 查询部门数据
```

```
        dept.getCompany().getCname() ;                    // 此处将查询公司数据
        JPAEntityFactory.close();                         // 关闭数据库连接
    }
```

程序执行结果	**【操作分析】find方法发出查询dept数据指令。** Hibernate: select dept0_.deptno as deptno1_1_0_, dept0_.cid as cid3_1_0_, dept0_.dname as dname2_1_0_ from Dept dept0_ where dept0_.deptno=? **【操作分析】dept.getCompany().getCname()操作发出company查询指令。** Hibernate: select company0_.cid as cid1_0_0_, company0_.cname as cname2_0_0_ from Company company0_ where company0_.cid=?

此时采用了两个查询语句获取了全部的数据，但是需要注意的是，如果现在执行的是 dept.getCompany().getCid()命令，则不会查询 company 表，因为 dept 表中保存有 cid 外键数据。

13.6.3　多对多数据关联

多对多数据关联可以理解为“一对多”与“多对一”的关联组合。在进行多对多配置时，通常会引入一张关联表，保存两张实体表的关系。例如，描述系统中用户所具有的角色信息，就会存在有如下关系：一个用户有多个角色，一个角色属于多个用户，其关联关系如图 13-10 所示。

图 13-10　多对多数据关联

1.【mldnspring-jpa 项目】编写数据库，创建脚本。

```
DROP DATABASE IF EXISTS mldn ;
CREATE DATABASE mldn CHARACTER SET UTF8 ;
USE mldn ;
--1. 用户数据表
CREATE TABLE member (
    mid                     VARCHAR(50),
    name                    VARCHAR(50),
    CONSTRAINT pk_mid PRIMARY KEY (mid)
) engine = innodb;
-- 2. 角色数据表
CREATE TABLE role (
    rid                     VARCHAR(50) ,
    title               VARCHAR(50),
    CONSTRAINT pk_rid PRIMARY KEY (rid)
) engine = innodb;
-- 3. 用户-角色关系表
CREATE TABLE member_role (
    mid                     VARCHAR(50),
```

```
    rid                    VARCHAR(50) ,
    CONSTRAINT fk_mid1 FOREIGN KEY(mid) REFERENCES member(mid) ,
    CONSTRAINT fk_rid1 FOREIGN KEY(rid) REFERENCES role(rid)
) engine = innodb;
INSERT INTO role(rid,title) VALUES ('company','公司管理') ;
INSERT INTO role(rid,title) VALUES ('dept','部门管理') ;
INSERT INTO role(rid,title) VALUES ('emp','雇员管理') ;
INSERT INTO role(rid,title) VALUES ('salgrade','薪资管理') ;
INSERT INTO role(rid,title) VALUES ('sale','销售管理') ;
INSERT INTO role(rid,title) VALUES ('market','市场管理') ;
INSERT INTO role(rid,title) VALUES ('project','项目管理') ;
```

在给出的数据库脚本中，核心的操作流程是通过用户找到对应的角色；同时，维护一个用户信息时，也一定会进行 member_role 这个中间关系表的数据更新。

2.【mldnspring-jpa 项目】定义 Member 实体类，一个 Member（用户）包含多个 Role（角色）对象。

```
@SuppressWarnings("serial")
@Entity
public class Member implements Serializable {
    @Id
    private String mid;
    private String name;
    @ManyToMany(fetch = FetchType.LAZY)                         // 启用延迟加载
    @JoinTable(                                                 // 描述的是一个关联表
        name="member_role" ,                                    // 定义中间表名称
        joinColumns = { @JoinColumn(name = "mid") }    ,        // member与member_role表的连接
        inverseJoinColumns = { @JoinColumn(name = "rid") })     // 通过Member找到Role中的rid的数据
    private List<Role> roles;
    // setter、getter、toString略
}
```

3.【mldnspring-jpa 项目】定义 Role 实体类，一个 Role（角色）对应多个 Member（用户）。

```
@Entity
@SuppressWarnings("serial")
public class Role implements Serializable {
    @Id
    private String rid;
    private String title;
    @ManyToMany(mappedBy = "roles", fetch = FetchType.LAZY)     // 多对多关联
    private List<Member> members;
    // setter、getter、toString略
}
```

上述程序定义了两个表实体类：Member 和 Role，但没有为 member_role 中间关系表定义实体类。这张表的数据维护是由 Member 类负责的，当用户进行增加、修改或查询操作时，将自动维护此表数据。

4.【mldnspring-jpa 项目】编写测试类，实现用户数据增加。

```
@Test
public void testMemberAdd() {
    // member_role关联表里是rid信息，下面定义的都是rid内容
    String rids[] = new String[] { "company", "dept", "emp" };      // 定义角色编号
    // 设置了一堆的Role对象只是为了获取一个rid属性内容，但是关联关系必须要求产生Role集合对象
    List<Role> allRoles = new ArrayList<Role>();                    // 保存角色信息
    Member member = new Member();                                   // 实例化持久类对象
    member.setMid("mldnjava");                                      // 设置数据
    member.setName("李兴华老师");                                    // 设置数据
    for (int x = 0; x < rids.length; x++) {                         // 配置用户角色
        Role role = new Role();                                     // 实例化持久类对象
        role.setRid(rids[x]);                                       // 设置角色
        allRoles.add(role);                                         // 向集合保存角色信息
    }
    member.setRoles(allRoles);                                      // 一个用户有多个角色
    JPAEntityFactory.getEntityManager().getTransaction().begin();   // 开启事务
    JPAEntityFactory.getEntityManager().persist(member);            // 持久化数据
    JPAEntityFactory.getEntityManager().getTransaction().commit();  // 提交事务
    JPAEntityFactory.close();                                       // 关闭连接
}
```

程序执行结果	**【操作分析】向member表中增加新数据。** Hibernate: insert into Member (name, mid) values (?, ?) **【操作分析】由于配置了Member与List<Role>关联，自动向member_role表中追加数据。** Hibernate: insert into member_role (mid, rid) values (?, ?) Hibernate: insert into member_role (mid, rid) values (?, ?) Hibernate: insert into member_role (mid, rid) values (?, ?)

上述程序实现了用户数据的创建，创建用户的同时为其分配角色。虽然 member_role 表中只需要一个 rid 数据即可，但是在 JPA 开发框架中必须依靠持久类来进行数据设置，所以准备了一个 List<Role>集合保存角色信息（主要保存的是 rid）。在持久化 Member 对象时，会自动根据关联关系维护 member_role 表中的数据。

5.【mldnspring-jpa 项目】在查询 member 表数据时，也可以利用集合中的方法来实现对应角色的查询。在进行用户角色信息查询时，必须使用 member_role 关系表和 role 数据表一起进行查询。

```
@Test
public void testMemberFind() {
```

```
        Member member = JPAEntityFactory.getEntityManager()
                .find(Member.class, "mldnjava") ;                    // 获取指定ID数据
        member.getRoles().size() ;                                   // 发出查询全部角色SQL命令
        JPAEntityFactory.close();                                    // 关闭数据库连接
    }
```

程序执行结果

【操作分析】根据mid查询出Member对象信息。

Hibernate: select member0_.mid as mid1_0_0_, member0_.name as name2_0_0_ from Member member0_ where member0_.mid=?

【操作分析】根据role_member表与role表查询出原本角色数据。

Hibernate: select roles0_.mid as mid1_1_0_, roles0_.rid as rid2_1_0_, role1_.rid as rid1_2_1_, role1_.title as title2_2_1_ from member_role roles0_ inner join Role role1_ on roles0_.rid=role1_.rid where roles0_.mid=?

上述程序在进行用户对应角色信息查询时（member.getRoles().size()语句发出查询），由于需要通过 member_role 数据表才可以确定 member 表与 role 表的数据关联关系，所以默认使用了内连接方式。

提示：多对多建议使用原生 SQL 查询。

实际上对于此时的数据表结构，很多情况下开发者需要的可能只是一个角色编号信息（例如，在系统开发进行角色与权限认证时，需要的往往只是一个 ID 编号），所以在这样的状态下为了提升查询效率，可以使用原生 SQL 实现角色信息查询。

范例：【mldnspring-jpa 项目】使用原生 SQL 查询用户角色信息。

```
    @Test
    public void testMemberFindSQL() {
        Member member = JPAEntityFactory.getEntityManager()
                .find(Member.class, "mldnjava") ;                    // 获取指定ID数据
        String sql = "SELECT rid FROM member_role WHERE mid=:mid" ;// 原生SQL
        System.out.println("*** 用户ID：" + member.getMid() + "、真实姓名：" +
member.getName());
        Query query = JPAEntityFactory.getEntityManager().createNativeQuery(sql) ;
        query.setParameter("mid", member.getMid()) ;
        List<Object> allRids = query.getResultList() ;               // 返回单列数据
        System.out.println("*** 用户角色：" + allRids);
        JPAEntityFactory.close();                                    // 关闭数据库连接
    }
```

程序执行结果

【操作分析】利用JPA提供的find方法查询出member表数据。

Hibernate: select member0_.mid as mid1_0_0_, member0_.name as name2_0_0_ from Member member0_ where member0_.mid=?

*** 用户ID：mldnjava、真实姓名：李兴华老师

【操作分析】利用原生SQL实现用户对应角色编号查询。

Hibernate: SELECT rid FROM member_role WHERE mid=?

*** 用户角色：[market, project, sale]

> 虽然本程序没有采用级联关系实现数据加载，但是由于不再采用多表关联的形式查询，整体的执行效率将获得极大提升，所以对于数据关联操作，笔者强烈建议：如果不是必须的情况下不建议使用，更多时候单表处理性能会更佳，而且也更加灵活。

6.【mldnspring-jpa 项目】多对多关联中，还有一个麻烦的问题，就是 member_role 关系表的数据维护处理。例如，用户角色更新时应先将对应的所有 member_role 中的数据删除，而后再重新增加 member_role 表数据。这种操作可以直接利用 JPA 自动来完成维护。

```
@Test
public void testMemberEdit() {
    // member_role关联表里需要rid信息，所以下面定义的都是rid内容
    String rids[] = new String[] { "market", "project", "sale" };       // 定义角色编号
    // 设置了一堆Role对象只是为了获取一个rid属性内容，但是关联关系必须要求产生Role集合
对象
    List<Role> allRoles = new ArrayList<Role>();                        // 保存角色信息
    Member member = new Member();                                       // 实例化持久类对象
    member.setMid("mldnjava");                                          // 设置数据
    member.setName("李兴华老师");                                       // 设置数据
    for (int x = 0; x < rids.length; x++) {                             // 配置用户角色
        Role role = new Role();                                         // 实例化持久类对象
        role.setRid(rids[x]);                                           // 设置角色
        allRoles.add(role);                                             // 向集合保存角色信息
    }
    member.setRoles(allRoles);                                          // 一个用户有多个角色
    JPAEntityFactory.getEntityManager().getTransaction().begin();       // 开启事务
    JPAEntityFactory.getEntityManager().merge(member);                  // 更新数据
    JPAEntityFactory.getEntityManager().getTransaction().commit();      // 提交事务
    JPAEntityFactory.close();                                           // 关闭连接
}
```

程序执行结果	**【操作分析】由于Member属于新的实体类所以合并前需要进行持久化数据确认。** Hibernate: select member0_.mid as mid1_0_0_, member0_.name as name2_0_0_ from Member member0_ where member0_.mid=? **【操作分析】确认Member与Role（List集合）是否为持久化数据。** Hibernate: select roles0_.mid as mid1_1_0_, roles0_.rid as rid2_1_0_, role1_.rid as rid1_2_1_, role1_.title as title2_2_1_ from member_role roles0_ inner join Role role1_ on roles0_.rid=role1_.rid where roles0_.mid=? **【操作分析】确认所设置的Role对象是否为持久化数据。** Hibernate: select role0_.rid as rid1_2_0_, role0_.title as title2_2_0_ from Role role0_ where role0_.rid=? Hibernate: select role0_.rid as rid1_2_0_, role0_.title as title2_2_0_ from Role role0_ where role0_.rid=? **【操作分析】更新member表数据。** Hibernate: update Member set name=? where mid=?

```
【操作分析】删除member_role表中的原始关联数据。
Hibernate: delete from member_role where mid=?
【操作分析】向member_role表中重新保存关联数据（rid）。
Hibernate: insert into member_role (mid, rid) values (?, ?)
Hibernate: insert into member_role (mid, rid) values (?, ?)
Hibernate: insert into member_role (mid, rid) values (?, ?)
```

本程序涉及的数据表较多，同时维持对象持久化状态，执行了许多次查询后才开始进行相应数据的更新处理。

13.7 Spring 整合 JPA 开发框架

JPA 框架虽然提供了合理的数据层处理逻辑，但重复度依然很高。为了更方便进行数据库连接管理、EntityManagerFactory 管理和业务层事务控制，最好的做法是将其与 Spring 进行整合，利用 Spring 简化 JPA 的开发复杂度。下面将使用之前的数据库连接池、AOP 事务控制等配置文件，进行二者的整合。

提示：关于本次整合操作。

为了方便理解，这里将使用最初的 dept 单表进行 Spring 与 JPA 的整合。进行环境配置时，需要引入Spring的相关依赖库，同时需要C3P0数据源支持以及AOP事务控制(mldnspring-jdbc 项目中提供)。许多代码在之前章节中已经讲解过了，所以本处只显示核心配置操作。完整的整合代码，读者可从本例的源代码中获取。

1.【mldnspring 项目】修改父 pom.xml 配置文件，追加 spring-orm 依赖库，以实现 Spring 对 ORMapping 开发框架管理。

```
<dependency>
    <groupId>org.springframework</groupId>
    <artifactId>spring-orm</artifactId>
    <version>${spring.version}</version>
</dependency>
```

2.【mldnspring-jpa 项目】在子 pom.xml 配置文件中引入 spring-orm 依赖库。

```
<dependency>
    <groupId>org.springframework</groupId>
    <artifactId>spring-orm</artifactId>
</dependency>
```

3.【mldnspring-jpa 项目】建立 src/main/resources/spring/spring-jpa.xml 配置文件，进行 EntityManagerFactory 配置。

```
<bean id="entityManagerFactory"
```

```
        class="org.springframework.orm.jpa.LocalContainerEntityManagerFactoryBean">
        <property name="dataSource" ref="dataSource"/>                    <!-- 数据源 -->
        <property name="persistenceXmlLocation"
            value="classpath:META-INF/persistence.xml" />                 <!-- JPA核心配置文件 -->
        <property name="persistenceUnitName" value="MLDNJPA"/>            <!-- 持久化单元名称 -->
        <property name="packagesToScan" value="cn.mldn.mldnspring.po"/><!-- PO类扫描包 -->
        <property name="persistenceProvider">            <!-- 持久化提供类，本次为Hibernate -->
            <bean class="org.hibernate.jpa.HibernatePersistenceProvider"/>
        </property>
        <property name="jpaVendorAdapter">                               <!-- JPA操作实现者 -->
            <bean class="org.springframework.orm.jpa.vendor.HibernateJpaVendorAdapter"/>
        </property>
        <property name="jpaDialect">                                     <!-- JPA实现方言 -->
            <bean class="org.springframework.orm.jpa.vendor.HibernateJpaDialect"/>
        </property>
    </bean>
```

由于所有的数据库连接都已经在 spring-database.xml 配置文件中定义了，所以 persistence.xml 配置文件不需要再进行任何的数据库连接配置，只需要定义相关的 JPA 操作属性即可。

4．【mldnspring-jpa 项目】由于本次操作整合 JPA 开发框架，所以要修改 spring/spring-transaction.xml 配置文件，更换事务管理器为 JpaTransactionManager，此类可以直接针对 EntityManagerFactory 实现事务控制。

```
    <!-- 定义事务管理的配置，必须配置PlatformTransactionManager接口子类 -->
    <bean id="transactionManager" class="org.springframework.orm.jpa.JpaTransactionManager">
        <property name="entityManagerFactory" ref="entityManagerFactory" />   <!-- 事务控制 -->
    </bean>
```

5．【mldnspring-jpa 项目】建立 IDeptDAO 数据层接口。

```
package cn.mldn.mldnspring.dao;
import java.util.List;
import cn.mldn.mldnspring.po.Dept;
public interface IDeptDAO {                                          // 最原始的纯粹接口
    /**
     * 增加新的部门数据
     * @param vo 部门持久化对象
     * @return 增加成功返回true
     */
    public boolean doCreate(Dept vo) ;
    /**
     * 查询全部部门数据
     * @return 部门持久化对象集合
     */
```

```
    public List<Dept> findAll() ;
}
```

6.【mldnspring-jpa 项目】定义 IDeptDAO 接口的子类 DeptDAOImpl。

```
package cn.mldn.mldnspring.dao.impl;
import java.util.List;
import javax.persistence.EntityManager;
import javax.persistence.PersistenceContext;
import javax.persistence.TypedQuery;
import org.springframework.stereotype.Repository;
import cn.mldn.mldnspring.dao.IDeptDAO;
import cn.mldn.mldnspring.po.Dept;
@Repository
public class DeptDAOImpl implements IDeptDAO {
    @PersistenceContext  // 获得EntityManager（配置文件只配置了EntityManagerFactory）
    private EntityManager entityManager ;                              // JPA操作对象
    @Override
    public boolean doCreate(Dept vo) {
        return this.entityManager.merge(vo) != null ;                  // 数据增加
    }
    @Override
    public List<Dept> findAll() {
        String jpql = "SELECT d FROM Dept AS d" ;                      // JPQL查询
        TypedQuery<Dept> query = this.entityManager.createQuery(jpql,Dept.class) ;
        return query.getResultList() ;
    }
}
```

7.【mldnspring-jpa 项目】定义 IDeptService 业务接口。

```
package cn.mldn.mldnspring.service;
import java.util.List;
import cn.mldn.mldnspring.po.Dept;
public interface IDeptService {
    /**
     * 增加部门数据，调用IDeptDAO.doCreate方法处理
     * @param vo 持久化类对象，没有设置主键数据
     * @return 增加成功返回true，否则返回false
     */
    public boolean add(Dept vo) ;
    /**
     * 查询dept表中的全部数据
     * @return Dept持久化对象集合
     */
```

```
	public List<Dept> list() ;
}
```

8.【mldnspring-jpa 项目】定义 IDeptService 接口实现子类 DeptServiceImpl。

```
package cn.mldn.mldnspring.service.impl;
import java.util.List;
import org.springframework.beans.factory.annotation.Autowired;
import org.springframework.stereotype.Service;
import cn.mldn.mldnspring.dao.IDeptDAO;
import cn.mldn.mldnspring.po.Dept;
import cn.mldn.mldnspring.service.IDeptService;
@Service
public class DeptServiceImpl implements IDeptService {
	@Autowired
	private IDeptDAO deptDAO ;                              // 注入IDeptDAO接口实例
	@Override
	public boolean add(Dept vo) {
		return this.deptDAO.doCreate(vo);                    // 数据增加
	}
	@Override
	public List<Dept> list() {
		return this.deptDAO.findAll() ;                      // 数据查询
	}
}
```

9.【mldnspring-jpa 项目】建立测试类，实现 IDeptService 增加业务测试。

```
@ContextConfiguration(locations = { "classpath:spring/spring-*.xml" })
@RunWith(SpringJUnit4ClassRunner.class)
public class TestDeptService {
	@Autowired
	private IDeptService deptService ;                      // 注入IDeptService业务接口实例
	@Test
	public void testAdd() {
		Dept po = new Dept() ;                               // 实例化持久类对象
		po.setDname("MLDN教学部");                           // 设置数据
		po.setCreatedate(new Date());                        // 设置数据
		po.setNum(55);                                       // 设置数据
		po.setAvgsal(89998.00);                              // 设置数据
		System.out.println(this.deptService.add(po));        // 数据持久化
	}
}
```

程序执行结果	Hibernate: insert into Dept (avgsal, createdate, dname, num) values (?, ?, ?, ?) true（数据成功保存，IDeptDAO.doCreate方法返回结果）

10．【mldnspring-jpa 项目】测试 IDeptService 接口查询功能。

```
@Test
public void testList() {
    List<Dept> allDepts = this.deptService.list() ;                // 查询全部数据
    allDepts.forEach((dept)->{                                     // 迭代输出查询结果
        System.out.println(dept);
    });
}
```

程序执行结果	Hibernate: select dept0_.deptno as deptno1_0_, dept0_.avgsal as avgsal2_0_, dept0_.createdate as createda3_0_, dept0_.dname as dname4_0_, dept0_.num as num5_0_ from Dept dept0_ 部门数据输出略

此时，程序利用 Spring 实现了 JPA 开发框架的管理，并且从整体实现效果来讲已经帮助用户简化了事务控制、连接管理等操作，用户可以只关注于核心业务实现过程。

13.8 SpringDataJPA

Spring 开发框架虽然可以整合 JPA 开发框架，更加便于管理，但在实际开发中仍然存在一定的问题。项目的核心在于业务层，业务层的处理是围绕数据层展开的，而数据层在进行 SQL 数据库操作时需要使用 JDBC，于是对整个项目开发而言，只要定义了 DAO 数据接口，就必须去定义 DAO 接口的子类，即便有 JPA 支持依然需要手动使用 EntityManager 接口进行操作。最为关键的是，这些操作还几乎都是相似的。为了简化 JPA 数据层的开发操作，Spring 开发框架提供了 SpringDataJPA 技术支持，以帮助用户解决重复定义 DAO 接口子类这一问题：只要项目中提供了 DAO 接口，SpringDataJPA 就会自动提供子类实例。这样，开发者就可以不再关注数据层的重复实现，而只关注业务层的设计与开发即可。

在项目中使用 SpringDataJPA 进行持久层开发，一般需要 3 个步骤。

☑ 声明持久层的接口。该接口继承了 Repository（或 Repository 的子接口，其中定义了一些常用的增、删、改、查，以及分页相关的方法）。

☑ 在接口中声明需要的业务方法，SpringDataJPA 将根据给定的策略生成实现代码。

☑ 在配置文件中定义<jpa:repositories>配置项。这样 Spring 初始化容器时将扫描 base-package 指定的包目录及其子目录，为继承 Repository 或其子接口的接口创建代理对象，并将代理对象注册为 Spring 所管理的 Bean 对象，业务层便可以通过 Spring 自动封装的特性来直接使用该对象。

在整个 SpringDataJPA 中，核心的关键在于 org.springframework.data.repository.Repository 接口以及其子接口（关系见图 13-11），这些接口的名称以及具体作用如下。

☑ Repository：一个标识，不包含任何方法，主要是为了方便 Spring 自动扫描识别。

☑ CrudRepository：继承于 Repository，实现一组 CRUD 相关的方法。

☑ PagingAndSortingRepository：继承 CrudRepository，实现一组分页排序相关的方法。

☑ JpaRepository：继承 PagingAndSortingRepository，实现一组 JPA 规范相关的方法。

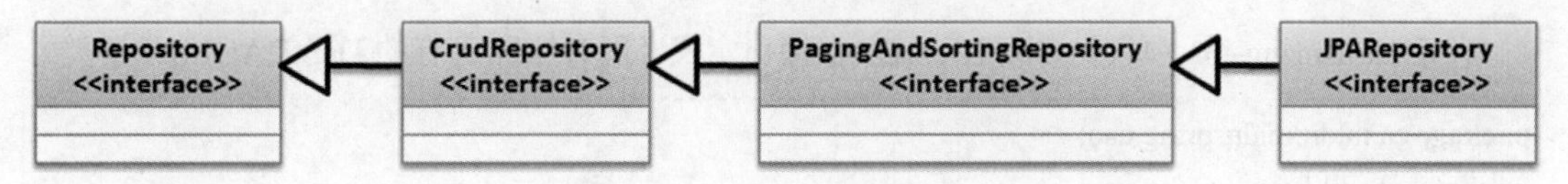

图 13-11　Repository 继承关系

在讲解具体的 SpringDataJPA 与 Repository 接口操作之前，需要先对项目环境进行相关配置。

1.【mldnspring 项目】在父 pom.xml 中定义 spring-data-jpa 依赖库的管理。

属性配置	`<spring-data-jpa.version>2.0.4.RELEASE</spring-data-jpa.version>`

依赖库配置	`<dependency>`
	`<groupId>org.springframework.data</groupId>`
	`<artifactId>spring-data-jpa</artifactId>`
	`<version>${spring-data-jpa.version}</version>`
	`</dependency>`

2.【mldnspring-jpa 项目】在子 pom.xml 中引入 spring-data-jpa 依赖库。

```
<dependency>
    <groupId>org.springframework.data</groupId>
    <artifactId>spring-data-jpa</artifactId>
</dependency>
```

3.【mldnspring-jpa 项目】在 spring-jpa.xml 配置文件中追加 jpa 命名空间，如图 13-12 所示。

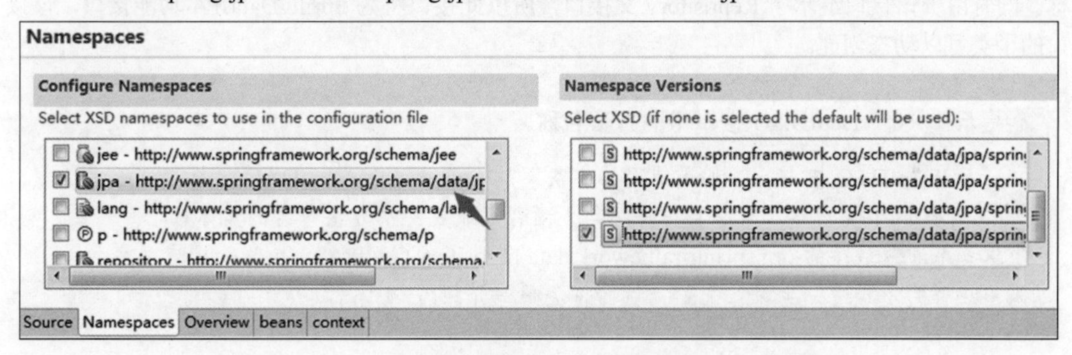

图 13-12　追加 jpa 命名空间

4.【mldnspring-jpa 项目】在 spring-jpa.xml 配置文件中追加配置项。

```
<!-- 定义SpringDataJPA的数据层接口所在包，该包中的接口一定是Repository子接口 -->
<jpa:repositories base-package="cn.mldn.mldnspring.dao"/>
```

配置完成后，只需要继承 org.springframework.data.repository.Repository 接口，就可以基于 SpringDataJPA 处理标准简化数据层定义。

13.8.1　Repository 基本使用

SpringDataJPA 开发中，所有的数据层接口都要求继承于 Repository 接口。Repository 的主要作用是可以直接使用注解或方法名称定义规则的形式来实现数据层功能。

1.【mldnspring-jpa 项目】修改 IDeptDAO 接口定义（同时删除已有的 DeptDAOImpl 子类）。

```
package cn.mldn.mldnspring.dao;
import java.util.List;
import org.springframework.data.repository.Repository;
import cn.mldn.mldnspring.po.Dept;
public interface IDeptDAO extends Repository<Dept, Long> {              // SpringData数据接口
    /**
     * 增加新的部门数据，方法名称定义为save，以自动实现merge()功能
     * @param vo 部门持久化对象
     * @return 增加成功，返回持久化对象
     */
    public Dept save(Dept vo) ;
    /**
     * 查询全部部门数据
     * @return 部门持久化对象集合
     */
    public List<Dept> findAll() ;
}
```

IDeptDAO 数据层接口与原始定义上有一些差别，主要是在数据保存上使用了 save 方法名称。同时由于该接口继承了 Repository 父接口，所以此接口将为 SpringDataJPA 功能接口，该接口的子类可以动态创建。

提示：关于 IDeptDAO 接口中的方法名称。

定义 IDeptDAO 接口时，如果使用的方法不是 save 和 findAll，DAO 层将无法正常工作，因为 SpringDataJPA 规定了数据层方法命名。读者可能更关心方法命名的具体标准是什么，实际上这些可通过后续的 org.springframework.data.repository.CrudRepository 子接口观察到，也可以采用注解的形式进一步深入配置，后面将会一一介绍。

2.【mldnspring-jpa 项目】由于 IDeptDAO 接口的方法名称发生了改变，所以此时修改 DeptServiceImpl 子类定义。不管方法名称如何改变，在进行子类定义时依然可以通过 Spring 容器自动注入 IDeptDAO 接口实例。

```
@Service
public class DeptServiceImpl implements IDeptService {
    @Autowired
    private IDeptDAO deptDAO ;                                          // 注入IDeptDAO接口实例
    @Override
    public boolean add(Dept vo) {
        return this.deptDAO.save(vo).getDeptno() != null;               // 数据增加
    }
    @Override
```

```
    public List<Dept> list() {
        return this.deptDAO.findAll() ;                    // 数据查询
    }
}
```

此时程序实现了一个最简单的 SpringDataJPA 操作，可以发现在此技术下，开发者只需要定义 DAO 接口标准就可以通过容器动态生成实现子类，使得开发者的注意力不需要再放到数据层上，而只关心业务层即可。

提示：可以使用注解代替继承的 Repository 父接口。

本程序所定义的 IDeptDAO 接口必须强制性继承 Repository 父接口，才可以实现 SpringDataJPA 操作。Spring 设计中考虑到不同开发者的需求，也可以利用注解来实现。

范例：使用@RepositoryDefinition 注解。

```
@RepositoryDefinition(domainClass=Dept.class,idClass=Long.class)
public interface IDeptDAO {                    // SprintData接口
    public Dept save(Dept vo) ;                // 实现数据增加，数据增加需要发生改变
    public List<Dept> findAll() ;              // 实现数据的列表显示
}
```

这里 DAO 接口上使用的@RepositoryDefinition 注解功能，与直接继承 Repository 父接口的效果相同。两种 DAO 层定义模式，读者可以根据习惯自行选择。

13.8.2 Repository 实现 CRUD

在使用 Repository 定义数据层接口操作方法时，也可以使用@Query 注解手动绑定要执行的 JPQL 语句（只允许绑定查询、修改、删除）。该注解的功能与 Query 接口使用形式类似，只是具体的操作实现将由 Spring 处理完成。

提示：SpringDataJPA 中如果不依靠方法名称绑定操作，则可以通过@Query 绑定操作类型。

以之前程序中的 findAll 方法为例，为其绑定查询语句也可以采用如下形式定义。

范例：【 mldnspring-jpa 项目 】为 findAll 绑定 JPQL 语句。

```
    @Query("SELECT d FROM Dept AS d")             // 写与不写此类语句，效果相同
    public List<Dept> findAll() ;
```

这里为 findAll 方法绑定了 JPQL 查询语句，即便方法名称不是 findAll，也可以实现全部数据查询。在 SpringDataJPA 中，数据层可以通过方法名称绑定操作，也可以通过@Query 接口绑定操作。

下面将通过几个具体的程序来进行程序功能实现，所有的方法绑定处理全部需要在 DAO 数据层接口上定义，同时如果需要测试，也需要增加相应的业务方法。

范例：【mldnspring-jpa 项目】为 IDeptDAO 接口设置一个根据 ID 查询方法。

☑ 修改 IDeptDAO 接口，增加 findById 查询方法。

```
/**
 * 根据ID进行查询，在JPQL获取参数时使用SpEl表达式获取第一个参数
 * @param id 要查询的部门编号
 * @return 部门持久化对象
 */
@Query("SELECT d FROM Dept AS d WHERE d.deptno=?#{[0]}")        // 使用SpEL语法获得参数
public Dept findById(Long id) ;
```

☑ 在 IDeptService 接口定义一个 get 方法，用于调用 findById 方法，同时实现子类。

```
@Override
public Dept get(long id) {
    return this.deptDAO.findById(id);
}
```

☑ 编写测试类，测试 get 方法功能。

```
@Test
public void testGet() {
    Dept dept = this.deptService.get(1L) ;                          // 查询指定ID数据
}
```

程序执行结果	Hibernate: select dept0_.deptno as deptno1_0_, dept0_.avgsal as avgsal2_0_, dept0_.createdate as createda3_0_, dept0_.dname as dname4_0_, dept0_.num as num5_0_ from Dept dept0_ where dept0_.deptno=? Dept [deptno=1, avgsal=89998.0, createdate=2018-02-27, dname=MLDN教学部, num=55]

在本程序中定义的 IDeptDAO 接口可以使用 SpEL 表达式“?#{[参数索引]}”访问传递到 findById 中的第一个参数，由于该方法属于自定义的数据操作方法，所以必须要提供@Query 声明要操作的 JPQL 语句，否则程序将出现错误。

范例：【mldnspring-jpa 项目】根据指定的主键范围查询全部部门信息（IN 操作）。

☑ 在 IDeptDAO 接口中增加新的数据查询方法。

```
/**
 * 根据指定ID范围实现部门信息查询
 * @param ids 全部部门ID
 * @return Dept持久化类集合
 */
@Query("SELECT d FROM Dept AS d WHERE d.deptno IN :pids")        // 使用pids访问参数
public List<Dept> findByIds(@Param(value="pids") Set<Long> ids) ;  // 根据设置的ID范围查询
```

☑ 在 IDeptService 方法中追加新的业务方法 gets 方法，并在 DeptServiceImpl 子类中实现此方法。

```
@Override
public List<Dept> gets(Set<Long> ids) {
    return this.deptDAO.findByIds(ids);
}
```

☑ 编写测试类，进行 gets 方法功能测试。

```
@Test
public void testGets() {
    Set<Long> allIds = new HashSet<Long>() ;                    // 实例化Set集合，用于保存查询ID
    allIds.addAll(Arrays.asList(1L, 3L, 5L));                   // 增加ID数据
    List<Dept> allDepts = this.deptService.gets(allIds) ;       // 数据查询
    allDepts.forEach((dept)->{                                  // 迭代输出查询结果
        System.out.println(dept);
    }) ;
}
```

程序执行结果	Hibernate: select dept0_.deptno as deptno1_0_, dept0_.avgsal as avgsal2_0_, dept0_.createdate as createda3_0_, dept0_.dname as dname4_0_, dept0_.num as num5_0_ from Dept dept0_ where dept0_.deptno in (? , ? , ?) 部门数据输出略

本程序在 IDeptDAO 定义新方法时，使用了@Param(value="pids")注解定义了@Query 注解可以访问的参数名称，这种形式要比使用 SpEL 访问更加直观，同时该操作实现的 IN 处理只需要传递 Set 集合即可自动转换为查询参数。

范例：【mldnspring-jpa 项目】在 DAO 方法中使用@Query 注解时，也可以利用 SpEL 获取方法参数类中的属性。

☑ 在 IDeptDAO 接口中定义方法，可以根据部门 ID 与部门名称进行数据查询，此时将使用 Dept 作为参数类型。

```
/**
 * 根据指定的部门ID与部门名称获取部门数据
 * @param dept 包含部门数据信息
 * @return Dept持久化类对象
 */
@Query("SELECT d FROM Dept AS d WHERE d.deptno=:#{#mydept.deptno} AND
dname=:#{#mydept.dname}")
public Dept findByIdAndDname(@Param(value="mydept") Dept dept) ; // 多条件查询，传递类对象
```

☑ 在 IDeptService 接口中定义一个 getIdAndDname 业务方法，并在子类进行实现。

```
@Override
public Dept getIdAndDname(Dept po) {
```

```
        return this.deptDAO.findByIdAndDname(po);
    }
```

☑ 编写测试类进行业务测试。

```
    @Test
    public void testGetIdAndDname() {
        Dept po = new Dept() ;                                    // 定义部门对象，用于传递参数
        po.setDeptno(1L);                                         // 设置参数内容
        po.setDname("MLDN教学部");                                 // 设置参数内容
        System.out.println(this.deptService.getIdAndDname(po));
    }
```

程序执行结果	Hibernate: select dept0_.deptno as deptno1_0_, dept0_.avgsal as avgsal2_0_, dept0_.createdate as createda3_0_, dept0_.dname as dname4_0_, dept0_.num as num5_0_ from Dept dept0_ where dept0_.deptno=? and dept0_.dname=? Dept [deptno=1, avgsal=89998.0, createdate=2018-02-27, dname=MLDN教学部, num=55]

本程序实现了多个参数的使用，并且为了方便将多个参数绑定在了 Dept 对象中，这样就可以在 DAO 层中利用 SpEL 访问对象的属性作为查询参数配置。

在 JPA 中，对于数据的修改操作如果使用 EntityManager 类中提供的方法，需要始终维护对象持久化状态，所以可以利用 Query 接口实现数据更新，而对于修改与删除的操作也可以利用@Query 注解进行定义。

范例：【mldnspring-jpa 项目】实现部门数据修改。

☑ 在 IDeptDAO 接口中定义数据修改方法。

```
    /**
     * 修改指定编号的部门信息
     * @param po 传递要修改的部门数据
     * @return 数据库更新影响的数据行数
     */
    @Modifying(clearAutomatically=true)                       // 追加缓存的清除与更新
    @Query("UPDATE Dept AS d SET d.dname=:#{#mydept.dname},d.num=:#{#mydept.num} WHERE
d.deptno=:#{#mydept.deptno}")
    public int doEdit(@Param(value="mydept") Dept po) ;
```

☑ 在 IDeptService 接口定义 edit 方法，并且在子类实现此更新业务方法。

```
    @Override
    public boolean edit(Dept po) {
        return this.deptDAO.doEdit(po) > 0;
    }
```

☑ 编写测试类，进行业务方法功能测试。

```
    @Test
    public void testEdit() {
        Dept po = new Dept();                                  // 定义对象，用于保存更新数据
```

```
        po.setDeptno(3L);                                    // 设置更新数据
        po.setNum(99);                                       // 设置更新数据
        po.setDname("魔乐科技教学研发中心");                   // 设置更新数据
        System.out.println(this.deptService.edit(po));
    }
```

程序执行结果	Hibernate: update Dept set dname=?, num=? where deptno=?

在进行更新数据时有可能需要进行相应缓存数据的清除，所以在定义 doEdit 方法时使用了@Modifying 注解进行缓存更新，由于使用的是 JPQL 更新操作，所以不需要维护对象状态，可以直接更新。

范例：【mldnspring-jpa 项目】实现数据删除。

☑　在 IDeptDAO 接口定义数据删除方法。

```
/**
 * 根据指定编号删除部门数据
 * @param deptno  部门编号
 * @return  数据库更新影响的数据行数
 */
@Modifying(clearAutomatically = true)                  // 追加缓存的清除与更新
@Query("DELETE FROM Dept AS d WHERE d.deptno=:dno")
public int doRemove(@Param("dno") Long deptno);
```

☑　在 IDeptService 业务层中定义 remove 方法，并且在子类实现此删除业务方法。

```
@Override
public boolean remove(long dno) {
    return this.deptDAO.doRemove(dno) > 0;
}
```

☑　编写测试类，进行业务方法功能测试。

```
@Test
public void testRemove() {
    System.out.println(this.deptService.remove(9L));
}
```

程序执行结果	Hibernate: delete from Dept where deptno=?

本程序利用 JPQL 实现了数据删除处理操作，删除时由于可能会存在数据缓存问题，所以依然配置@Modifying 注解，可以发现在 Repository 整体设计中都只需要开发者定义很简单的部分，即可方便地实现自定义数据层操作。

13.8.3　Repository 方法映射

SpringDataJPA 中，除了可以使用@Query 定义指定查询方法语句外，也可以采用方法映射形式利用容器自动生成查询语句定义。此时就需要使用一些特定的关键字进行方法名称的声明，在 SpringDataJPA 官方文档中给出了如表 13-5 所示的关键字和参考操作。

表 13-5 SpringDataJPA 关键字

编号	Keyword	Sample	JPQL snippet
1	And	findByLastnameAndFirstname	… where x.lastname = ?1 and x.firstname = ?2
2	Or	findByLastnameOrFirstname	… where x.lastname = ?1 or x.firstname = ?2
3	Is,Equals	findByFirstname, findByFirstnameIs, findByFirstnameEquals	… where x.firstname = ?1
4	Between	findByStartDateBetween	… where x.startDate between ?1 and ?2
5	LessThan	findByAgeLessThan	… where x.age < ?1
6	LessThanEqual	findByAgeLessThanEqual	… where x.age <= ?1
7	GreaterThan	findByAgeGreaterThan	… where x.age > ?1
8	GreaterThanEqual	findByAgeGreaterThanEqual	… where x.age >= ?1
9	After	findByStartDateAfter	… where x.startDate > ?1
10	Before	findByStartDateBefore	… where x.startDate < ?1
11	IsNull	findByAgeIsNull	… where x.age is null
12	IsNotNull,NotNull	findByAge(Is)NotNull	… where x.age not null
13	Like	findByFirstnameLike	… where x.firstname like ?1
14	NotLike	findByFirstnameNotLike	… where x.firstname not like ?1
15	StartingWith	findByFirstnameStartingWith	… where x.firstname like ?1(parameter bound with appended %)
16	EndingWith	findByFirstnameEndingWith	… where x.firstname like ?1(parameter bound with prepended %)
17	Containing	findByFirstnameContaining	… where x.firstname like ?1(parameter bound wrapped in %)
18	OrderBy	findByAgeOrderByLastnameDesc	… where x.age = ?1 order by x.lastname desc
19	Not	findByLastnameNot	… where x.lastname <> ?1
20	In	findByAgeIn(Collection<Age> ages)	… where x.age in ?1
21	NotIn	findByAgeNotIn(Collection<Age> ages)	… where x.age not in ?1
22	True	findByActiveTrue()	… where x.active = true
23	False	findByActiveFalse()	… where x.active = false
24	IgnoreCase	findByFirstnameIgnoreCase	… where UPPER(x.firstame) = UPPER(?1)

SpringDataJPA 框架在解析方法名时，会把方法名多余的前缀截取掉（如 find、findBy、read、readBy、get、getBy 等），然后对剩下部分进行解析。下面通过几个具体范例讲解方法映射处理。

范例：【mldnspring-jpa 项目】根据部门人数与部门名称进行查询。

☑　在 IDeptDAO 接口中定义新的查询方法，该查询方法将进行方法名称映射。

```
/**
 * 根据部门编号与部门名称进行查询
 * @param num  部门人数
 * @param dname  部门名称
 * @return  部门持久化对象集合
 */
public List<Dept> findByNumAndDname(Integer num, String dname);
```

☑　在 IDeptService 接口中定义 getNumAndDname 方法，并且在子类实现此业务方法。

```
@Override
public List<Dept> getNumAndDname(int num, String dname) {
    return this.deptDAO.findByNumAndDname(num, dname);
}
```

☑　编写测试类，测试业务方法。

```
@Test
public void testGetNumAndDname() {
    List<Dept> allDepts = this.deptService.getNumAndDname(55, "MLDN教学部") ;
    allDepts.forEach((dept)->{                                    // 迭代输出查询结果
        System.out.println(dept);
    }) ;
}
```

程序执行结果	Hibernate: select dept0_.deptno as deptno1_0_, dept0_.avgsal as avgsal2_0_, dept0_.createdate as createda3_0_, dept0_.dname as dname4_0_, dept0_.num as num5_0_ from Dept dept0_ where dept0_.num=? and dept0_.dname=? 部门数据输出略

本程序利用方法映射处理，从而省略了@Query 注解定义查询语句。

范例：【mldnspring-jpa 项目】查询指定范围 ID 的部门数据。

☑　在 IDeptDAO 接口中定义数据查询。

```
/**
 * 根据指定ID进行部门信息查询
 * @param ids 要查询的部门编号
 * @return  部门持久化对象集合
 */
public List<Dept> findByDeptnoIn(Set<Long> ids) ;
```

☑ 在 IDeptService 业务接口中定义 getIn 方法，并且在子类实现此方法。

```
@Override
public List<Dept> getIn(Set<Long> ids) {
    return this.deptDAO.findByDeptnoIn(ids);
}
```

☑ 编写测试类，进行业务测试。

```
@Test
public void testGetIn() {
    Set<Long> allIds = new HashSet<Long>() ;                // 实例化Set集合，用于保存查询ID
    allIds.addAll(Arrays.asList(1L, 3L, 5L));               // 增加ID数据
    List<Dept> allDepts = this.deptService.getIn(allIds) ;  // 数据查询
    allDepts.forEach((dept)->{                              // 迭代输出查询结果
        System.out.println(dept);
    }) ;
}
```

程序执行结果	Hibernate: select dept0_.deptno as deptno1_0_, dept0_.avgsal as avgsal2_0_, dept0_.createdate as createda3_0_, dept0_.dname as dname4_0_, dept0_.num as num5_0_ from Dept dept0_ where dept0_.deptno in (? , ? , ?) 部门数据输出略

在使用 IN 操作符时，方法映射需要接收集合数据，所以本程序在数据层方法中定义了 Set 集合，保存所有要查询的部门编号，这样会自动将参数转换为 IN 使用的参数。

范例：【mldnspring-jpa 项目】实现数据模糊查询与排序。

☑ 在 IDeptDAO 接口中定义查询方法，根据部门名称进行模糊查询与排序处理。

```
/**
 * 根据部门名称进行模糊查询（Containing表示前后都追加%），而后将查询结果按照deptno降序
排列
 * @param keyWord  查询关键字
 * @return  部门持久化对象集合
 */
public List<Dept> findByDnameContainingOrderByDeptnoDesc(String keyWord);
```

☑ 在 IDeptService 接口中追加查询业务方法，并且在子类进行实现。

```
@Override
public List<Dept> listSearch(String keyWord) {
    return this.deptDAO.findByDnameContainingOrderByDeptnoDesc(keyWord);
}
```

☑ 在测试类中进行业务方法测试。

```
@Test
public void testListSearch() {
```

```
		List<Dept> allDepts = this.deptService.listSearch("MLDN") ;	// 模糊查询
		allDepts.forEach((dept)->{						// 迭代输出查询结果
			System.out.println(dept);
		});
	}
```

程序执行结果	Hibernate: select dept0_.deptno as deptno1_0_, dept0_.avgsal as avgsal2_0_, dept0_.createdate as createda3_0_, dept0_.dname as dname4_0_, dept0_.num as num5_0_ from Dept dept0_ where dept0_.dname like ? order by dept0_.deptno desc 部门数据输出略

本程序实现了数据模糊查询。在方法映射中，模糊查询可以使用 StartingWith（开头追加%）、EndingWith（结尾追加%）和 Containing（前后都追加%）3 种边界匹配方法。在数据查询后使用了 ORDER BY 排序定义，将部门编号由高到底降序排列。

13.8.4　CrudRepository 数据接口

Repository 是整个 JPA 执行自动处理的核心接口，其下还有很多子接口。如果开发者针对单表 CRUD 需要重复进行方法定义，显然费时费力。实际上只需要继承 CrudRepository（Repository 子接口）父接口，就可以自动支持基础 CRUD 处理能力。在 CrudRepository 接口中定义的方法如表 13-6 所示。

表 13-6　CrudRepository 接口定义方法

编号	方法	类型	描述
1	public <S extends T> S save(S entity);	普通	追加单个的实体对象
2	public <S extends T> Iterable<S> saveAll(Iterable<S> entities)	普通	实现数据批量增加，需要传递 List、Set 集合
3	public Optional<T> findById(ID id)	普通	根据 ID 进行查询
4	public boolean existsById(ID id)	普通	判断当前的 ID 是否存在
5	public Iterable<T> findAll();	普通	返回全部的查询结果
6	public Iterable<T> findAllById(Iterable<ID> ids);	普通	得到部分的查询结果
7	public long count();	普通	取得全部数据量
8	public void deleteById(ID id);	普通	根据 id 删除
9	public void delete(T entity);	普通	直接根据实体对象删除
10	public void deleteAll(Iterable<? extends T> entities);	普通	删除一堆的实体
11	public void deleteAll();	普通	清空数据表

通过表 13-6 可以发现，里面提供了大部分常用的 CRUD 处理方法，开发者可根据需求采用方法映射或@Query 接口扩充查询方法。

1.【mldnspring-jpa 项目】定义 IDeptDAO 接口，继承 CrudRepository 父接口。

```
package cn.mldn.mldnspring.dao;
import org.springframework.data.repository.CrudRepository;
```

```
import cn.mldn.mldnspring.po.Dept;
public interface IDeptDAO extends CrudRepository<Dept, Long> {        // SpringData数据接口
}
```

2.【mldnspring-jpa 项目】由于此数据层提供的方法较多，且与之前相比方法名称有所改变，所以将在业务层中测试部分方法，重新定义 IDeptService 业务接口。

```
public interface IDeptService {
    public boolean add(Dept vo) ;                                      // 增加部门数据
    public List<Dept> list() ;                                         // 查询dept表中的全部数据
    public Dept get(long id) ;                                         // 根据部门编号查询部门信息
}
```

3.【mldnspring-jpa 项目】定义 IDeptService 业务接口实现子类。

```
@Service
public class DeptServiceImpl implements IDeptService {
    @Autowired
    private IDeptDAO deptDAO ;                                         // 注入IDeptDAO接口实例
    @Override
    public boolean add(Dept vo) {
        return this.deptDAO.save(vo).getDeptno() != null ;
    }
    @Override
    public List<Dept> list() {
        return (List<Dept>) this.deptDAO.findAll();                    // Iterable转为List
    }
    @Override
    public Dept get(long id) {
        // 使用Optional接收返回数据，可以避免null数据
        Optional<Dept> result = this.deptDAO.findById(id) ;
        return result.get() ;
    }
}
```

4.【mldnspring-jpa 项目】编写测试类调用业务方法。

```
@ContextConfiguration(locations = { "classpath:spring/spring-*.xml" })
@RunWith(SpringJUnit4ClassRunner.class)
public class TestDeptService {
    @Autowired
    private IDeptService deptService ;                                 // 注入IDeptService业务接口实例
    @Test
    public void testAdd() {
        Dept po = new Dept() ;                                         // 实例化持久类对象
        po.setDname("MLDN教学部");                                      // 设置数据
```

```
        po.setCreatedate(new Date());                                   // 设置数据
        po.setNum(55);                                                  // 设置数据
        po.setAvgsal(89998.00);                                         // 设置数据
        System.out.println(this.deptService.add(po));                   // 数据持久化
    }
    @Test
    public void testList() {
        this.deptService.list().forEach((dept)->{                       // 迭代输出查询结果
            System.out.println(dept);
        });
    }
    @Test
    public void testGet() {
        System.out.println(this.deptService.get(1L));
    }
}
```

本程序的测试结果与之前相同，区别在于用户自定义 DAO 层中的方法可以减少重复功能定义。

13.8.5　PagingAndSortingRepository 数据接口

CrudRepository 实现的只是基础的 CRUD 处理功能，其本身不支持数据的分页显示及排序处理，所以在 SpringDataJPA 中还提供了 CrudRepository 子接口 PagingAndSortingRepository，以实现分页与排序。此接口中定义的方法如表 13-7 所示。

表 13-7　PagingAndSortingRepository 接口方法

编号	方法	类型	描述
1	public Iterable<T> findAll(Sort sort);	普通	需要设置一个排序的类
2	public Page<T> findAll(Pageable pageable);	普通	需要将所有的内容设置到 Pageable 类

PagingAndSortingRepository 接口提供的 findAll 方法可以实现分页与排序查询，此方法需要接收一个 Pageable 接口参数。该类型可以保存分页信息，如当前页（currentPage，从 0 开始）、每页显示数据行数（lineSize）与排序定义。

1.【mldnspring-jpa 项目】定义 IDeptDAO 接口，继承 PagingAndSortingRepository 父接口。

```
package cn.mldn.mldnspring.dao;
import org.springframework.data.repository.PagingAndSortingRepository;
import cn.mldn.mldnspring.po.Dept;
public interface IDeptDAO extends PagingAndSortingRepository<Dept, Long> {
}
```

此时的 IDeptDAO 接口将具有全部的单表 CRUD 处理方法，同时也支持数据分页操作。

2.【mldnspring-jpa 项目】修改 IDeptService 业务层定义，增加新的方法。

```
/**
 * 实现数据的分页处理
 * @param currentPage 当前页
 * @param lineSize 每页的数据行
 * @return 包含以下的返回结果
 * 1. key = allDepts，value =所有的部门信息
 * 2. key = deptCount，value =统计结果
 * 3. key = deptPage，value =总页数
 */
public Map<String,Object> listSplit(int currentPage,int lineSize) ;
```

3.【mldnspring-jpa 项目】修改 DeptServiceImpl 子类，为 list 方法追加排序，同时覆写 listSplit 分页查询。

```
@Override
public List<Dept> list() {
    Sort sort = new Sort(Sort.Direction.DESC,"deptno") ;          // 设置deptno字段为降序排列
    return (List<Dept>) this.deptDAO.findAll(sort) ;              // 排序查询
}
```

```
@Override
public Map<String, Object> listSplit(int currentPage, int lineSize) {
    Sort sort = new Sort(Sort.Direction.DESC, "deptno");          // 设置deptno字段为降序排列
    // 将分页与排序操作保存到Pageable接口对象中，通过DAO层进行方法调用，页数从0开始
    Pageable pageable = PageRequest.of(currentPage - 1, lineSize, sort);
    // Page会自动保存全部数据记录、总记录数，同时会计算总页数
    Page<Dept> pageDept = this.deptDAO.findAll(pageable);          // 数据查询
    Map<String, Object> map = new HashMap<String, Object>();       // 保存返回结果
    map.put("allDepts", pageDept.getContent());                    // 保存返回数据
    map.put("deptCount", pageDept.getTotalElements());             // 保存总记录数
    map.put("deptPage", pageDept.getTotalPages());                 // 保存总页数
    return map;
}
```

4.【mldnspring-jpa 项目】在测试类中编写代码，测试 listSplit。

```
@Test
public void testListSplit() {
    System.out.println(this.deptService.listSplit(1, 2));
}
```

程序执行结果	【操作分析】采用分页形式查询数据。 Hibernate: select dept0_.deptno as deptno1_0_, dept0_.avgsal as avgsal2_0_, dept0_.createdate as createda3_0_, dept0_.dname as dname4_0_, dept0_.num as num5_0_ from Dept dept0_ order by dept0_.deptno desc limit ?

【操作分析】统计表中数据记录。

```
Hibernate: select count(dept0_.deptno) as col_0_0_ from Dept dept0_
部门数据输出略
```

本程序实现了数据的分页查询处理。在进行分页查询（数据查询与统计查询）中，需要将所有分页的参数利用 PageRequest 类提供的方法进行创建，而返回的数据会通过 Page 接口进行包装。

13.8.6　JpaRepository 数据接口

Repository 数据接口支持 SpringData 相关处理，同时为了方便开发者进行项目编写还提供了 CrudRepository、PagingAndSortingRepository 子接口，但是考虑到 JPA 开发中的一些特性，SpringDataJPA 又在 PagingAndSortingRepository 子接口下继续定义了一个 JpaRepository 子接口，该接口扩充的主要方法如表 13-8 所示。

表 13-8　JpaRepository 扩充方法

编号	方法	类型	描述
1	public List<T> findAllById(Iterable<ID> ids)	普通	查询全部 ID 数据
2	public T getOne(ID id)	普通	根据 ID 查询，等价于 getReference 方法
3	public <S extends T> S saveAndFlush(S entity)	普通	数据保存并刷新缓存

下面为了验证 JpaRepository 接口功能，将按照如下步骤编写代码完成处理。

1.【mldnspring-jpa 项目】修改 IDeptDAO 接口定义。

```
package cn.mldn.mldnspring.dao;
import org.springframework.data.jpa.repository.JpaRepository;
import cn.mldn.mldnspring.po.Dept;
public interface IDeptDAO extends JpaRepository<Dept, Long> {          // SpringData数据接口
}
```

2.【mldnspring-jpa 项目】在 IDeptService 接口中定义新的业务方法。

```
    /**
     * 根据指定范围的ID进行数据查询
     * @param ids 要查询的部门编号
     * @return 全部部门信息
     */
    public List<Dept> list(Set<Long> ids) ;
```

3.【mldnspring-jpa 项目】在 IDeptService 子类中实现新定义业务方法。

```
    @Override
    public List<Dept> list(Set<Long> ids) {
        return this.deptDAO.findAllById(ids);                      // 此时直接返回List集合
    }
```

4．【mldnspring-jpa 项目】编写业务测试。

```
@Test
public void testListIn() {
    Set<Long> allIds = new HashSet<Long>() ;              // 实例化Set集合，用于保存查询ID
    allIds.addAll(Arrays.asList(1L, 3L, 5L));             // 增加ID数据
    List<Dept> allDepts = this.deptService.list(allIds) ; // 数据查询
    allDepts.forEach((dept)->{                            // 迭代输出查询结果
        System.out.println(dept);
    }) ;
}
```

程序执行结果	Hibernate: select dept0_.deptno as deptno1_0_, dept0_.avgsal as avgsal2_0_, dept0_.createdate as createda3_0_, dept0_.dname as dname4_0_, dept0_.num as num5_0_ from Dept dept0_ where dept0_.deptno in (? , ? , ?) 部门数据输出略

通过一系列分析可以发现，在实际项目开发中，使用 JpaRepository 实现的数据层开发操作是最接近于 JPA 开发框架使用环境的，同时也支持其他的 Repository 接口功能。利用此接口实现的 JPA 整合才是最方便的。

13.9　本章小结

1．JPA 是 Java 持久化操作的技术标准，Hibernate 是 JPA 最为常用的一种实现技术。

2．JPA 的所有操作都是通过 EntityManager 接口展开的，但是此接口进行数据操作需要维护数据状态，所以性能不高。在进行数据更新时，建议使用 JPQL 结合 Query 接口完成。

3．JPA 支持数据缓存，且一级缓存永远存在，二级缓存需要配置后生效。

4．JPA 中为了防止并发更新操作所造成的数据错误，提供了悲观锁机制与乐观锁机制。

5．JPA 中支持数据的一对一关联、一对多关联以及多对多关联。

6．传统的 Spring 整合 JPA 开发框架可以将数据库连接管理、EntityManagerFactory 管理、事务管理交由容器完成，但是需要开发者提供数据层实现子类。

7．SpringDataJPA 可以利用 Repository 接口简化数据层定义复杂度，利用容器可以帮助用户减少 DAO 接口子类定义。

第 14 章

SpringMVC

通过本章学习，可以达到以下目标：

1．掌握 SpringMVC 运行原理与设计架构。
2．掌握 SpringMVC 数据处理与跳转配置。
3．掌握 SpringMVC 文件上传技术。
4．掌握 SpringMVC 拦截器定义与数据验证操作。
5．掌握 SpringMVC 与 Druid 和 SpringDataJPA 技术整合。

除了基本功能外，为了方便开发者编写代码，Spring 还提供了 SpringMVC 支持。利用 SpringMVC，可以与 Spring 开发框架进行整合，更容易实现控制层处理。本章将为读者详细讲解 SpringMVC 中的各种配置，最后会通过一个实际案例进行应用说明。

14.1 SpringMVC 简介

MVC（Model View Controller）是 Java 项目开发中应用最广泛的一种设计模式，也是 Java EE 唯一提倡的整体设计模式。在传统 Web 开发中（使用 Servlet、JSP），要想搭建一个便于维护、可动态扩充又无明确依赖的 MVC 设计模式，需要利用反射与配置文件进行大量的编码设计实现。Spring 中，为了简化 MVC 的设计难度，提出了 SpringMVC 技术。利用 SpringMVC 提供的 DispatcherServlet 程序类，可轻松实现用户请求与具体 Action 控制类之间的连接。

SpringMVC 的整体架构如图 14-1 所示。

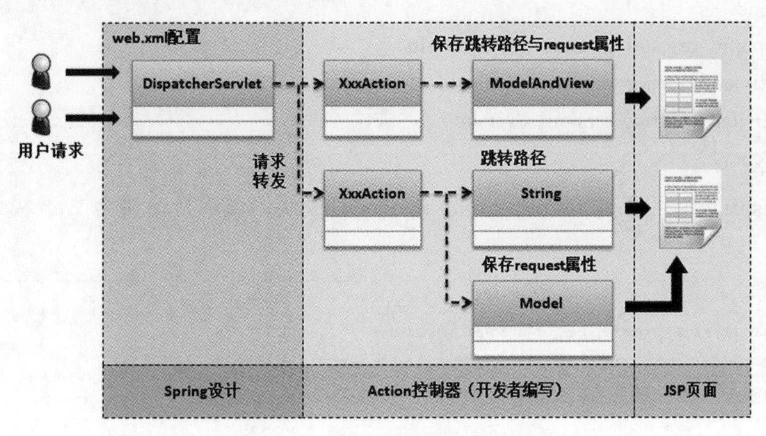

图 14-1　SpringMVC 核心架构

SpringMVC 中，用户的所有请求都会被 DispatcherServlet 程序类接收。该类是利用反射与 Spring 提供的工具类实现的一个动态请求处理，它会根据用户访问的请求路径，将请求转交给指定的 Action 类（开发者定义的控制器）进行处理，处理完成后再利用字符串进行路径跳转（使用 Model 类保存属性）。也可以利用 ModelAndView 类进行跳转路径的配置与传递 request 属性定义。

 **提示：SpringMVC 与 Struts。**

Struts 是一款优秀的 MVC 开发设计框架，从 Struts 1.x 到 Struts 2.x，一直在行业中应用广泛。本质上来说，SpringMVC 与 Struts 这两个开发框架在整体设计上并无太大差别。但 Struts 路径直接基于反射进行操作，产生的漏洞曾导致过数据泄露。从那以后，SpringMVC 迎来了发展契机。SpringMVC 使用注解配置访问路径，更加安全，对于参数的接收与处理过程也更加简单。

除了如图 14-1 所示的核心架构外，在 SpringMVC 中还支持视图解析器（以实现页面安全访问）、请求参数与类对象转换、Restful 风格展示、文件上传、拦截器等核心技术。

14.2 搭建 SpringMVC 项目开发环境

SpringMVC 是在 Spring 开发框架基础上构建的 MVC 技术，所以需要在项目中先配置好相应的 Spring 与 SpringMVC 依赖支持。同时，还需要在 web.xml 配置文件中配置 Spring 监听器与定义 DispatcherServlet 路径。下面将通过具体的步骤进行讲解。

1.【mldnspring 项目】修改父 pom.xml 配置文件，追加 SpringMVC 相关依赖配置。

```
<dependency>
    <groupId>org.springframework</groupId>
    <artifactId>spring-web</artifactId>
    <version>${spring.version}</version>
</dependency>
<dependency>
    <groupId>org.springframework</groupId>
    <artifactId>spring-webmvc</artifactId>
    <version>${spring.version}</version>
</dependency>
```

2.【mldnspring 项目】建立 mldnspring-mvc 项目模块，该项目模块为 Web 模块，如图 14-2 所示。

Group Id	Artifact Id	Version
org.apache.maven.archetypes	maven-archetype-site	1.1
org.apache.maven.archetypes	maven-archetype-site-simple	1.1
org.apache.maven.archetypes	maven-archetype-webapp	1.0

图 14-2　创建 Web 模块

3.【mldnspring-mvc 项目】修改子 pom.xml 配置文件，除了引入之前配置的依赖库外，还需要导入 Web 依赖库。

```
<dependency>
    <groupId>org.springframework</groupId>
    <artifactId>spring-web</artifactId>
</dependency>
<dependency>
    <groupId>org.springframework</groupId>
    <artifactId>spring-webmvc</artifactId>
</dependency>
```

除了要引入 spring-mvc、spring-webmvc 相关依赖外，因为这是一个 Web 项目模块，所以还需要导入 servlet、jsp 等依赖支持库。

4.【mldnspring-mvc 项目】要在 Web 容器中配置 SpringMVC，首先需要配置一个 Spring 监听器，这样才可以在 Web 容器启动时自动启动 Spring 容器，维护 Spring 中的对象。修改 web.xml 配置文件，追加控制器。

```
<!-- 配置Spring容器启动的监听器，以加载Spring中的核心配置文件 -->
<listener>
    <listener-class>
        org.springframework.web.context.ContextLoaderListener
    </listener-class>
</listener>
```

5.【mldnspring-mvc 项目】要启动 Spring 容器，还需要指明 Spring 配置文件路径。

```
<!-- 设置上下文参数，实际上就是设置application属性，等价于setAttribute() -->
<context-param>
    <param-name>contextConfigLocation</param-name>              <!-- Spring中定义的属性名称 -->
    <param-value>classpath:spring/spring-*.xml</param-value>    <!-- 资源文件 -->
</context-param>
```

6.【mldnspring-mvc 项目】SpringMVC 的核心是 DispatcherServlet 转发处理类，该类需要在 web.xml 中配置。

```
<!-- 设置SpringMVC的核心控制器类，利用此类实现所有的请求分发处理（Action） -->
<servlet>
    <servlet-name>springmvc</servlet-name>
    <servlet-class>
        org.springframework.web.servlet.DispatcherServlet
    </servlet-class>
    <init-param>                                <!-- 定义SpringMVC的配置文件 -->
        <param-name>contextConfigLocation</param-name>
```

```
            <param-value>classpath:spring/spring-mvc.xml</param-value>
        </init-param>
    </servlet>
    <servlet-mapping>                      <!-- SpringMVC设计中的路径都是以action结尾的 -->
        <servlet-name>springmvc</servlet-name>
        <url-pattern>*.action</url-pattern>
    </servlet-mapping>
```

本程序中配置的 DispatcherServlet 类的访问路径名称为*.action，此路径也为所有 Action 执行路径的默认后缀。

7.【mldnspring-mvc 项目】为了杜绝程序开发中的乱码问题，还需要统一配置编码过滤器。

```
    <!-- 配置编码过滤器，以解决数据传输乱码问题 -->
    <filter>
        <filter-name>encoding</filter-name>
        <filter-class>
            org.springframework.web.filter.CharacterEncodingFilter
        </filter-class>
        <init-param>                                   <!-- 设置要使用的编码 -->
            <param-name>encoding</param-name>
            <param-value>UTF-8</param-value>
        </init-param>
    </filter>
    <filter-mapping>                                   <!-- 所有路径都必须经过此过滤器 -->
        <filter-name>encoding</filter-name>
        <url-pattern>/*</url-pattern>
    </filter-mapping>
```

8.【mldnspring-mvc 项目】建立 spring-mvc.xml 配置文件，追加 mvc 命名空间，如图 14-3 所示。

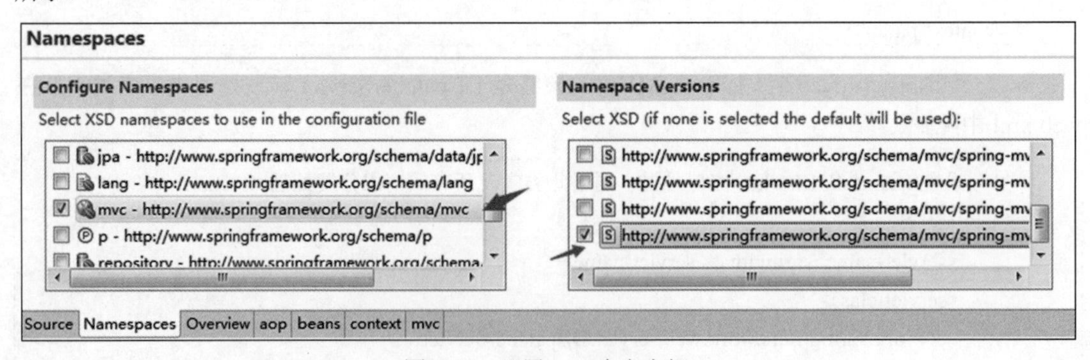

图 14-3　配置 mvc 命名空间

9.【mldnspring-mvc 项目】修改 spring-mvc.xml 配置文件，追加 mvc 配置。

```
    <context:component-scan base-package="cn.mldn.mldnspring.action" />
```

```
<mvc:annotation-driven/>                    <!-- 启用SpringMVC注解配置支持 -->
<mvc:default-servlet-handler/>              <!-- 启动Servlet的配置处理 -->
```

至此，SpringMVC 开发环境搭建完成，而后就可以在此基础上编写相应代码了。

提示：为防止重复扫描配置，扫描路径可以定义为子目录。

整合 SpringMVC 开发框架时，需要使用<context:component-scan>注解定义扫描包。如果此时配置为父包（n.mldn.mldnspring），将与 Spring 监听器配置文件中的自动扫描包重复，造成重复扫描问题，产生未知的错误。

14.3　编写第一个 SpringMVC 程序

SpringMVC 程序符合标准 MVC 设计模式的要求，即前台通过表单或 URL 地址重写，进行请求参数的发送，而后通过指定路径访问 Action 类中的处理方法。数据处理完成后，通过 request 属性范围保存要显示的数据信息，并在 JSP 页面中显示。SpringMVC 为了简化内置对象的应用，所有的参数都可以通过 Action 方法进行自动接收。对于配置跳转路径与传递属性数据，建议直接使用 ModelAndView 类进行包装。

ModelAndView 类定义的方法如表 14-1 所示。

表 14-1　ModelAndView 类方法

编号	方法	类型	描述
1	public ModelAndView()	构造	实例化 ModelAndView 对象，但是不设置数据
2	public ModelAndView(String viewName)	构造	设置跳转视图名称（路径）
3	public ModelAndView(String viewName, Map<String, ?> model)	构造	设置跳转路径与传递数据（会自动迭代 Map 集合，将 key 作为参数名称，value 作为内容）
4	public void setViewName(String viewName)	普通	设置视图名称
5	public ModelAndView addObject(String attributeName, Object attributeValue)	普通	设置一个传递数据
6	public ModelAndView addAllObjects(Map<String, ?> modelMap)	普通	设置多个传递数据
7	public void clear()	普通	清空路径与数据信息

下面将根据前面配置好的 SpringMVC 开发环境，编写一个 ECHO 处理程序，即由页面通过表单输入数据，交由 Action 处理，而后将处理过的数据交由 JSP 页面显示。程序的执行流程如图 14-4 所示。

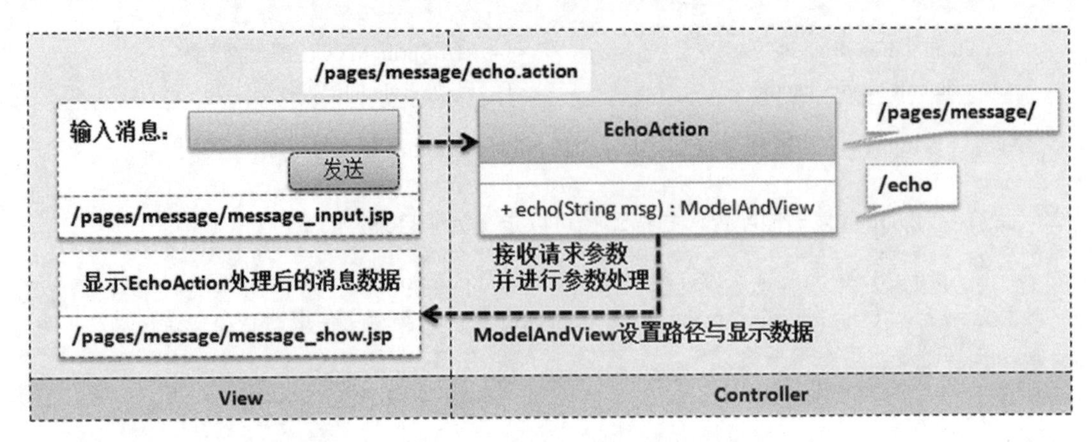

图 14-4 ECHO 程序处理流程

1.【mldnspring-mvc 项目】建立 EchoAction 程序类，负责请求处理。

```
package cn.mldn.mldnspring.action;
import org.slf4j.Logger;
import org.slf4j.LoggerFactory;
import org.springframework.stereotype.Controller;
import org.springframework.web.bind.annotation.RequestMapping;
import org.springframework.web.servlet.ModelAndView;
@Controller                                                     // 定义控制器
@RequestMapping("/pages/message/*")          // 定义访问父路径，与方法中的路径组合为完整路径
public class EchoAction {                                       // 自定义Action程序类
    private Logger log = LoggerFactory.getLogger(EchoAction.class) ; // 日志记录
    @RequestMapping("/echo")                                    // 访问的路径为echo.action
    public ModelAndView echo(String msg) {                      // 根据参数名称自动进行匹配处理
        this.log.info("*** EchoAction接收到请求参数，msg = " + msg);    // 日志输出
        // 日志输出ModelAndView主要功能是设置跳转路径以及保存request属性
        ModelAndView mav = new ModelAndView("/pages/message/message_show.jsp") ;
        mav.addObject("echoMessage", "【ECHO】msg = " + msg) ; // 设置request数据
        return mav ;
    }
}
```

本程序完成了一个最基础的 Action 类定义。

☑ @Controller：SpringMVC 的控制器需要使用@Controller 注解声明（与@Service、@Repository 注解意义相同）。功能是向 Spring 中配置一个自定义的 Bean。

☑ @RequestMapping：控制器的访问路径。注解中如果直接定义访问路径，则表示支持全部的 HTTP 请求模式（GET、POST、PUT 等），可通过 RequestMethod 枚举类设置请求类型。例如：

GET访问	@RequestMapping(value="/echo",method=RequestMethod.GET)
POST访问	@RequestMapping(value="/echo",method=RequestMethod.POST)

|- 本程序利用此注解定义了两个访问路径，最终的访问路径是这两个路径的叠加。

|- 在 EchoAction 类上定义了访问路径/pages/message/*，该路径属于父路径，所以使用通配符“*”。

|- 在 echo 方法上定义了子路径/echo，实际访问中需要将两个路径合并在一起使用。

|- 如果觉得此注解过于麻烦，可以采用@GetMapping("/echo") 或 @PostMapping("/echo") 之类的注解简化 HTTP 请求模式的定义。

☑ ModelAndView：主要功能是配置跳转路径与传递的 request 属性。

☑ echo(String msg)：方法的参数名称与请求参数名称匹配时，可以自动接收，不需要再编写 request.getParameter("msg")语句。

提示：访问路径可以返回字符串。

SpringMVC 设计模式中，推荐读者使用 ModelAndView 操作处理。Spring 中，同一种操作可以有多种处理方案，这里也可以在方法上使用 String 作为返回路径配置，通过方法参数定义一个 Model 对象，以实现属性传递。

范例：另一种定义形式。

```
@Controller                                                        // 定义控制器
@RequestMapping("/pages/message/*")
public class EchoAction {                                           // 自定义Action程序类
    @RequestMapping("/pages/message/echo")                         // 访问的路径
    public String echo(String msg,Model model) {                   // Model设置属性
        model.addAttribute("echoMessage", "【ECHO】msg = " + msg) ;   // 设置属性
        return "/pages/message/message_show.jsp" ;
    }
}
```

本程序并没有定义父路径，而是在 echo 方法上直接定义了完整路径，同时在配置 echo 参数时配置了 Model 对象，利用该对象实现了属性传递。这样 echo 方法执行后的跳转路径就可以通过字符串进行描述了。

2.【mldnspring-mvc 项目】在 Web 目录中创建/pages/message/message_show.jsp 页面，显示参数内容。

```
<%@ page pageEncoding="UTF-8"%>
<%
    request.setCharacterEncoding("UTF-8") ;
    String basePath = request.getScheme() + "://" + request.getServerName() + ":" +
            request.getServerPort() + request.getContextPath() + "/";
%>
<base href="<%=basePath%>" />
<h1>ECHO消息显示：${echoMessage}</h1>
```

消息显示页面定义完成后，可以采用地址重写的方式处理 msg 参数传递。将项目发布到

Tomcat 服务器上并启动，随后输入如下访问地址。

```
http://localhost:8080/mldnspring-mvc/pages/message/echo.action?msg=www.mldn.cn
```

程序运行后，可以在后台观察 EchoAction 程序类提供的日志信息，也可以在浏览器中观察到如图 14-5 所示的最终信息显示页面。

图 14-5　ECHO 数据处理

3.【mldnspring-mvc 项目】参数传递方式过于烦琐，最好的做法是建立一个 message_input.jsp 页面，利用表单输入参数。

```
<%@ page pageEncoding="UTF-8"%>
<%
    request.setCharacterEncoding("UTF-8") ;
    String basePath = request.getScheme() + "://" + request.getServerName() + ":" +
            request.getServerPort() + request.getContextPath() + "/";
    String message_input_url = basePath + "pages/message/echo.action" ;
%>
<base href="<%=basePath%>" />
<form action="<%=message_input_url%>" method="post">
    请输入消息：<input type="text" name="msg" id="msg" value="www.mldn.cn">
    <input type="submit" value="发送">
</form>
```

本程序利用表单实现了 msg 参数的输入处理，同时在表单上设置了 EchoAction 完整的处理路径。

4.【mldnspring-mvc 项目】在标准 MVC 设计模式中，用户必须通过控制层才能够访问显示层，所以最好在 EchoAction 类中追加一个页面跳转处理。

```
@RequestMapping("/echo_pre")
public String echoPre() {
    return "/pages/message/message_input.jsp" ;
}
```

本程序只是一个跳转，并不传递任何显示属性，所以直接在方法上返回 String 即可。与父路径相结合后，访问路径为 http://localhost:8080/mldnspring-mvc/pages/message/echo_pre.action，显示效果如图 14-6 所示。

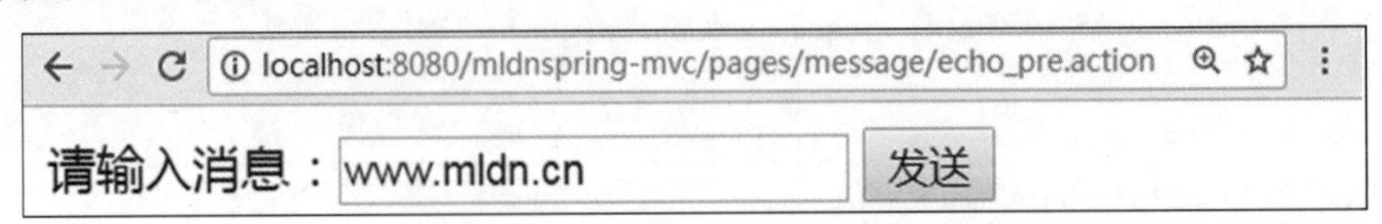

图 14-6　访问消息输入表单

14.4 接收请求参数

SpringMVC 设计中，为了简化参数的接收过程，可直接在控制层方法上通过方法参数名称匹配，实现请求参数的自动接收。如果请求参数的名称与方法中的名称不相同，可在定义方法参数上使用@RequestParam 注解进行配置。

@RequestParam 注解中有 3 个常用配置属性，如表 14-2 所示。

表 14-2 @RequestParam 注解属性

编号	属性	描述
1	name	定义要接收的参数名称
2	required	该参数是否必须传递
3	defaultValue	未传递此参数时设置的默认值

范例：【mldnspring-mvc 项目】修改 EchoAction 类中的 echo 方法。

```
public ModelAndView echo(
        @RequestParam(                              // 表示对请求参数的配置
            name = "msg",                           // 表示映射的请求参数的名称
            required = false,                       // 表示该参数是否必须传递
            defaultValue = "www.mldnjava.cn")       // 设置该参数为null时的默认值
            String str) {}                          // 接收请求参数
```

本程序中，echo 方法中的参数名称与请求参数名称并不相同，但由于使用了@RequestParam 注解，仍然可以实现指定参数的接收，并可在不传递参数时使用默认值进行设置。

提示：@RequestParam 注解。

如果只为了实现请求参数与方法接收参数的匹配，不需要设置默认值，使用@RequestParam("msg") 配置注解即可。这种配置较为复杂，实际开发中建议采用原始方式，保持请求参数名称与方法参数名称的一致。

进行参数接收时，可以实现数据类型的自动转换处理，也可以在一个方法中实现多个参数的接收。下面通过一个完整的程序进行演示。

1.【mldnspring-jpa 项目】修改 message_input.jsp 页面，实现多个参数的输入。

```
<form action="<%=message_input_url%>" method="post">
    消息内容：<input type="text" name="msg" id="msg" value="www.mldn.cn"><br>
    消息级别：<select id="level" name="level">
            <option value="0">紧急</option>
            <option value="1">普通</option>
            <option value="2">延迟</option>
        </select><br>
```

```
    发送日期：<input type="text" id="pubdate" name="pubdate" value="2262-01-21"><br>
    消息标签：<input type="checkbox" name="tags" id="tags" value="政治">政治
              <input type="checkbox" name="tags" id="tags" value="经济">经济
              <input type="checkbox" name="tags" id="tags" value="文化">文化<br>
    <input type="submit" value="发送">    <input type="reset" value="重置">
</form>
```

此时程序共定义了 3 个请求参数：msg（String）、level（Integer 或 int）和 tags（String[]），页面运行后的效果如图 14-7 所示。

图 14-7 多参数表单

2.【mldnspring-jpa 项目】修改 EchoAction 中的 echo 方法，进行参数接收与处理。

```
@PostMapping("/echo")                                              // 访问路径为echo.action
public ModelAndView echo(String msg, Integer level, String tags[], Date pubdate) {
    ModelAndView mav = new ModelAndView("/pages/message/message_show.jsp") ;
    mav.addObject("echoMessage", "【ECHO】msg = " + msg) ;            // 设置request数据
    mav.addObject("echoLevel", "【ECHO】level = " + level) ;          // 设置request数据
    mav.addObject("echoTags", "【ECHO】tags = " +
        Arrays.toString(tags)) ;                                     // 设置request数据
    mav.addObject("echoPubdate", "【ECHO】pubdate = " + pubdate) ;    // 设置request数据
    return mav ;
}
```

本程序中，echo 定义的方法参数与请求参数的方法名称相同，由于复选框是一个数组数据，所以使用 String[]进行接收。这里面最特殊的是 pubdate 参数，此类型为日期型数据。由于 SpringMVC 中没有定义日期接收的自动转换处理，所以需要开发者自行定义一个日期转换处理。

3.【mldnspring-mvc 项目】在 EchoAction 中扩充一个日期转换处理方法。

```
@InitBinder
public void initBinder(WebDataBinder binder) {                     // 设置一个Web数据的绑定转换
    SimpleDateFormat sdf = new SimpleDateFormat("yyyy-MM-dd") ;     // 定义转换处理
    // 在Web数据绑定中注册自定义规则绑定器，主要用于处理java.util.Date类型，允许为null
    binder.registerCustomEditor(java.util.Date.class, new CustomDateEditor(sdf, true));
}
```

本方法实现了一个数据绑定处理，这样每当发现需要接受 java.util.Date 类型参数时，都会先

按照指定格式进行转换。

提示：可以在父类中定义转换处理。

如果所有的 Action 程序类都需要重复定义，initBinder 方法将会非常麻烦。为了提高代码重用性，可以建立一个统一的抽象类 AbstractAction，在此类中定义 initBinder 方法，这样所有继承自 AbstractAction 的程序类都将具备日期数据处理支持。

4.【mldnspring-jpa 项目】修改 message_show.jsp 页面，实现数据输出。

```
<h1>ECHO消息显示：${echoMessage}</h1>
<h1>ECHO消息显示：${echoLevel}</h1>
<h1>ECHO消息显示：${echoTags}</h1>
```

本程序在 JSP 页面中追加了更多的 request 数据输出操作，表单提交后，页面显示效果如图 14-8 所示。

```
localhost:8080/mldnspring-mvc/pages/message/echo.action
ECHO消息显示：【ECHO】msg = www.mldn.cn
ECHO消息显示：【ECHO】level = 0
ECHO消息显示：【ECHO】pubdate = Tue Jan 21 00:00:00 CST 2262
ECHO消息显示：【ECHO】tags = [经济, 文化]
```

图 14-8　获取 ECHO 数据信息

14.5　参数与对象转换

SpringMVC 中除了可单独实现参数接收外，还可以实现参数与类对象的转换，以避免过多使用关键字 new 进行对象的实例化操作。同时，还支持多级属性设置，使程序开发更加方便。

1.【mldnspring-mvc 项目】建立描述部门与雇员的 VO 类。

Dept.java类

```
package cn.mldn.mldnspring.vo;
import java.io.Serializable;
import java.util.Date;
@SuppressWarnings("serial")
public class Dept implements Serializable {
	private Long deptno ;
	private String dname ;
	private Date createdate ;
	// setter、getter、toString略
}
```

Emp.java类

```
package cn.mldn.mldnspring.vo;
import java.io.Serializable;
import java.util.Date;
@SuppressWarnings("serial")
public class Emp implements Serializable {
	private Long empno ;
	private String ename ;
	private Double salary ;
	private Date hiredate ;
	private Integer level ;
```

```
	private Dept dept ;
	// setter、getter、toString略
}
```

2.【mldnspring-mvc 项目】建立 EmpAction 程序类，在类中定义日期格式转换处理与控制方法。

```
@Controller                                                         // 定义控制器
@RequestMapping("/pages/emp/*")                                     // 访问父路径
public class EmpAction {                                            // 自定义Action程序类
	private Logger log = LoggerFactory.getLogger(EmpAction.class) ;  // 日志记录
	@PostMapping("/add")                                             // 访问的路径为echo.action
	public ModelAndView add(Emp emp) {                               // 接收请求参数
		this.log.info(emp.toString());                               // 信息输出
		ModelAndView mav = new ModelAndView() ;
		mav.setViewName("/pages/emp/emp_do.jsp");
		mav.addObject("myemp", emp) ;                                // 保存数据
		return mav ;                                                 // 不进行跳转
	}
	@RequestMapping("/add_pre")
	public String addPre() {                                         // 数据增加前跳转
		return "/pages/emp/emp_add.jsp" ;
	}
	// 其他重复定义方法略
}
```

本程序中，add 方法用于 Emp 对象接收，因此必须保证表单提交参数的名称与 Emp 类中的成员属性名称相同。数据接收后，内容会交给 emp_do.jsp 页面进行显示。

3.【mldnspring-mvc 项目】在 Web 目录中创建 pages/emp/emp_add.jsp 页面，定义数据增加表单。

```
<%@ page pageEncoding="UTF-8"%>
<%
	request.setCharacterEncoding("UTF-8") ;
	String basePath = request.getScheme() + "://" + request.getServerName() + ":" +
		request.getServerPort() + request.getContextPath() + "/";
	String emp_add_url = basePath + "pages/emp/add.action" ;
%>
<base href="<%=basePath%>" />
<form action="<%=emp_add_url%>" method="post">
	雇员编号：<input type="text" name="empno" id="empno" value="7369"><br>
	雇员姓名：<input type="text" name="ename" id="ename" value="李兴华"><br>
	基本工资：<input type="text" name="salary" id="salary" value="3000.33"><br>
	雇佣日期：<input type="text" name="hiredate" id="hiredate" value="1991-11-11"><br>
```

```
    雇员等级：<input type="text" name="level" id="level" value="0"><br>
    部门编号：<input type="text" name="dept.deptno" id="dept.deptno" value="10"><br>
    部门名称：<input type="text" name="dept.dname" id="dept.dname" value="财务部"><br>
    部门成立：<input type="text" name="dept.createdate" id="dept.createdate" value="1986-09-15"><br>
    <input type="submit" value="添加"><input type="reset" value="重置">
</form>
```

本程序实现了雇员增加表单定义。在进行雇员数据增加时，采用的参数名称也都是类中的属性名称，由于在 Emp 类中提供了 Dept 类对象，所以可以采用“. ”形式找到 Dept 类中的属性。页面运行效果如图 14-9 所示。

图 14-9　雇员增加表单

4.【mldnspring-mvc 项目】在 Web 目录中创建 pages/emp/emp_do.jsp 页面，显示部分数据。

```
<h3>雇员姓名：${myemp.ename}、基本工资：${myemp.salary}、所在部门名称：
${myemp.dept.dname}</h3>
```

本页面并没有显示全部的雇员信息，当表单提交后可以看见如图 14-10 所示的信息显示页面。

图 14-10　显示雇员信息

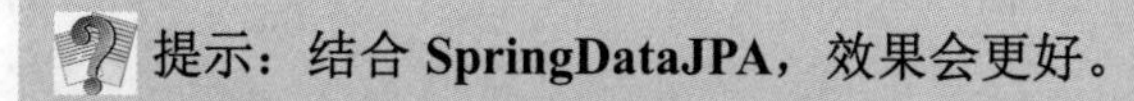

提示：结合 SpringDataJPA，效果会更好。

本程序提供了请求参数与类对象的自动转换，如果项目数据层采用 SpringDataJPA 实现，则可以直接利用持久化对象 PO 类进行参数接收，从而进一步简化程序代码。

14.6　Restful 展示风格

Restful（简称 Rest）是现在最为流行的一种数据交换格式，简单点理解就是可以直接在控制层的方法上将数据信息以 JSON 结构的形式进行返回，这样可以帮助开发者简化对象与 JSON

数据的转换处理。

提示：Restful 技术应用。

本书并不是讨论 Restful 设计的图书，在此处只是为了方便 Ajax 调用，而对于 Restful 开发的详细内容，可以参考笔者的《Java 微服务架构实战》一书进行完整学习。

要在 SpringMVC 中使用 Restful 风格进行返回数据处理，需要在项目中引入 jackson 相关依赖库，同时还需要使用@ResponseBody 注解定义控制层方法，下面通过具体的程序进行演示。

1.【mldnspring 项目】修改父 pom.xml 配置文件，追加 jackson 相关依赖库管理。

属性配置

```xml
<jackson.version>2.9.4</jackson.version>
```

依赖库配置

```xml
<dependency>
    <groupId>com.fasterxml.jackson.core</groupId>
    <artifactId>jackson-core</artifactId>
    <version>${jackson.version}</version>
</dependency>
<dependency>
    <groupId>com.fasterxml.jackson.core</groupId>
    <artifactId>jackson-databind</artifactId>
    <version>${jackson.version}</version>
</dependency>
<dependency>
    <groupId>com.fasterxml.jackson.core</groupId>
    <artifactId>jackson-annotations</artifactId>
    <version>${jackson.version}</version>
</dependency>
```

2.【mldnspring-mvc 项目】修改子 pom.xml 配置文件，引入依赖库。

```xml
<dependency>
    <groupId>com.fasterxml.jackson.core</groupId>
    <artifactId>jackson-core</artifactId>
</dependency>
<dependency>
    <groupId>com.fasterxml.jackson.core</groupId>
    <artifactId>jackson-databind</artifactId>
</dependency>
<dependency>
    <groupId>com.fasterxml.jackson.core</groupId>
    <artifactId>jackson-annotations</artifactId>
</dependency>
```

项目中引入 jackson 相关依赖库后，可以自动将类对象（List、Map 也包含）根据结构转为

JSON 数据格式。

3.【mldnspring-mvc 项目】在 EmpAction 类中追加返回 List 集合方法。

```
@GetMapping("/list")
@ResponseBody                          // 使用此注解就表示返回的对象自动变为JSON对象
public Object list() {
    List<Emp> all = new ArrayList<Emp>() ;
    for (int x = 0 ; x < 3 ; x ++) {
        Emp vo = new Emp() ;
        vo.setEmpno(7369L + x);
        vo.setEname("李兴华 - " + x);
        Dept dept = new Dept() ;
        dept.setDeptno(10L + x);
        dept.setDname("教学部 - " + x);
        vo.setDept(dept);
        all.add(vo) ;
    }
    return all ;
}
```

本程序通过方法返回了一个 List 集合，这个集合会自动转为 JSON 数组的形式进行定义，并且会将属性名称作为数据 key，属性内容作为数据 value，显示效果如图 14-11 所示。

localhost:8080/mldnspring-mvc/pages/emp/list.action

[{"empno":7369,"ename":"李兴华 - 0","salary":null,"hiredate":null,"level":null,"dept":{"deptno":10,"dname":"教学部 - 0","createdate":null}},{"empno":7370,"ename":"李兴华 - 1","salary":null,"hiredate":null,"level":null,"dept":{"deptno":11,"dname":"教学部 - 1","createdate":null}},{"empno":7371,"ename":"李兴华 - 2","salary":null,"hiredate":null,"level":null,"dept":{"deptno":12,"dname":"教学部 - 2","createdate":null}}]

图 14-11　Restful 显示

14.7　获取内置对象

虽然 SpringMVC 为数据接收与处理提供了自动化支持，但很多情况下，开发者依然需要直接进行内置对象处理。获得内置对象有如下两种途径。

获得方式 1：在控制层处理方法上直接定义要使用的内置对象（HttpServletRequest、HttpServletResponse）。

```
@Controller                                    // 定义控制器
@RequestMapping("/pages/web/*")                // 访问父路径，与方法中的路径组合为完整路径
public class WebObjectAction {                 // 自定义Action程序类
    private Logger log = LoggerFactory.getLogger(WebObjectAction.class) ; // 日志记录
    @RequestMapping("showOne")
```

```
    public ModelAndView showOne(HttpServletRequest request, HttpServletResponse response) {
        this.log.info("*** 【REQUEST内置对象】ContextPath = " + request.getContextPath()) ;
        this.log.info("*** 【APPLICATION内置对象】RealPath = " + request
                        .getServletContext().getRealPath("/")) ;
        this.log.info("*** 【RESPONSE内置对象】Locale = " + response.getLocale()) ;
        this.log.info("*** 【SESSION内置对象】SessionID = " + request.getSession().getId()) ;
        return null ;
    }
}
```

程序执行结果	*** 【REQUEST内置对象】ContextPath = /mldnspring-mvc *** 【APPLICATION内置对象】RealPath = F:\myeclipse\...\mldnspring-mvc\ *** 【RESPONSE内置对象】Locale = zh_CN *** 【SESSION内置对象】SessionID = CFCF99B111C7D39711FECC09F71FEB26

获得方式 2：利用 org.springframework.web.context.request.RequestContextHolder 请求上下文获得。

虽然可以在控制层方法上通过参数形式获取内置对象，但这样的做法不是很方便。开发中往往会自定义一些工具类，此时可以使用 Spring 提供的 RequestContextHolder 来获取内置对象信息。

```
    @RequestMapping("/show_two")
    public ModelAndView showTwo() {
        // 通过RequestContextHolder类可以获得HttpServletRequest与HttpServletResponse对象
        HttpServletRequest request = ((ServletRequestAttributes) RequestContextHolder
            .getRequestAttributes()).getRequest() ;
        HttpServletResponse response = ((ServletRequestAttributes) RequestContextHolder
            .getRequestAttributes()).getResponse() ;
        HttpSession session = request.getSession() ;
        ServletContext context = request.getServletContext() ;
        this.log.info("*** 【REQUEST内置对象】ContextPath = " + request.getContextPath()) ;
        this.log.info("*** 【APPLICATION内置对象】RealPath = " + context.getRealPath("/")) ;
        this.log.info("*** 【RESPONSE内置对象】Locale = " + response.getLocale()) ;
        this.log.info("*** 【SESSION内置对象】SessionID = " + session.getId()) ;
        return null ;
    }
```

程序执行结果	*** 【REQUEST内置对象】ContextPath = /mldnspring-mvc *** 【APPLICATION内置对象】RealPath = F:\myeclipse\...\mldnspring-mvc\ *** 【RESPONSE内置对象】Locale = zh_CN *** 【SESSION内置对象】SessionID = CFCF99B111C7D39711FECC09F71FEB26

本程序实现了与之前相同的功能，但是对比在方法上定义内置对象，此时通过 RequestContextHolder 实现的内置对象获取处理会更加灵活。

14.8　Web 资源安全访问

Web 项目开发中，安全永远是非常重要的问题。WEB-INF 是 Web 项目中安全级别最高的目录，此目录不允许直接进行访问，因此可以将所有的 Web 资源统一放置到此目录下（建议保留一个 index 首页），然后利用 MVC 设计模式，通过控制器跳转和 ViewResolver（视图解析器）实现资源访问。下面通过具体步骤进行演示。

1.【mldnspring-mvc 项目】将所有保存在 Web 目录下的资源（*.jsp、*.html、*.css、*.js、图片等）按照已有结构保存到 WEB-INF 目录中。

2.【mldnspring-mvc 项目】修改 spring-mvc.xml 配置文件，追加 ViewResolver 配置。

```
<!-- 启动一个视图访问的解析处理器，该操作会自动在容器中加载，不需要做任何的依赖配置 -->
<bean class="org.springframework.web.servlet.view.InternalResourceViewResolver">
    <property name="prefix" value="/WEB-INF/pages/"/>          <!-- 定义路径前缀 -->
    <property name="suffix" value=".jsp"/>                     <!-- 定义路径后缀 -->
</bean>
```

为了便于页面路径定义，本程序使用了视图解析器。在控制器中编写访问路径时，可以省略其前缀以及后缀内容。

3.【mldnspring-mvc 项目】以 EchoAction 程序类为例，修改其跳转路径。

```
@GetMapping("/echo_pre")
public String echoPre() {
    return "message/message_input";                    // 不写前缀或后缀
}
@PostMapping("/echo")                                  // 访问的路径为echo.action
public ModelAndView echo(String msg, Integer level, String tags[], Date pubdate) {
    ModelAndView mav = new ModelAndView("message/message_show") ; // 不写前缀与后缀
    // 其他语句略
    return mav ;
}
```

本程序通过控制器实现页面跳转时，由于视图解析器的配置存在，所以只需要编写访问路径的中间部分即可。

4.【mldnspring-mvc 项目】一个 Web 项目里，除了 JSP 页面，还会有*.js、*.css、图片等文件。为了安全，同样要将其保存在 WEB-INF 目录中，并在 Spring 中进行静态资源访问映射。修改 spring-mvc.xml 文件。

```
<!-- 进行静态Web资源的映射路径配置 -->
<mvc:resources location="/WEB-INF/js/" mapping="/mvcjs/**"/>
<mvc:resources location="/WEB-INF/css/" mapping="/mvccss/**"/>
<mvc:resources location="/WEB-INF/images/" mapping="/mvcimages/**"/>
```

5.【mldnspring-mvc 项目】修改 web.xml 配置文件，为 DispatcherServlet 增加一个新的访问路径，通过该路径实现静态资源的访问映射。

```
<servlet-mapping>                    <!-- SpringMVC设计中的路径都是以action结尾的 -->
    <servlet-name>springmvc</servlet-name>
    <url-pattern>*.action</url-pattern>
    <url-pattern>/</url-pattern>         <!-- 静态资源映射使用 -->
</servlet-mapping>
```

6.【mldnspring-springmvc 项目】配置完成后，如果要导入一些静态资源，可以通过映射名称完成。

访问CSS	<link rel="stylesheet" type="text/css" href="mvccss/style.css">
访问JS	<script type="text/javascript" src="mvcjs/pages/message/message_input.js"></script>
访问图片	<img src="mvcimages/diaosi.jpg">

读者在使用 SpringMVC 开发的时候，考虑到程序的标准化与安全性，所有的资源都应该进行上述配置。

14.9 读取资源文件

Web 项目里，通常有大量*.properties 资源文件。利用这类文件可以保存一些重要信息，如文字提示信息（messages.properties）、跳转路径（pages.properties）、验证规则（validations.properties）等。这些资源文件，用户可以直接在 Spring 配置文件中进行处理。

1.【mldnspring-mvc 项目】在 src/main/resources 源文件夹目录中创建一个新的包 cn.mldn.message。

2.【mldnspring-mvc 项目】在 cn.mldn.message 包中建立两个资源文件。

pages.properties	messages.properties
echo.input.page=message/message_input	welcome.info=欢迎{0}光临访问！

3.【mldnspring-mvc 项目】修改 spring-mvc.xml 配置文件，追加资源映射。

```
<!-- 配置SpringMVC里面要使用的各项资源 -->
<bean id="messageSource"
    class="org.springframework.context.support.ResourceBundleMessageSource">
    <property name="basenames">    <!-- 定义所有的BaseName名称 -->
        <array>
            <value>cn.mldn.message.pages</value>          <!-- 资源定位 -->
            <value>cn.mldn.message.messages</value>       <!-- 资源定位 -->
        </array>
    </property>
</bean>
```

资源文件读取，使用 ResourceBundleMessageSource 类完成配置。此类是 MessageSource 接口的子类，在程序中可以利用 MessageSource 接口提供的方法实现资源读取。操作方法如表 14-3 所示。

表 14-3　MessageSource 接口方法

编号	方法	类型	描述
1	public String getMessage(String code, Object[] args, String defaultMessage, Locale locale);	普通	读取指定资源，可以配置参数、默认返回数据、Local 国际化处理
2	public String getMessage(String code, Object[] args, Locale locale) throws NoSuchMessageException;	普通	读取指定资源

4.【mldnspring-mvc 项目】在 EchoAction 程序类中注入 MessageSource 接口对象。

```
@Autowired
private MessageSource messageSource ;                // 资源读取
```

5.【mldnspring-mvc 项目】在 EchoAction 类中建立一个方法，进行资源读取测试。

```
@GetMapping("/message")
public ModelAndView message() {
        System.out.println("【echo.input.page】" + this.messageSource
                .getMessage("echo.input.page", null, Locale.getDefault()));
        System.out.println("【welcome.info】" + this.messageSource
                .getMessage("welcome.info", new Object[] { "李兴华" }, Locale.getDefault()));
        System.out.println("【nothing】" + this.messageSource
                .getMessage("nothing", null, Locale.getDefault()));
        return null;
}
```

程序执行结果	【echo.input.page】message/message_input 【welcome.info】欢迎李兴华光临访问！

本程序根据 key 实现了资源文件中数据的读取。读取时可利用对象数组，设置资源中占位符的数据。资源不存在时，会出现 org.springframework.context.NoSuchMessageException 异常。

14.10　文件上传

文件上传是 Web 开发中的重要环节，也是 MVC 开发框架必须支持的功能。SpringMVC 中主要使用 Apache FileUpload 组件实现了文件的上传管理。下面通过具体步骤讲解 SpringMVC 的操作实现。

1.【mldnspring 项目】在父 pom.xml 文件中配置上传组件依赖管理。

属性配置	<fileupload.version>1.3.3</fileupload.version>

	<commons-io.version>2.6</commons-io.version>
依赖库配置	<dependency> <groupId>commons-fileupload</groupId> <artifactId>commons-fileupload</artifactId> <version>${fileupload.version}</version> </dependency> <dependency> <groupId>commons-io</groupId> <artifactId>commons-io</artifactId> <version>${commons-io.version}</version> </dependency>

2.【mldnspring-mvc 项目】修改子模块 pom.xml 配置文件，引入上传依赖库。

```
<dependency>
    <groupId>commons-fileupload</groupId>
    <artifactId>commons-fileupload</artifactId>
</dependency>
<dependency>
    <groupId>commons-io</groupId>
    <artifactId>commons-io</artifactId>
</dependency>
```

3.【mldnspirng-mvc 项目】SpringMVC 中，上传依赖 CommonsMultipartResolver 类进行处理，所以首先要在 spring-mvc.xml 配置文件中进行定义。

```
<!-- 配置FileUpload组件在Spring中整合的处理类 -->
<bean id="multipartResolver"
    class="org.springframework.web.multipart.commons.CommonsMultipartResolver">
    <!-- 定义上传文件的最大的字节大小，这里为5MB -->
    <property name="maxUploadSize" value="5242880"/>
    <!-- 定义上传文件允许的最大字节数，这里为2MB -->
    <property name="maxUploadSizePerFile" value="2097152"/>
    <!-- 定义内存中允许占用的最大字节数 -->
    <property name="maxInMemorySize" value="10485760"/>
</bean>
```

4.【mldnspirng-mvc 项目】由于定义了上传限制，所以一旦上传出现问题，应跳转到错误页。

```
<!-- 配置错误的映射处理，出现了指定的异常后，可以让其跳转到错误页
（/WEB-INF/pages/errors.jsp） -->
<bean class="org.springframework.web.servlet.handler.SimpleMappingExceptionResolver">
    <property name="exceptionMappings"><!-- 进行整体的错误映射 -->
        <props>   <!-- 定义错误类型 -->
            <prop key="org.springframework.web.multipart.MaxUploadSizeExceededException">
                errors
```

```
                </prop>
            </props>
        </property>
    </bean>
```

项目中有视图解析机制，所以错误页/WEB-INF/pages/errors.jsp 只需要配置 errors 即可（不需要编写前缀与后缀）。

5.【mldnspirng-mvc 项目】定义文件上传表单页面/pages/photo/photo_upload.jsp。

```
<%@ page pageEncoding="UTF-8"%>
<%
    request.setCharacterEncoding("UTF-8") ;
    String basePath = request.getScheme() + "://" + request.getServerName() + ":" +
            request.getServerPort() + request.getContextPath() + "/";
    String photo_input_url = basePath + "pages/photo/upload.action" ;
%>
<html>
<head>
    <base href="<%=basePath%>" />
</head>
<body>
<form action="<%=photo_input_url%>" method="post" enctype="multipart/form-data">
    输入消息：<input type="text" name="msg" id="msg" value="www.mldn.cn"><br>
    上传图片：<input type="file" name="photo" id="photo"><br>
    <input type="submit" value="发送">
</form>
</body>
</html>
```

6.【mldnspirng-mvc 项目】定义 UploadAction 类，实现上传处理。文件的保存路径为/WEB-INF/upload。

```
package cn.mldn.mldnspring.action;
import java.io.File;
import java.util.HashMap;
import java.util.Map;
import java.util.UUID;
import org.springframework.stereotype.Controller;
import org.springframework.web.bind.annotation.GetMapping;
import org.springframework.web.bind.annotation.PostMapping;
import org.springframework.web.bind.annotation.RequestMapping;
import org.springframework.web.bind.annotation.ResponseBody;
import org.springframework.web.context.ContextLoader;
import org.springframework.web.multipart.MultipartFile;
```

```
@Controller                                                    // 定义控制器
@RequestMapping("/pages/photo/*")                  // 访问父路径，与方法中的路径进行组合为完整路径
public class UploadAction {                                    // 自定义Action程序类
    @PostMapping("/upload")
    @ResponseBody                                              // 使用Restful风格返回
    public Object upload(String msg, MultipartFile photo) throws Exception {
        Map<String,Object> result = new HashMap<String,Object>() ;        // 保存结果
        result.put("msg", msg) ;                                          // 保存信息
        if (photo == null || photo.isEmpty()) {
            result.put("photo-name", "nophoto") ;                         // 没有文件，保存默认名称
        } else {    // 现在有上传文件
            String photoName = UUID.randomUUID() + "." + photo.getContentType()
                .substring(photo.getContentType().lastIndexOf("/") + 1) ; // 创建保存文件名称
            String photoPath = ContextLoader.getCurrentWebApplicationContext()
                .getServletContext().getRealPath("/WEB-INF/upload/") + photoName;// 保存路径
            result.put("photo-size", photo.getSize()) ;                   // 保存上传文件信息
            result.put("photo-mime", photo.getContentType()) ;            // 保存上传文件信息
            result.put("photo-name", photoName) ;                         // 保存上传文件信息
            result.put("photo-path", photoPath) ;                         // 保存上传文件信息
            photo.transferTo(new File(photoPath));                        // 进行文件转存
        }
        return result ;                                                   // 返回文件信息
    }
    @GetMapping("/upload_pre")
    public String uploadPre() {
        return "photo/photo_upload";                                      // 上传表单
    }
}
```

本程序使用 MultipartFile 实现了上传文件接收，并利用此类中提供的方法实现了上传文件的相关信息获取。为了简化理解，对于上传的处理结果，使用 Restful 风格返回了相应的文件信息与保存路径。

提示：关于上传错误处理。

本节程序实现了对上传的限制，并将项目发布到了 Tomcat 服务器中。如果此时开发者上传了一个非常大的文件，执行程序仍会出现错误（不跳转到错误页），同时会在控制台打印 org.apache.commons.fileupload.FileUploadBase$SizeLimitExceededException 异常信息。这是由 Tomcat 自身的上传限制导致的。要解决这个问题，需要取消 Tomcat 的上传限制。

范例：修改 server.xml 配置文件，取消上传限制。

```
<Connector connectionTimeout="20000" port="80"
    protocol="HTTP/1.1" redirectPort="8443" maxSwallowSize="-1"/>
```

本程序配置了 **maxSwallowSize="-1"**属性，表示不再受上传限制。此时，SpringMVC 的上传检测可以正常执行。

14.11　拦截器

拦截器是 AOP 概念的一种应用，即在用户发送请求与控制层接收请求之间追加一些拦截程序，以保证提交到 Action 中的请求数据的合法性。拦截器的运行过程如图 14-12 所示。

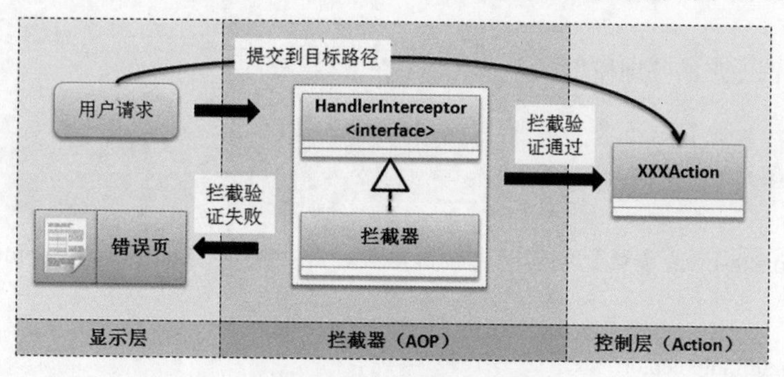

图 14-12　拦截器运行

通过图 14-12 可以发现，用户发送请求到控制层的处理过程中，所有的请求路径依然只关心控制层的目标路径。拦截器就像 Web 过滤器一样，会根据配置自动执行处理。

在 SpringMVC 中，HandlerInterceptor 是拦截器处理的标准父接口，子类只有实现此接口才可以实现拦截器功能。HandlerInterceptor 接口定义的方法如表 14-4 所示。

表 14-4　HandlerInterceptor 接口方法

编号	方法名称	类型	描述
1	public default boolean preHandle(HttpServletRequest request, HttpServletResponse response, Object handler) throws Exception	普通	在控制层处理之前执行拦截处理
2	public default void postHandle(HttpServletRequest request, HttpServletResponse response, Object handler, ModelAndView modelAndView) throws Exception	普通	控制层处理完毕但是还未回应请求，此时已经返回 ModelAndView 对象
3	public default void afterCompletion(HttpServletRequest request, HttpServletResponse response, Object handler, @Nullable Exception ex) throws Exception	普通	在请求处理完成后进行拦截

在实际项目开发中，往往需要在请求处理前进行拦截，所以 HandlerInterceptor 接口的 preHandle 方法使用得较多。

提示：Spring 版本不同，HandlerInterceptor 接口也不同。

本书使用的是 Spring 5.x 开发版本，所以 HandlerInterceptor 接口中的方法全部使用 default 进行定义。在这之前，该接口的全部方法都是抽象方法，即子类必须要全部实现。

14.11.1 定义基础拦截器

要定义拦截器，首先需要有一个完整的请求处理操作，然后需要配置拦截器的执行路径，之后才可以观察到最终的执行结果。

提示：本例的程序基础模型。

本例需要进行完整的请求与回应处理，这里使用 14.5 节的程序，即雇员数据发送处理，通过此程序观察拦截器的运用。

1.【mldnspring-mvc 项目】创建一个拦截器类，此类实现 HandlerInterceptor 接口，并覆写接口全部方法。

```
package cn.mldn.util.interceptor;
import javax.servlet.http.HttpServletRequest;
import javax.servlet.http.HttpServletResponse;
import org.slf4j.Logger;
import org.slf4j.LoggerFactory;
import org.springframework.web.servlet.HandlerInterceptor;
import org.springframework.web.servlet.ModelAndView;
public class ValidationInterceptor implements HandlerInterceptor {          // 定义拦截器
    private Logger logger = LoggerFactory.getLogger(ValidationInterceptor.class) ;
    @Override
    public boolean preHandle(HttpServletRequest request,
            HttpServletResponse response, Object handler) throws Exception {
        this.logger.info("1、preHandle方法执行，" + handler.getClass());
        return true ;    // 返回true，表示请求继续；返回false，表示不执行后续的Action或拦截器
    }
    @Override
    public void postHandle(HttpServletRequest request,   HttpServletResponse response,
            Object handler, ModelAndView modelAndView) throws Exception {
        this.logger.info("2、postHandle方法执行，" + handler.getClass() +
                "、ModelAndView = " + modelAndView);
    }
    @Override
    public void afterCompletion(HttpServletRequest request,
            HttpServletResponse response, Object handler, Exception ex) throws Exception {
```

```
        this.logger.info("3、afterCompletion方法执行，" +    handler.getClass());
    }
}
```

本程序中拦截器定义了 3 个方法，并在这 3 个方法中提供了一个重要的对象 Object handler，为了方便观察，利用反射进行了此对象所属类的打印。

2.【mldnspring-mvc 项目】拦截器需要配置拦截路径后才可以正常执行。修改 spring-mvc.xml 文件，进行配置。

```
<mvc:interceptors>                                              <!-- 配置拦截器，可以定义多个 -->
    <mvc:interceptor>
        <mvc:mapping path="/pages/**/*.action" />               <!-- 拦截路径 -->
        <bean class="cn.mldn.util.ValidationInterceptor" />     <!-- 拦截器处理类 -->
    </mvc:interceptor>
</mvc:interceptors>
```

本程序配置了拦截路径，只要访问此路径，就会自动触发 ValidationInterceptor 拦截器的执行。

3.【mldnspring-mvc 项目】启动 Web 项目，访问/pages/emp/add_pre.action 路径，拦截器的执行结果如下。

【操作分析】用户发出请求，访问**add_pre.action**之前会触发**preHandle**方法。

1. preHandle方法执行，class org.springframework.web.method.HandlerMethod

【操作分析】add_pre.action对应的控制器方法处理完毕，并且返回了**ModelAndView**，会触发**postHandle**方法。

2. postHandle方法执行，class org.springframework.web.method.HandlerMethod、ModelAndView = ModelAndView: reference to view with name 'emp/emp_add'; model is {}

【操作分析】服务器已经成功的回应了客户端请求，会触发**afterCompletion**方法。

3. afterCompletion方法执行，class org.springframework.web.method.HandlerMethod

分析可以发现，拦截器中每个拦截方法里都有一个 HandlerMethod 类对象，该类是进行拦截处理的核心。

14.11.2　HandlerMethod 类

实际开发中，为了确保进入控制器中的请求都是合法请求，定义最多的拦截器方法就是 preHandle()。该方法中提供了 3 个对象：HttpServletRequest、HttpServletResponse 和 Object（HandlerMethod）。利用 HandlerMethod 类对象，可以获取请求访问的目标程序类的相关信息。

HandlerMethod 类的常用方法如表 14-5 所示。

表 14-5　HandlerMethod 类常用方法

编号	方法名称	类型	描述
1	public Object getBean()	普通	获取处理方法的 Bean 对象（XxxAction）
2	public Method getMethod()	普通	返回指定的访问路径对应的具体方法

续表

编号	方法名称	类型	描述
3	public Class<?> getBeanType()	普通	获取处理方法所在 Bean 的 Class 对象
4	public MethodParameter[] getMethodParameters()	普通	获得要调用方法的所有参数信息

为了更好地观察表 14-5 中的方法，下面修改之前的拦截器定义。

范例：【mldnspring-mvc 项目】在拦截器中对 HandlerMethod 类对象进行信息输出。

```
package cn.mldn.util.interceptor;
import javax.servlet.http.HttpServletRequest;
import javax.servlet.http.HttpServletResponse;
import org.slf4j.Logger;
import org.slf4j.LoggerFactory;
import org.springframework.web.servlet.HandlerInterceptor;
import org.springframework.web.servlet.ModelAndView;
public class ValidationInterceptor implements HandlerInterceptor {              // 定义拦截器
    private Logger logger = LoggerFactory.getLogger(ValidationInterceptor.class) ;
    @Override
    public boolean preHandle(HttpServletRequest request,
        HttpServletResponse response, Object handler) throws Exception {
        if (handler instanceof HandlerMethod) {                                   // 执行向下转型
            HandlerMethod handlerMethod = (HandlerMethod) handler;                // 强制转换
            this.logger.info("Action对象：" + handlerMethod.getBean());
            this.logger.info("Action类：" + handlerMethod.getBeanType());
            this.logger.info("Action方法：" + handlerMethod.getMethod());
        }
        return true ;     // 返回true，表示请求继续；返回false，表示不执行后续Action或拦截器
    }
}
```

本程序定义的拦截器中只使用了 preHandle 方法。为了更好地观察 HandlerMethod 类的作用，将 emp_add.jsp 提交表单到 EmpAction.add 方法的拦截结果，如下所示。

```
【操作分析】执行的Action程序类对象信息。
Action对象：cn.mldn.mldnspring.action.EmpAction@6c55dc79
【操作分析】用户发送请求的目标Action程序类（Class对象）。
Action类：class cn.mldn.mldnspring.action.EmpAction
【操作分析】用户发送请求的目标Action程序类的方法。
Action方法：public org.springframework.web.servlet.ModelAndView
            cn.mldn.mldnspring.action.EmpAction.add(cn.mldn.mldnspring.vo.Emp)
```

可以发现，开发者使用 HandlerMethod 类处理时，可以取得目标处理 Action 类的相关反射信息。如果想实现拦截控制，可以利用这些反射对象进行处理。

14.11.3　使用拦截器实现服务端请求验证

实际开发中，拦截器最大的作用是保证用户请求的正确性。例如，在某些控制器进行参数接收时，必须保证参数的格式正确。开发中最常用的参数类型有以下几种：字符串（string）、整数（int、long）、小数（double）、日期（date）、日期时间（datetime）、验证码（rand）、字符串数组（string[]）、整数数组（int[]、long[]）。抽象出常用的数据类型后，就可以实现验证规则的定义，而有了验证规则，就可以利用拦截器实现一个可重用的服务端数据验证组件的定义。

下面将充分利用资源文件与反射的特点进行实现，拦截器验证的基本类结构关系如图 14-13 所示。

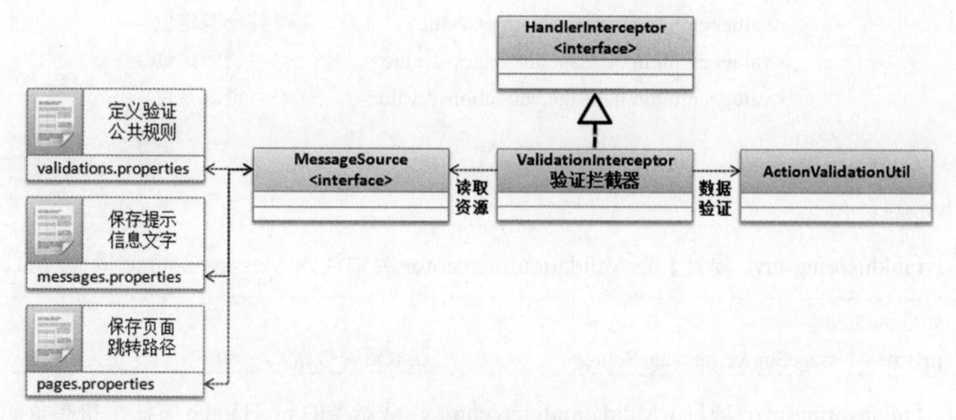

图 14-13　数据拦截器类结构关系

图 14-13 给出的设计结构中，实现的关键是 MessageSource 接口。Spring 中可以利用该接口获取资源信息。

本程序一共使用 3 个资源资源文件。

☑　validations.properties：采用“Action 名称.方法名称”的形式保存执行本方法时所使用的验证规则结构。

☑　messages.properties：保存提示信息以及验证出错后的信息。

☑　pages.properties：保存所有的跳转路径与某一个控制器拦截器验证失败后的跳转路径。

14.11.3.1　验证规则的定义与获取

拦截器中可以利用 HandlerMethod 类获取目标 Action 的反射对象（Class、Method），利用反射对象可以获取类名称与方法名称。要进行服务端数据验证，最好的做法是按照如下格式进行定义。

```
Action类名称.控制器方法名称=属性:规则|属性:规则|...
```

对于采用的验证规则，可以在程序中使用字符串（string）、整数（int、long）、小数（double）、日期（date）、日期时间（datetime）、验证码（rand）、字符串数组（string[]）、整数数组（int[]、long[]）进行处理。

1.【mldnspring-mvc 项目】创建 cn.mldn.messages.validations.properties 文件，保存验证规则。

```
cn.mldn.mldnspring.action.EmpAction.add=empno:long|ename:string|salary:double|
hiredate:date|dept.deptno:long|dept.dname:string|dept.createdate:date
```

由于在 Emp 程序处理时，addPre 方法不需要执行验证，所以这里只定义了 add 方法的请求验证规则。

2.【mldnspring-mvc 项目】修改 spring-mvc.xml 配置文件，配置资源文件。

```
<bean id="messageSource"
class="org.springframework.context.support.ResourceBundleMessageSource">
    <property name="basenames">    <!-- 定义所有的BaseName名称 -->
        <array>
            <value>cn.mldn.message.pages</value>            <!-- 资源定位 -->
            <value>cn.mldn.message.messages</value>         <!-- 资源定位 -->
            <value>cn.mldn.message.validations</value>      <!-- 资源定位 -->
        </array>
    </property>
</bean>
```

3.【mldnspring-mvc 项目】在 ValidationInterceptor 类中注入 MessageSource 接口对象。

```
@Autowired
private MessageSource messageSource ;              // 获取资源数据
```

4.【mldnspring-mvc 项目】ValidationInterceptor 拦截器类的 preHandle 方法里拼凑资源 key，以获取资源数据。

```
@Override
public boolean preHandle(HttpServletRequest request, HttpServletResponse response,
        Object handler) throws Exception {
    if (handler instanceof HandlerMethod) {      // 执行向下转型前应先判断是否是指定类的实例
        HandlerMethod handlerMethod = (HandlerMethod) handler; // 强制转换
        String validationRuleKey = handlerMethod.getBeanType().getName() + "." +
            handlerMethod.getMethod().getName() ;
        String validationRule = null ;           // 保存要读取指定的资源key对应的验证规则
        try {                                    // 如果指定key不存在，表示现在不需要验证
            validationRule = this.messageSource.getMessage(validationRuleKey, null, null) ;
        } catch (Exception e) {}
        if (validationRule != null) {            // 验证处理操作，则需要进行验证处理
            this.logger.info("【验证规则 - {"+request.getRequestURI()+"}】" +
                validationRule);
        }
    }
    return true ;    // 返回true表示请求继续，返回false表示不执行后续Action或拦截器
}
```

5.【mldnspring-mvc 项目】启动 Web 程序，由于只针对 add 方法绑定了验证规则，所以只

有在访问 add.action 路径时才会出现如下提示信息。

```
【验证规则 - {/mldnspring-mvc/pages/emp/add.action}】
empno:long|ename:string|salary:double|hiredate:date|dept.deptno:long|dept.dname:string|dept.createdate:date
```

这样就实现了资源文件保存验证规则的处理逻辑，而程序编写完成后，开发者在以后的开发过程中只需要清楚验证规则的配置，即可轻松实现服务端数据验证处理。

14.11.3.2 数据验证处理

数据验证处理主要是针对用户的请求参数发出的内容进行检测，在 validations.properties 文件中已经定义了所有可能提交到控制器中的参数信息，而此时程序只需要基于正则表达式实现验证处理即可。

1.【mldnspring-mvc 项目】在 messages.properties 配置文件中定义验证错误提示信息。

```
# ***************【验证规则信息开始】ValiodationInterceptor拦截器使用 ***************
validation.string.msg=该请求参数的内容不允许为空！
validation.int.msg=该请求参数的内容必须为整数！
validation.long.msg=该请求参数的内容必须为整数！
validation.double.msg=该请求参数的内容必须为小数！
validation.date.msg=该请求参数的内容必须为日期格式（yyyy-MM-dd）！
validation.datetime.msg=该请求参数的内容必须为日期时间格式（yyyy-MM-dd HH:mm:ss）！
validation.rand.msg=验证码输入错误，请核实后重新输入！
validation.string[].msg=该请求参数的内容不允许为空！
validation.int[].msg=该请求参数的内容必须为整数！
validation.long[].msg=该请求参数的内容必须为整数！
# ***************【验证规则信息结束】ValiodationInterceptor拦截器使用 ***************
```

2.【mldnspring-mvc 项目】验证错误后，需要指定错误页。修改 pages.properties 资源文件，定义错误页路径。

```
# ********* 定义公共错误页的跳转路径 **********
error.page=/pages/error.action
# ********* EmpAction相关跳转页面路径（验证失败路径） **********
cn.mldn.mldnspring.action.EmpAction.add.error.page=/pages/emp/add_pre.action
```

在本配置文件中，定义了两个错误页的信息。

☑ 某一控制层处理指定错误页，格式为“控制器类名称.方法名称.error.page=路径”，优先级高。

☑ 公共错误页，格式为“error.page=路径”，优先级低（在不配置具体错误页时生效）。

3.【mldnspring-mvc 项目】建立一个公共处理的 Action 控制器类，实现公共页面跳转配置。

```
package cn.mldn.mldnspring.action;
import org.springframework.stereotype.Controller;
import org.springframework.web.bind.annotation.RequestMapping;
@Controller                                                    // 定义控制器
```

```
public class CommonAction {                              // 公共Action
    @RequestMapping("/pages/error")                      // 访问路径
    public String error() {
        return "errors";                                 // 跳转到/WEB-INF/pages/errors.jsp
    }
}
```

此时开发者可以自行定义 errors.jsp 页面的错误提示信息。

4.【mldnspring-mvc 项目】为了减少拦截器中的操作代码，可以建立一个 ActionValidationUtil 验证类，该类主要针对各种给定的验证规则进行正则检测，同时将所有错误的验证信息保存在 Map 集合中。Map 集合的 key 为参数名称，value 就是 messages.properties 中保存的信息。

```
package cn.mldn.util.validate;
import java.util.HashMap;
import java.util.Map;
import javax.servlet.http.HttpServletRequest;
import org.slf4j.Logger;
import org.slf4j.LoggerFactory;
import org.springframework.context.MessageSource;
/**
 * 实现Action数据的验证处理类
 * @author 李兴华
 */
public class ActionValidationUtil {
    private Logger logger = LoggerFactory.getLogger(ActionValidationUtil.class); // 日志组件
    private Map<String,String> errors = new HashMap<String,String>() ;           // 保存错误信息
    private String rule ;                                                         // 保存验证规则
    private HttpServletRequest request ;                                          // 请求对象
    private MessageSource messageSource ;                                         // 读取资源文件
    /**
     * 实例化Action数据验证工具类对象，在此类中可以直接实现数据验证以及错误信息保存
     * @param rule 要执行的数据验证规则
     * @param request 通过该参数可以取得用户的请求参数
     * @param messageSource 所有的消息资源的文字提示信息
     */
    public ActionValidationUtil(String rule, HttpServletRequest request,
            MessageSource messageSource) {
        this.rule = rule ;                                        // 保存规则
        this.request = request ;                                  // 保存request对象
        this.messageSource = messageSource ;                      // 保存MessageSource对象
        this.handleValidator();                                   // 构造方法进行验证操作
    }
    /**
```

```
 * 实现验证的具体操作，根据指定的验证规则来获取验证数据以实现各个数据的检测处理
 */
private void handleValidator() {
    String rulesResult [] = this.rule.split("\\|") ;                    // 验证规则拆分
    for (int x = 0 ; x < rulesResult.length ; x ++) {                     // 获取全部参数
        // 第一个元素为请求参数、第二个元素为验证规则
        String temp [] = rulesResult[x].split(":") ;                      // 获取每一个规则
        try {
            String paramterValue = this.request.getParameter(temp[0]) ;   // 根据参数获取数据
            switch(temp[1]) {                                             // 验证处理操作
                case "int" : {                                            // int规则
                    if (!this.validateInt(paramterValue)) {               // 没有验证通过
                        this.errors.put(temp[0], this.messageSource
                                .getMessage("validation.int.msg",null,null)) ;
                    }
                    break ;
                }
                case "string" : {                                         // string规则
                    if (!this.validateString(paramterValue)) {            // 没有验证通过
                        this.errors.put(temp[0], this.messageSource
                                .getMessage("validation.string.msg",null,null)) ;
                    }
                    break ;
                }
                case "double" : {                                         // double规则
                    if (!this.validateDouble(paramterValue)) {            // 没有验证通过
                        this.errors.put(temp[0], this.messageSource
                                .getMessage("validation.double.msg",null,null)) ;
                    }
                    break ;
                }
                case "long" : {                                           // long规则
                    if (!this.validateLong(paramterValue)) {              // 没有验证通过
                        this.errors.put(temp[0], this.messageSource
                                .getMessage("validation.long.msg",null,null)) ;
                    }
                    break ;
                }
                case "date" : {                                           // date规则
                    if (!this.validateDate(paramterValue)) {              // 没有验证通过
                        this.errors.put(temp[0], this.messageSource
                                .getMessage("validation.date.msg",null,null)) ;
                    }
```

```
                    break ;
                }
                case "datetime" : {                            // datetime规则
                    if (!this.validateDatetime(paramterValue)) {    // 没有验证通过
                        this.errors.put(temp[0], this.messageSource
                                .getMessage("validation.datetime.msg",null,null)) ;
                    }
                    break ;
                }
                case "rand" : {                                // rand规则
                    if (!this.validateRand(paramterValue)) {        // 没有验证通过
                        this.errors.put(temp[0], this.messageSource
                                .getMessage("validation.rand.msg",null,null)) ;
                    }
                    break ;
                }
                case "string[]" : {                            // string[]规则
                    if (!this.validateStringArray(
                        this.request.getParameterValues(temp[0]))) { // 没有验证通过
                        this.errors.put(temp[0], this.messageSource
                                .getMessage("validation.string[].msg",null,null)) ;
                    }
                    break ;
                }
                case "long[]" : {                              // long[]规则
                    if (!this.validateLongArray(
                        this.request.getParameterValues(temp[0]))) { // 没有验证通过
                        this.errors.put(temp[0], this.messageSource
                                .getMessage("validation.long[].msg",null,null)) ;
                    }
                    break ;
                }
                case "int[]" : {                               // int[]规则
                    if (!this.validateIntArray(
                        this.request.getParameterValues(temp[0]))) { // 没有验证通过
                        this.errors.put(temp[0], this.messageSource
                                .getMessage("validation.int[].msg",null,null)) ;
                    }
                    break ;
                }
            }
        } catch (Exception e) {
            this.logger.error(e.toString());
```

```
            }
        }
    }
    /**
     * 验证字符串的数据是否为空（null和""）
     * @param str 要验证的字符串数据
     * @return 如果不为空返回true，为空返回false
     */
    private boolean validateString(String str) {
        if (str == null || "".equals(str)) {                              // 数据是否为空
            return false ;                                                // 验证失败
        }
        return true ;                                                     // 验证成功
    }
    /**
     * 验证字符串的数据是否为空（null和""）
     * @param str 要验证的字符串数据
     * @return 如果不为空返回true，为空返回false
     */
    private boolean validateStringArray(String str[]) {
        if (str == null || str.length == 0) {                             // 数据是否为空
            return false ;                                                // 验证失败
        } else {                                                          // 验证内容是否为空
            for (int x = 0 ; x < str.length ; x ++) {
                if (str[x] == null || "".equals(str[x])) {                // 检测每一个数据
                    return false ;
                }
            }
        }
        return true ;                                                     // 验证成功
    }

    /**
     * 验证指定的字符串是否由数字所组成
     * @param str 字符串
     * @return 如果全部由数字所组成返回true
     */
    private boolean validateInt(String str) {
        if (this.validateString(str)) {                                   // 检验是否为空
            return str.matches("\\d+") ;                                  // 检测是否为数字
        }
        return false ;                                                    // 验证失败
    }
```

```
/**
 * 验证指定的字符串是否由数字所组成
 * @param str 字符串
 * @return 如果全部由数字所组成返回true
 */
private boolean validateLong(String str) {
    if (this.validateString(str)) {                              // 检验是否为空
        return str.matches("\\d+") ;                             // 检测是否为数字
    }
    return false ;                                               // 验证失败
}
/**
 * 验证指定的字符串是否由数字所组成
 * @param str 字符串
 * @return 如果全部由数字所组成返回true
 */
private boolean validateLongArray(String str[]) {
    if (this.validateStringArray(str)) {                         // 检验是否为空
        for (int x = 0 ; x < str.length ; x ++) {
            if (this.validateString(str[x])) {
                if (!str[x].matches("\\d+")) {                   // 没有验证通过
                    return false ;                               // 验证失败
                }
            } else {                                             // 有内容为空
                return false ;
            }
        }
    }
    return false ;                                               // 验证失败
}
/**
 * 验证指定的字符串是否由数字所组成
 * @param str 字符串
 * @return 如果全部由数字所组成返回true
 */
private boolean validateIntArray(String str[]) {
    if (this.validateStringArray(str)) {                         // 检验是否为空
        for (int x = 0 ; x < str.length ; x ++) {
            if (this.validateString(str[x])) {
                if (!str[x].matches("\\d+")) {                   // 未通过验证
                    return false ;                               // 验证失败
                }
            } else {                                             // 有内容为空
```

```
                    return false ;                          // 验证失败
                }
            }
        }
        return false ;
    }
    /**
     * 验证指定的字符串是否由数字所组成
     * @param str 字符串
     * @return 如果全部由数字所组成返回true
     */
    private boolean validateDouble(String str) {
        if (this.validateString(str)) {                     // 检验是否为空
            return str.matches("\\d+(\\.\\d+)") ;           // 正则验证
        }
        return false ;                                      // 验证失败
    }
    /**
     * 验证指定的字符串是否为日期格式
     * @param str 字符串
     * @return 如果全部由数字所组成返回true
     */
    private boolean validateDate(String str) {
        if (this.validateString(str)) {                     // 检验是否为空
            return str.matches("\\d{4}-\\d{2}-\\d{2}") ;    // 正则验证
        }
        return false ;
    }
    /**
     * 验证指定的字符串是否为日期时间格式
     * @param str 字符串
     * @return 如果全部由数字所组成返回true
     */
    private boolean validateDatetime(String str) {
        if (this.validateString(str)) {                     // 检验是否为空
            return str.matches("\\d{4}-\\d{2}-\\d{2} \\d{2}:\\d{2}:\\d{2}") ;// 正则验证
        }
        return false ;                                      // 验证失败
    }
    /**
     * 验证指定的字符串是与指定的验证码相符合
     * @param str 字符串
     * @return 如果全部由数字所组成返回true
```

```
     */
    private boolean validateRand(String str) {
        String rand = (String) this.request.getSession().getAttribute("rand") ; // 获取验证码
        if (this.validateString(str) && this.validateString(rand)) {        // 检验是否为空
            return str.equalsIgnoreCase(rand) ;                             // 验证码检测
        }
        return false ;                                                      // 验证失败
    }
    /**
     * 获取全部的错误信息，如果没有错误则集合的长度为0
     * @return 错误内容
     */
    public Map<String, String> getErrors() {                                // 返回错误信息
        return errors;
    }
}
```

5.【mldnspring-mvc 项目】在 ValidationInterceptor 拦截器中使用 ActionValidationUtil 进行验证处理。

```
    @Override
    public boolean preHandle(HttpServletRequest request, HttpServletResponse response,
        Object handler) throws Exception {
        if (handler instanceof HandlerMethod) { // 向下转型前应先判断是否是指定类的实例
            HandlerMethod handlerMethod = (HandlerMethod) handler;   // 强制转换
            String validationRuleKey = handlerMethod.getBeanType().getName() + "." +
                    handlerMethod.getMethod().getName() ;
            String validationRule = null ; // 保存要读取指定资源key对应的验证规则
            try {                         // 指定key不存在，则不需要验证
                validationRule = this.messageSource.getMessage(validationRuleKey, null, null) ;
            } catch (Exception e) {}
            if (validationRule != null) { // 验证处理操作，需要进行验证处理
                this.logger.info("【验证规则 - {"+request.getRequestURI()+"}】" +
                    validationRule);
                ActionValidationUtil avu = new ActionValidationUtil(
                    validationRule, request, this.messageSource) ;
                if (avu.getErrors().size() > 0) {                    // 如果有错误信息
                    request.setAttribute("errors", avu.getErrors());  // 保存错误信息
                    String errorPage = null ;                          // 错误页
                    try {                                              // 获取当前访问错误页
                        errorPage = this.messageSource.getMessage(validationRuleKey +
                            ".error.page", null, null) ;
                    } catch (Exception e) {        // 如果无指定路径，跳转到公共errorPage
                        errorPage = this.messageSource.getMessage("error.page", null, null) ;
```

```
                    }                                           // 跳转到错误页
                    request.getRequestDispatcher(errorPage).forward(request, response);
                    return false ;                              // 请求拦截
                }
            }
        }
        return true ;   // 返回true，表示请求继续；返回false，表示不执行后续Action或拦截器
    }
```

本程序实现了服务端验证操作，利用获取的验证规则与 ActionValidationUtil 类检测后，如果发现没有错误信息则表示验证通过，如果发现有错误信息则会按照配置跳转到错误页（可以是具体的某一个错误页或公共错误页）。

14.11.3.3　验证上传文件类型

在进行验证时，除了要针对提交参数验证，也有可能需要针对上传文件进行验证处理。例如，在进行上传时只允许接收图片文件，对于此类的操作也应该在拦截器中进行处理。

在使用 preHandle 方法处理 Action 请求前，可以在此方法中获取 HttpServletRequest 接口对象。但是对于这时的接口对象，如果用户请求没有进行表单封装，那么采用的是 Tomcat 实现类 org.apache.catalina.connector.RequestFacade 实例化。而如果有表单封装，则使用的是 org.springframework.web.multipart.support.DefaultMultipartHttpServletRequest 类进行实例化。此时可以在拦截器中利用当前 request 对象是否为 DefaultMultipartHttpServletRequest 类型，来实现是否有表单封装的判断，从而可以进一步获取上传文件的 MIME 类型数据与既定的规则进行验证。

1.【mldnspring-mvc 项目】修改 validations.properties 文件，定义公共 MIME 与 Action 指定 MIME 规则。

公共MIME规则	# 定义公共的验证规则配置
	mime.rule=image/bmp|image/png|image/jpg|image/jpeg|image/gif

Action指定MIME规则	# 为某一个Action处理请求单独定义MIME规则
	cn.mldn.mldnspring.action.EchoAction.add.mime.rule=image/png|image/bmp

在定义的两个 MIME 验证规则里面，公共的 MIME 规则优先级会比具体的 MIME 规则低。

2.【mldnspring-mvc 项目】当上传文件违反了 MIME 规则后应该显示错误，修改 messages.properties 资源文件，追加 MIME 验证失败的提示信息。

```
validation.mime.msg=该上传文件不符合上传规则，所以无法进行接收处理！
```

3.【mldnspring-mvc 项目】建立 ActionMIMEValidationUtil 类型验证工具类，检测上传文件类型。

```
package cn.mldn.util.validate;
import java.util.HashMap;
import java.util.Iterator;
import java.util.Map;
import javax.servlet.http.HttpServletRequest;
```

```
import org.slf4j.Logger;
import org.slf4j.LoggerFactory;
import org.springframework.context.MessageSource;
import org.springframework.web.multipart.MultipartFile;
import org.springframework.web.multipart.MultipartHttpServletRequest;
import org.springframework.web.multipart.support.DefaultMultipartHttpServletRequest;
/**
 * 实现Action上传文件验证的处理规则
 * @author 李兴华
 */
public class ActionMIMEValidationUtil {
    private Logger logger = LoggerFactory.getLogger(ActionMIMEValidationUtil.class);
    private Map<String,String> errors = new HashMap<String,String>() ;    // 保存错误信息
    private String rule ;                                                 // 保存验证规则
    private MultipartHttpServletRequest request ;                         // 请求对象
    private MessageSource messageSource ;
    /**
     * 实例化Action数据验证工具类对象，实现数据验证及错误信息保存
     * @param rule 要执行的数据验证规则
     * @param request 用户的请求参数
     * @param messageSource 消息资源的文字提示信息
     */
    public ActionMIMEValidationUtil(String rule, HttpServletRequest request,
        MessageSource messageSource) {
        this.rule = rule ;
        this.messageSource = messageSource ;
        if (request instanceof DefaultMultipartHttpServletRequest) {
            this.request = (MultipartHttpServletRequest) request ;        // 包含所有上传信息
            this.handleValidator();                                       // 构造方法里验证
        }
    }
    /**
     * 实现验证的具体操作，根据指定验证规则获取验证数据，实现各个数据的检测处理
     */
    private void handleValidator() {
        String rulesResult [] = this.rule.split("\\|") ;                  // 验证规则拆分
        for (int x = 0 ; x < rulesResult.length ; x ++) {
            try {
                Map<String, MultipartFile> fileMap = this.request.getFileMap() ;   // 接收文件
                if (fileMap.size() > 0) {                                 // 确定有上传内容
                    Iterator<Map.Entry<String, MultipartFile>> iter =
                        fileMap.entrySet().iterator() ;
                    while (iter.hasNext()) {                              // 迭代上传文件
```

```
                    Map.Entry<String, MultipartFile> me = iter.next() ;
                    if (me.getValue().getSize() > 0) {                      // 有文件上传
                        if (!this.validateMime(me.getValue()
                                .getContentType(), this.rule)) {            // 不符合规则
                            this.errors.put(me.getKey(), this.messageSource
                                    .getMessage("validation.mime.msg",null,null)) ;
                        }
                    }
                }
            }
        } catch (Exception e) {
            this.logger.error(e.toString());
        }
    }
}
/**
 * 如果当前传递MIME类型符合定义范围，则表示允许上传
 * @param mime 当前传递文件的规则
 * @param mimeRule 所有满足的验证规则
 * @return 如果验证通过返回true，否则返回false
 */
private boolean validateMime(String mime,String mimeRule) {
    if (mime == null || "".equals(mime)) {                           // 上传文件是否有MIME类型
        return false ;                                               // 验证失败
    }
    String rules [] = mimeRule.split("\\|") ;                        // 拆分规则
    for (int x = 0 ; x < rules.length ; x ++) {                      // 检测规则
        if (mime.equals(rules[x])) {
            return true ;                                            // 验证通过
        }
    }
    return false ;                                                   // 验证失败
}
/**
 * 获取全部的错误信息，如果没有错误则集合的长度为0
 * @return 错误内容
 */
public Map<String, String> getErrors() {                             // 获取错误信息
    return errors;
}
}
```

4.【mldnspring-mvc 项目】修改 ValidationInterceptor 程序类，在基本数据验证通过后再进行

上传文件类型验证。

```
@Override
public boolean preHandle(HttpServletRequest request, HttpServletResponse response,
        Object handler) throws Exception {
    if (handler instanceof HandlerMethod) { // 向下转型前，应先判断是否是指定类的实例
        HandlerMethod handlerMethod = (HandlerMethod) handler;   // 强制转换
        String validationRuleKey = handlerMethod.getBeanType().getName() + "." +
            handlerMethod.getMethod().getName() ;
        String validationRule = null ;       // 保存要读取指定资源key对应的验证规则
        try {                                // 如果指定key不存在，表示暂时不需要验证
            validationRule = this.messageSource.getMessage(validationRuleKey, null, null) ;
        } catch (Exception e) {}
        if (validationRule != null) {        // 验证处理操作，则需要进行验证处理
            this.logger.info("【验证规则 - {"+request.getRequestURI()+"}】" +
                validationRule);
            ActionValidationUtil avu = new ActionValidationUtil(validationRule,
                request, this.messageSource) ;
            String errorPage = null ;        // 错误页
            try {                            // 获取错误页
                errorPage = this.messageSource.getMessage(validationRuleKey +
                        ".error.page", null, null) ;
            } catch (Exception e) {          // 如果没有指定路径，则跳转到公共的errorPage
                errorPage = this.messageSource.getMessage("error.page", null, null) ;
            }
            if (avu.getErrors().size() > 0) {                    // 现在有错误信息
                request.setAttribute("errors", avu.getErrors());   // 保存错误信息
                request.getRequestDispatcher(errorPage).forward(request, response);
                return false ;                                   // 请求拦截
            } else {
                if (request instanceof DefaultMultipartHttpServletRequest) {// 有上传
                    String mimeRule = null; ;
                    try {            // 获取文件规则，如果没有，则使用公共规则
                        mimeRule = this.messageSource.getMessage(validationRuleKey +
                            ".mime.rule", null,null) ;
                    } catch (Exception e) {
                        mimeRule = this.messageSource.getMessage("mime.rule", null,null) ;
                    }
                    ActionMIMEValidationUtil amvu = new ActionMIMEValidationUtil(
                            mimeRule,request,this.messageSource) ;
                    if (amvu.getErrors().size() > 0) {                 // 有错误信息
                        request.setAttribute("errors", amvu.getErrors());
                        request.getRequestDispatcher(errorPage)
```

```
                                    .forward(request, response);
                            return false ;                                    // 请求拦截
                        }
                    }
                }
            }
        }
        return true ;     // 返回true表示请求继续，返回false表示不执行后续的Action或拦截器
    }
```

本程序在之前数据验证的基础上追加了上传文件检测，在上传文件格式出现问题时将跳转到已有的错误页上，同时继续使用 errors 保存错误信息。

14.12　Spring 综合案例

清楚了 SpringMVC 的核心组成后，下面将通过一个完整案例介绍 SpringMVC 与 SpringDataJPA 的整合开发。

提示：关于本案例的学习。

本例是一个完整的综合性案例，将使用之前讲解的操作处理。由于代码较多，考虑到篇幅问题，并未做过多的统一管理设计。这里只为读者分析核心代码，完整的程序读者可通过源代码自行学习。

本例所给出的前端程序是基于 Bootstrap 完成的，因此这里已经配置好了 JS 客户端表单验证，即此时前端可以依靠 JS 验证，后端程序可以使用拦截器进行服务端验证。

本例将实现一个商品管理程序，其数据库结构如图 14-14 所示。

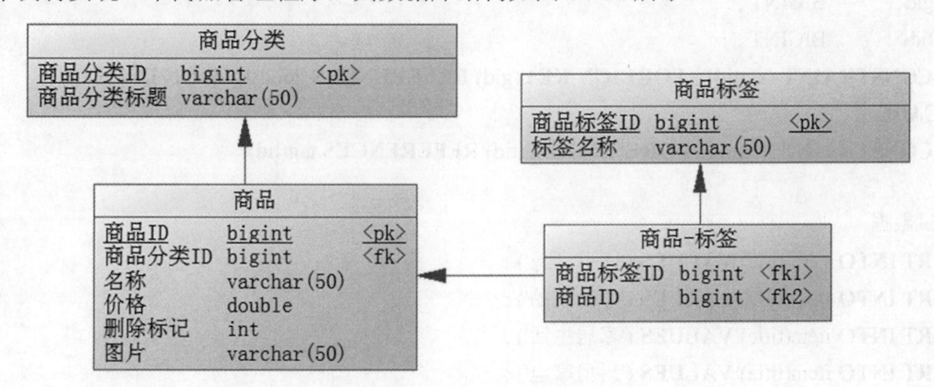

图 14-14　案例数据库结构

本程序中将存在一对多、多对多两种数据关系。

☑　一对多关联：所有的商品都对应有商品分类，一个商品分类可以保存有多个商品信息。

☑ 多对多关联：每个商品都拥有商品标签，一个标签可以属于多个商品，一个商品可以有多个标签。

范例：【mldnspring-mvc】数据库创建脚本（包含有表结构与测试数据）。

```
DROP DATABASE IF EXISTS mldn ;
CREATE DATABASE mldn CHARACTER SET UTF8 ;
USE mldn ;
CREATE TABLE item(
    iid         BIGINT AUTO_INCREMENT ,
    title       VARCHAR(50) ,
    CONSTRAINT pk_iid PRIMARY KEY(iid)
) ;
CREATE TABLE tag(
    tid         BIGINT AUTO_INCREMENT ,
    title       VARCHAR(50) ,
    CONSTRAINT pk_tid PRIMARY KEY(tid)
) ;
CREATE TABLE goods (
    gid         BIGINT AUTO_INCREMENT ,
    name        VARCHAR(50) ,
    price       DOUBLE ,
    photo       VARCHAR(100) ,
    dflag       INT ,
    iid         BIGINT ,
    CONSTRAINT pk_gid10 PRIMARY KEY(gid) ,
    CONSTRAINT fk_iid FOREIGN KEY(iid) REFERENCES item(iid)
) ;
CREATE TABLE goods_tag(
    gid         BIGINT ,
    tid         BIGINT ,
    CONSTRAINT fk_gid11 FOREIGN KEY(gid) REFERENCES goods(gid) ON DELETE
CASCADE ,
    CONSTRAINT fk_tid11 FOREIGN KEY(tid) REFERENCES tag(tid)
) ;
-- 测试数据
INSERT INTO item(title) VALUES ('图书音像') ;
INSERT INTO item(title) VALUES ('办公用品') ;
INSERT INTO item(title) VALUES ('家居生活') ;
INSERT INTO item(title) VALUES ('厨房家电') ;
INSERT INTO item(title) VALUES ('电子设备') ;
INSERT INTO tag(title) VALUES ('高端') ;
INSERT INTO tag(title) VALUES ('奢华') ;
INSERT INTO tag(title) VALUES ('性价比高') ;
```

```
INSERT INTO tag(title) VALUES ('免费') ;
INSERT INTO tag(title) VALUES ('耐用') ;
-- 提交事务
COMMIT ;
```

本数据表在进行商品信息删除时，使用了逻辑删除。如果商品表（goods 表）中的 dflag 内容为 1（表示为 true），则不再进行商品信息显示。

14.12.1 搭建项目开发环境

本例将按照标准 MVC 设计结构进行项目开发，核心接口与类定义如图 14-15 所示。

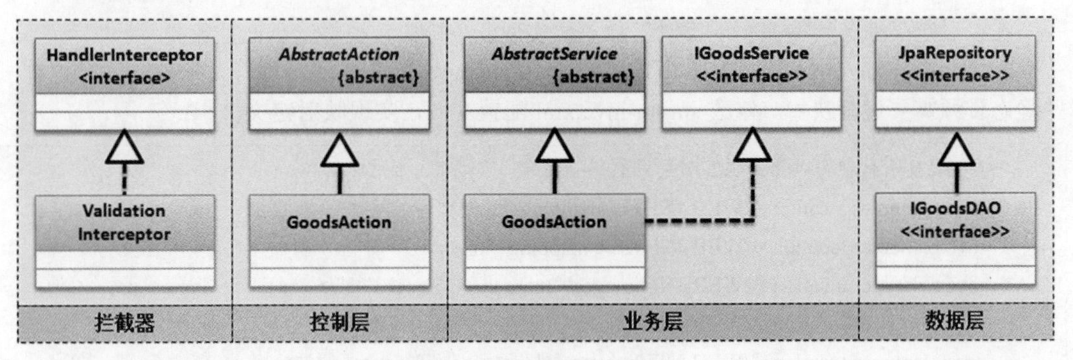

图 14-15 案例设计结构

本程序将使用之前讲解的 JPA 配置。在定义 Action 时，为了减少重复代码，创建了一个 AbstractAction 抽象类，利用此类可实现控制层方法重用。下面讲解项目环境搭建的核心过程。

1.【mldnspring-mvc 项目】定义抽象 Action 程序类。

```
package cn.mldn.util.action.abs;
import java.text.SimpleDateFormat;
import java.util.Locale;
import org.springframework.beans.factory.annotation.Autowired;
import org.springframework.beans.propertyeditors.CustomDateEditor;
import org.springframework.context.MessageSource;
import org.springframework.web.bind.WebDataBinder;
import org.springframework.web.bind.annotation.InitBinder;
/**
 * 定义公共的Action抽象类，定义Action可重用方法
 * @author 李兴华
 */
public abstract class AbstractAction {                              // 该类要被子类继承
    @InitBinder
    public void initBinder(WebDataBinder binder) {                  // 设置Web数据绑定转换
        SimpleDateFormat sdf = new SimpleDateFormat("yyyy-MM-dd") ;  // 定义转换处理
        // 在Web数据绑定中注册自定义规则绑定器，处理java.util.Date类型，允许为null
```

```
            binder.registerCustomEditor(java.util.Date.class, new CustomDateEditor(sdf, true));
        }
}
```

考虑到业务层也可能提供重复处理操作，所以本处提供了一个 AbstractService 抽象类，此类中暂不定义任何方法。

2.【mldnspring-mvc 项目】定义抽象业务实现类。

```
package cn.mldn.mldnspring.service.abs;
public abstract class AbstractService {
}
```

在以后定义业务层接口实现类时，除了要实现业务接口，也将继承此抽象类。

3.【mldnspring-mvc 项目】本项目用到了 Bootstrap、jQuery 等前端开发框架，同时也有一些静态资源需要进行映射。修改 spring-mvc.xml 配置文件，实现映射定义。

```
    <!-- 配置所有静态Web资源的映射路径 -->
    <mvc:resources location="/WEB-INF/js/" mapping="/js/**"/>
    <mvc:resources location="/WEB-INF/css/" mapping="/css/**"/>
    <mvc:resources location="/WEB-INF/images/" mapping="/images/**"/>
    <mvc:resources location="/WEB-INF/jquery/" mapping="/jquery/**"/>
    <mvc:resources location="/WEB-INF/bootstrap/" mapping="/bootstrap/**"/>
    <mvc:resources location="/WEB-INF/upload/" mapping="/upload/**"/>
```

4.【mldnspring-mvc 项目】定义 3 个持久化类，同时进行映射关系配置。

☑ 定义 Item.java 持久化类。

```
@SuppressWarnings("serial")
@Entity
public class Item implements Serializable {
    @Id
    @GeneratedValue(strategy=GenerationType.IDENTITY)
    private Long iid ;
    private String title ;
    @OneToMany(mappedBy="item")                 // 一对多关联
    private List<Goods> goodses ;
    // setter、getter、toString略
}
```

☑ 定义 Tag.java 持久化类。

```
@SuppressWarnings("serial")
@Entity
public class Tag implements Serializable {
    @Id
    @GeneratedValue(strategy=GenerationType.IDENTITY)
```

```
    private Long tid ;
    private String title ;
    @ManyToMany(mappedBy = "tags", fetch = FetchType.LAZY)
    private List<Goods> goodses ;
    // setter、getter、toString略
}
```

☑　定义 Goods.java 持久化类。

```
@SuppressWarnings("serial")
@Entity
public class Goods implements Serializable {
    @Id
    @GeneratedValue(strategy=GenerationType.IDENTITY)
    private Long gid ;
    private String name ;
    private Double price ;
    private String photo ;
    private Integer dflag ;
    @ManyToOne(fetch=FetchType.LAZY)                                    // 延迟加载
    @JoinColumn(name="iid")                                         // 设置关联字段
    private Item item ;
    @ManyToMany(fetch = FetchType.LAZY)                             // 启用延迟加载
    @JoinTable(                                                     // 描述的是一个关联表
            name="goods_tag" ,                                      // 定义中间表名称
            joinColumns = { @JoinColumn(name = "gid") }             // goods与goods_tag表的连接
            inverseJoinColumns = {
                        @JoinColumn(name = "tid") })                // 通过Goods类找到Tag类中的tid数据
    private List<Tag> tags ;                                        // 商品对应标签信息
    // setter、getter、toString略
}
```

5.【mldnspring-mvc 项目】定义数据层接口，所有的数据层接口继承 JpaRepository 父接口。

☑　定义 IItemDAO 数据层接口。

```
public interface IItemDAO extends JpaRepository<Item, Long> {}
```

☑　定义 ITagDAO 数据层接口。

```
public interface ITagDAO extends JpaRepository<Tag, Long> {}
```

☑　定义 IGoodsDAO 数据层接口。

```
public interface IGoodsDAO extends JpaRepository<Goods, Long> {}
```

基本环境与程序配置搭建完成后，就可以依据此结构逐步实现各项功能了。

14.12.2 商品信息增加页面

商品信息增加时需要提供一个表单页面，提供商品所属分类编号与商品标签信息。其中，商品分类信息要通过 item 表获取，商品标签信息要通过 tag 表获取。

1.【mldnspring-mvc 项目】定义 IGoodsService 业务接口，提供商品增加前的数据查询操作。

```
package cn.mldn.mldnspring.service;
import java.util.Map;
public interface IGoodsService {
    /**
     * 进行商品添加前的数据查询操作
     * @return 返回的数据包含有如下内容：
     *    key = allItems、value = 所有的商品分类
     *    key = allTags、value = 所有的商品标签
     */
    public Map<String,Object> preAdd() ;
}
```

2.【mldnspring-mvc 项目】建立 IGoodsService 业务接口子类 GoodsServiceImpl，该类继承自 AbstractService 父类。

```
package cn.mldn.mldnspring.service.impl;
import java.util.HashMap;
import java.util.Map;
import org.springframework.beans.factory.annotation.Autowired;
import org.springframework.stereotype.Service;
import cn.mldn.mldnspring.dao.IGoodsDAO;
import cn.mldn.mldnspring.dao.IItemDAO;
import cn.mldn.mldnspring.dao.ITagDAO;
import cn.mldn.mldnspring.service.IGoodsService;
import cn.mldn.mldnspring.service.abs.AbstractService;
@Service
public class GoodsServiceImpl extends AbstractService implements IGoodsService {
    @Autowired
    private IItemDAO itemDAO ;                              // 注入IItemDAO数据接口实例
    @Autowired
    private ITagDAO tagDAO ;                                // 注入ITagDAO数据接口实例
    @Autowired
    private IGoodsDAO goodsDAO ;                            // 注入IGoodsDAO数据接口实例
    @Override
    public Map<String, Object> preAdd() {
        Map<String, Object> map = new HashMap<String, Object>();
        map.put("allItems", this.itemDAO.findAll()) ;       // 查询全部分类信息
        map.put("allTags", this.tagDAO.findAll()) ;         // 查询全部标签信息
```

```
        return map;
    }
}
```

3.【mldnspring-mvc 项目】建立 GoodsAction 程序类，注入 IGoodsService 业务接口实例，定义商品增加前的相关数据查询，并将查询结果保存到 request 属性范围中。

```
@Controller                                                     // 定义控制器
@RequestMapping("/pages/back/admin/goods/*")                    // 父路径
public class GoodsAction extends AbstractAction {               // 自定义Action程序类
    private Logger logger = LoggerFactory.getLogger(GoodsAction.class) ;   // 日志记录
    @Autowired
    private IGoodsService goodsService ;
    public ModelAndView addPre() {
        Map<String,Object> map = this.goodsService.preAdd() ;
        this.logger.info("商品信息增加前信息查询：" + map);
        ModelAndView mav = new ModelAndView("back/admin/goods/goods_add") ;
        mav.addAllObjects(map) ;                                // 保存商品分类与标签数据
        return mav ;
    }
}
```

4.【mldnspring-mvc 项目】定义/pages/back/admin/goods/goods_add.jsp 页面，进行商品分类与商品标签信息迭代输出。

☑ 迭代输出商品分类信息。

```
<div class="form-group" id="iidDiv">
    <label class="col-md-2 control-label" for="iid">商品分类：</label>
    <div class="col-md-5">
        <select id="iid" name="iid" class="form-control">
            <option value="">========= 请选择商品所属分类 =========</option>
            <c:forEach items="${allItems}" var="item">
                <option value="${item.iid}">${item.title}</option>
            </c:forEach>
        </select>
    </div>
    <span class="col-md-5" id="iidSpan">*</span>
</div>
```

☑ 迭代输出商品标签信息。

```
<div class="form-group" id="tagDiv">
    <label class="col-md-2 control-label" for="tag">商品标签：</label>
    <div class="col-md-5">
        <c:forEach items="${allTags}" var="tag">
```

```
                <div class="col-md-3">
                        <div class="checkbox">
                                <label><input type="checkbox" id="tid" name="tid"
                                        value="${tag.tid}">${tag.title}</label>
                        </div>
                </div>
        </c:forEach>
    </div>
    <span class="col-md-5" id="tidSpan">*</span>
</div>
```

本页面依靠 IGoodsService 业务层中的 preAdd 方法进行了相关信息查询，页面运行效果如图 14-16 所示。

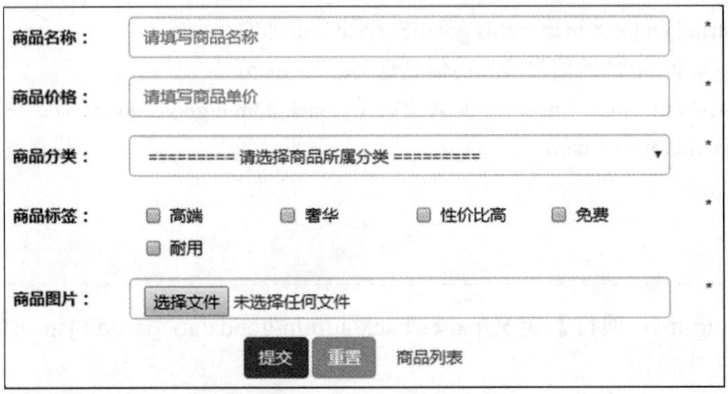

图 14-16　商品信息添加页

14.12.3　商品信息保存

商品信息需要通过表单进行输入，添加商品信息时需要上传图片，此时需要对图片类型进行验证。同时，所有上传的文件需要保存在/WEB-INF/upload 目录中，且上传的图片需要进行自动更名处理。

1.【mldnspring-mvc 项目】修改 validations.properties 资源文件，追加服务端验证规则。

```
cn.mldn.mldnspring.action.GoodsAction.add=name:string|price:double|iid:int|tid:string[]
# 定义公共的验证规则配置
mime.rule=image/bmp|image/png|image/jpg|image/jpeg|image/gif
```

2.【mldnspring-mvc 项目】在 IGoodsService 业务接口中追加商品增加方法。

```
    /**
     * 实现商品数据的追加处理，新添加商品的dflag内容为0
     * @param vo 要追加的商品信息（配置好关联关系）
     * @return 追加成功返回true，否则返回false
     */
    public boolean add(Goods vo) ;
```

3.【mldnspring-mvc 项目】在 GoodsServiceImpl 子类中覆写 add 业务方法。

```
@Override
public boolean add(Goods vo) {
    vo.setDflag(0);                                    // 新增商品信息删除标记为0
    return this.goodsDAO.save(vo).getGid() != null;    // 商品保存
}
```

4.【mldnspring-mvc 项目】表单提交数据时，分类信息只传递了 iid，标签信息只传递 tid（数组），所以需要在 GoodsAction 中进行手动处理。

```
@RequestMapping("add")
public ModelAndView add(Goods goods, long iid, String tid[],
    MultipartFile pic) throws Exception {
    ModelAndView mav = new ModelAndView(super.getMessage("forward.page")) ;   // 获取路径
    String msg = "商品数据增加失败！" ;
    String url = "/pages/back/admin/goods/add_pre.action" ;     // 显示后的跳转路径
    List<Tag> allTags = new ArrayList<Tag>() ;
    for (int x = 0 ; x < tid.length ; x ++) {                   // 将标签信息保存到集合中
        Tag tag = new Tag() ;
        tag.setTid(Long.parseLong(tid[x]));                     // 保存tid编号
        allTags.add(tag) ;
    }
    Item item = new Item() ;
    item.setIid(iid);                                           // 保存iid编号
    goods.setItem(item);                                        // 保存商品与分类关系
    goods.setTags(allTags);                                     // 保存商品与标签关系
    if (pic == null && pic.isEmpty()) {
        goods.setPhoto("nophoto.png");                          // 默认图片名称
    } else {                                                    // 创建新的图片名称
        goods.setPhoto(UUID.randomUUID() + "." + pic.getContentType()
                .substring(pic.getContentType().lastIndexOf("/") + 1) );
    }
    if (this.goodsService.add(goods)) {                         // 保存成功
        if (!(pic == null && pic.isEmpty())) {                  // 有文件上传
            String photoPath = ContextLoader.getCurrentWebApplicationContext()
                    .getServletContext().getRealPath("/WEB-INF/upload/") + goods.getPhoto();
            pic.transferTo(new File(photoPath));                // 保存上传文件
        }
        msg = "商品数据增加成功！" ;
    }
    mav.addObject("msg", msg) ;                                 // 保存提示信息
    mav.addObject("url", url) ;                                 // 保存跳转路径
    return mav ;
```

```
}
```

商品信息保存成功后，将通过 forward.jsp 页面进行信息提示，同时跳转到商品添加页面。

14.12.4 商品信息列表

保存后的商品信息需要列表显示，也就是说数据要进行分页处理，所以需要扩充 IGoodsDAO 接口中的业务方法。

1.【mldnspring-mvc 项目】修改 IGoodsDAO 接口，增加新的方法。

```
public interface IGoodsDAO extends JpaRepository<Goods, Long> {
    @Query("SELECT g FROM Goods AS g WHERE g.name LIKE :#{'%' + #keyWord + '%'}")
    public List<Goods> findSplit(@Param(value = "keyWord") String keyWord, Pageable page);
    @Query("SELECT COUNT(g) FROM Goods AS g WHERE g.name LIKE :#{'%' + #keyWord + '%'}")
    public Long getSplitCount(@Param(value = "keyWord") String keyWord) ;
}
```

由于需要分页处理，所以在定义 findSplit 方法时传入了 Pageable 接口对象。

2.【mldnspring-mvc 项目】在 IGoodsService 接口中定义分页查询方法。

```
/**
 * 查询商品信息的分页数据，如果没有查询列或查询关键字，则进行整体查询
 * @param keyWord 查询关键字
 * @param currentPage 当前页
 * @param lineSize 每页行
 * @return 返回的内容包含有如下信息：
 * 1. key = allGoods、value = 全部商品信息
 * 2. key = allRecorders、value = 统计结果
 * 3. key = allItems、value = 全部的分类信息（Map集合）
 */
public Map<String, Object> list(String keyWord, int currentPage, int lineSize) ;
```

3.【mldnspring-mvc 项目】在 GoodsServiceImpl 子类中覆写 list 方法。

```
@Override
public Map<String, Object> list(String keyWord, int currentPage, int lineSize) {
    Map<String, Object> map = new HashMap<String, Object>();
    Sort sort = new Sort(Sort.Direction.DESC, "gid");                          // 降序排列
    // 将分页与排序操作保存到Pageable接口对象中，后续可以通过DAO层调用方法，页数从0开始
    Pageable pageable = PageRequest.of(currentPage - 1, lineSize, sort);
    if (keyWord == null || "".equals(keyWord)) {                               // 查询全部操作
        Page<Goods> pageGoods = this.goodsDAO.findAll(pageable);               // 数据查询
        map.put("allRecorders", pageGoods.getTotalElements());                 // 数据统计
        map.put("allGoods", pageGoods.getContent());                           // 数据信息
    } else {                                                                   // 模糊查询
        map.put("allRecorders", this.goodsDAO.getSplitCount(keyWord));         // 数据统计
```

```
        map.put("allGoods", this.goodsDAO.findSplit(keyWord,pageable)); // 数据信息
    }
    Map<Long, String> itemMap = new HashMap<Long, String>();          // 保存分类信息
    this.itemDAO.findAll().forEach((item)->{
        itemMap.put(item.getIid(), item.getTitle()) ;                  // 保存item信息
    });
    map.put("allItems", itemMap) ;                                     // 商品分类
    return map;
}
```

4.【mldnspring-mvc 项目】为了方便处理分页，可以定义一个分页管理类，利用此类处理分页参数并可通过 request 属性将分页信息保存到 JSP 页面。

```
package cn.mldn.util.web;
import javax.servlet.http.HttpServletRequest;
import org.springframework.web.context.request.RequestContextHolder;
import org.springframework.web.context.request.ServletRequestAttributes;
/**
 * 分页的参数处理操作
 * @author 李兴华
 */
public class SplitPageUtil {
    private int currentPage = 1;                               // 参数：cp
    private int lineSize = 5;                                  // 参数：ls
    private String keyWord;                                    // 参数：kw
    private HttpServletRequest request;                        // request对象
    /**
     * 将需要进行模糊查询的columnData（下拉框）传递到组件中，目的是为了属性操作
     * @param handleUrl 设置分页路径
     */
    public SplitPageUtil(String handleUrl) {
        this.request = ((ServletRequestAttributes) RequestContextHolder
                            .getRequestAttributes()).getRequest();
        this.request.setAttribute("handleUrl", handleUrlKey);
        try {                                                  // 接收当前页码
            this.currentPage = Integer.parseInt(this.request.getParameter("cp"));
        } catch (Exception e) { }
        try {                                                  // 接收每页显示的数据行数
            this.lineSize = Integer.parseInt(this.request.getParameter("ls"));
        } catch (Exception e) { }
        this.keyWord = this.request.getParameter("kw");        // 接收关键字
        if (this.keyWord == null) {
            this.keyWord = "";
        }
```

```
        this.request.setAttribute("currentPage", this.currentPage);
        this.request.setAttribute("lineSize", this.lineSize);
        this.request.setAttribute("keyWord", this.keyWord);
    }
    public int getCurrentPage() {                                   // 得到当前页码
        return currentPage;
    }
    public int getLineSize() {                                      // 得到每页显示的数据行数
        return lineSize;
    }
    public String getKeyWord() {                                    // 得到关键字
        return keyWord;
    }
}
```

5.【mldnspring-mvc 项目】扩充 GoodsAction 类中的方法，进行分页处理。

```
    @RequestMapping("list")
    public ModelAndView list() {
        ModelAndView mav = new ModelAndView("back/admin/goods/goods_list") ;
        SplitPageUtil spu = new SplitPageUtil("/pages/back/admin/goods/list.action") ;
        mav.addAllObjects(this.goodsService.list(spu.getKeyWord(),
                spu.getCurrentPage(), spu.getLineSize())) ;
        return mav ;
    }
```

6.【mldnspring-mvc 项目】在/pages/back/admin/goods/goods_list.jsp 页面中迭代输出商品数据。

```
    <table class="table table-striped table-bordered table-hover">
        <tr>
            <td style="width:5%"><input type="checkbox" id="selectall"/></td>
            <td>商品名称</td>
            <td>商品单价</td>
            <td>商品分类</td>
            <td>操作</td>
        </tr>
        <c:forEach items="${allGoods}" var="goods">
        <tr>
            <td><input type="checkbox" id="gid" value="${goods.gid}"/></td>
            <td>${goods.name}</td>
            <td><fmt:formatNumber value="${goods.price}"/></td>
            <td>${allItems[goods.item.iid]}</td>
            <td><a href="<%=goods_edit_url%>?gid=${goods.gid}" class="btn btn-warning btn-xs">
                <span class="glyphicon glyphicon-pencil"></span> 编辑</a></td>
```

```
            </tr>
        </c:forEach>
    </table>
```

此时分页操作已被封装到了组件中，页面上直接利用<jsp:include>语句导入即可，页面执行效果如图 14-17 所示。

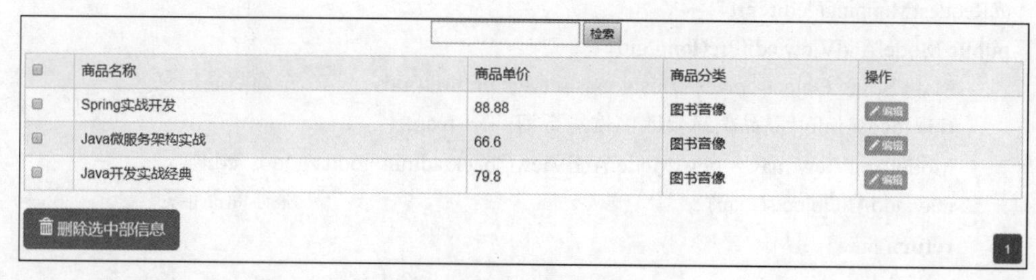

图 14-17　商品信息列表

14.12.5　商品信息编辑页面

编辑商品信息前，要先进行商品信息、分类信息、标签信息查询，并将查询结果回填到 JSP 页面。

1.【mldnspring-mvc 项目】编辑商品时，除了要查询商品相关信息外，还需要通过 goods_tag 表查询指定商品编号对应的标签编号，所以在 IGoodsDAO 接口中需要扩充一个原生 SQL 查询的数据层方法。

```
@Query(nativeQuery = true, value = "SELECT tid FROM goods_tag WHERE gid=:gid")
public List<Long> findTidByGoods(@Param(value = "gid") Long gid) ;
```

2.【mldnspring-mvc 项目】在 IGoodsService 接口中追加一个编辑前的数据查询业务方法。

```
/**
 * 商品修改前的数据查询操作
 * @return 返回的数据包含有如下内容：
 * key = allItems、value = 所有的商品分类
 * key = allTags、value = 所有的商品标签
 * key = goods、value = 要修改的商品信息
 * key = allTids、value = 商品标签
 */
public Map<String,Object> preEdit(long id) ;
```

3.【mldnspring-mvc 项目】在 GoodsServiceImpl 子类中实现 preEdit 方法。

```
@Override
public Map<String, Object> preEdit(long id) {
    Map<String, Object> map = new HashMap<String, Object>();
    map.put("allItems", this.itemDAO.findAll());                // 查询所有商品分类
    map.put("allTags", this.tagDAO.findAll());                  // 查询所有商品标签
    map.put("goods", this.goodsDAO.findById(id).get());         // 查询指定商品信息
```

```
        map.put("goodsTags", this.goodsDAO.findTidByGoods(id));   // 查询指定商品拥有的标签编号
        return map;
    }
```

4．【mldnspring-mvc 项目】在 GoodsAction 类中定义 editPre 方法，查询商品编辑前的数据。

```
@RequestMapping("edit_pre")
public ModelAndView editPre(long gid) {
    Map<String,Object> map = this.goodsService.preEdit(gid) ;     // 查询商品信息
    this.logger.info("商品信息修改前信息查询：" + map);
    ModelAndView mav = new ModelAndView("back/admin/goods/goods_edit") ;
    mav.addAllObjects(map) ;                                      // 保存商品信息
    return mav ;
}
```

5．【mldnspring-mvc 项目】在/pages/back/admin/goods/goods_edit.jsp 页面中回填表单数据，表单回填后的效果如图 14-18 所示。

图 14-18　商品信息编辑页面

本页面中，最为重要的两项数据处理是商品分类回填和商品标签回填（利用 JSTL 函数标签处理），核心代码如下。

```
<div class="form-group" id="iidDiv">
    <label class="col-md-2 control-label" for="iid">商品分类：</label>
    <div class="col-md-5">
        <select id="iid" name="iid" class="form-control">
            <option value="">========= 请选择商品所属分类 =========</option>
            <c:forEach items="${allItems}" var="item">
                <option value="${item.iid}"
                        ${item.iid==goods.item.iid?"selected":""}>${item.title}</option>
            </c:forEach>
```

```
                </select>
            </div>
            <span class="col-md-5" id="iidSpan">*</span>
        </div>
        <div class="form-group" id="tagDiv">
            <label class="col-md-2 control-label" for="tag">商品标签：</label>
            <div class="col-md-5">
                <c:forEach items="${allTags}" var="tag">
                    <div class="col-md-3">
                        <div class="checkbox">
                            <label><input type="checkbox" id="tid" name="tid" value="${tag.tid}"
                            ${fn:contains(goodsTags,tag.tid) ? "checked" : ""}>${tag.title}</label>
                        </div>
                    </div>
                </c:forEach>
            </div>
            <span class="col-md-5" id="tidSpan">*</span>
        </div>
```

14.12.6　商品信息更新

在商品编辑页面修改完商品信息后，需要更新商品信息。利用 JPA 中提供的数据关联关系，可以实现 goods_tag 表中的关联数据自动更新。商品图片如果不需要更新，则不需要重新上传图片；如果需要更新，可使用原始文件名称进行文件覆盖。

1.【mldnspring-mvc 项目】在 IGoodsService 业务接口中追加数据修改方法。

```
/**
 * 商品数据的修改处理
 * @param vo 要追加的商品信息
 * @return 修改成功返回true，否则返回false
 */
public boolean edit(Goods vo) ;
```

2.【mldnspring-mvc 项目】在 GoodsServiceImpl 子类中覆写 edit 方法。

```
@Override
public boolean edit(Goods vo) {
    this.goodsDAO.save(vo) ;
    return true ;
}
```

3.【mldnspring-mvc 项目】修改 validations.properties 配置文件，追加验证规则。

```
cn.mldn.mldnspring.action.GoodsAction.edit=
gid:long|name:string|price:double|iid:int|tid:string[]
```

4.【mldnspring-mvc 项目】在 GoodsAction 类中定义商品更新方法。

```
@RequestMapping("edit")
public ModelAndView edit(Goods goods, long iid, String tid[],
        MultipartFile pic) throws Exception {
    ModelAndView mav = new ModelAndView(super.getMessage("forward.page")) ;      // 跳转路径
    String msg = "商品数据修改失败！" ;
    String url = "/pages/back/admin/goods/list.action" ;                  // 显示跳转后的路径
    List<Tag> allTags = new ArrayList<Tag>() ;
    for (int x = 0 ; x < tid.length ; x ++) {                              // 将标签信息保存到集合中
        Tag tag = new Tag() ;
        tag.setTid(Long.parseLong(tid[x]));                                // 保存tid编号
        allTags.add(tag) ;
    }
    Item item = new Item() ;
    item.setIid(iid);                                                      // 保存iid编号
    goods.setItem(item);                                                   // 保存商品与分类关系
    goods.setTags(allTags);                                                // 保存商品与标签关系
    if (!pic.isEmpty()) {                                                  // 有新的文件上传
        if ("nophoto.png".equals(goods.getPhoto())) {
            goods.setPhoto(UUID.randomUUID() + "." + pic.getContentType()
                    .substring(pic.getContentType().lastIndexOf("/") + 1));
        }
    }
    if (this.goodsService.edit(goods)) {                                   // 保存成功
        if (!(pic == null && pic.isEmpty())) {                             // 有文件上传
            String photoPath = ContextLoader.getCurrentWebApplicationContext()
                    .getServletContext().getRealPath("/WEB-INF/upload/") + goods.getPhoto();
            pic.transferTo(new File(photoPath));                           // 保存上传文件
        }
        msg = "商品数据修改成功！" ;
    }
    mav.addObject("msg", msg) ;                                            // 保存提示信息
    mav.addObject("url", url) ;                                            // 保存跳转路径
    return mav ;
}
```

本程序的基本逻辑业务与商品增加方式类似，唯一的区别在于上传文件的覆盖更新处理。

14.12.7 商品信息删除

商品信息删除采用逻辑处理，主要通过修改 goods 表中的 dflag 字段实现。删除后再对列表数据进行查询，需要追加一个查询条件（dflag=0），才可以实现删除逻辑。

1．【mldnspring-mvc 项目】在 IGoodsDAO 接口中追加一个修改 dflag 字段数据的方法。

```
@Modifying(clearAutomatically = true)
@Query("UPDATE Goods AS g SET g.dflag=:dflag WHERE gid IN :gids")
public Integer editDflag(@Param(value = "gids") Set<Long> gids,
        @Param(value = "dflag") Integer dflag) ;
```

2．【mldnspring-mvc 项目】在 IGoodsService 业务层增加删除业务方法。

```
/**
 * 商品信息的删除处理
 * @param gids 要删除的商品编号
 * @return 如果没有商品或商品删除失败，返回false
 */
public boolean remove(Set<Long> gids) ;
```

3．【mldnspring-mvc 项目】GoodsServiceImpl 子类覆写 remove 方法。

```
@Override
public boolean remove(Set<Long> gids) {
    return this.goodsDAO.editDflag(gids, 1) > 0;          // 更新dflag字段
}
```

4．【mldnspring-mvc 项目】数据删除操作是通过前端的 JS 事件传递的要删除数据 id，参数名称为 ids，所以修改 validations.properties 配置文件，定义验证规则。

```
cn.mldn.mldnspring.action.GoodsAction.delete=ids:string
```

5．【mldnspring-mvc 项目】在 GoodsAction 程序类中定义删除方法。

```
@RequestMapping("delete")
public ModelAndView delete(String ids) {
    String msg = "商品数据删除失败！" ;
    String url = "/pages/back/admin/goods/list.action" ;          // 显示跳转后的路径
    ModelAndView mav = new ModelAndView("plugins/forward") ;
    // 传递要删除的所有商品ID，多个商品ID间用“,”分隔
    String idData [] = ids.split(",") ;                           // 根据“,”拆分数据
    Set<Long> gids = new HashSet<Long>() ;                        // 保存要删除的商品ID
    for (int x = 0 ; x < idData.length ; x ++) {
        gids.add(Long.parseLong(idData[x])) ;                     // 保存商品ID
    }
    if (this.goodsService.remove(gids)) {
        msg = "商品数据删除成功！" ;
    }
    mav.addObject("msg", msg) ;                                   // 保存提示信息
    mav.addObject("url", url) ;                                   // 保存跳转路径
    return mav ;
```

```
}
```

此时程序可以实现 goods 表中的 dflag 字段更新，但是逻辑更新处理还需要考虑数据查询问题。

提示：关于 TransactionRequiredException 异常。

本程序对业务操作使用了更新处理。如果配置不当，有可能出现事务处理异常（javax.persistence.TransactionRequiredException: Executing an update/delete query）。造成此问题的原因有两个：事务配置错误，即 spring-transaction.xml 配置有问题；或是重复进行了类扫描（Spring 中使用 context 自动扫描配置<context:component-scan>）。

6.【mldnspring-mvc 项目】修改 IGoodsDAO 数据接口，定义新的数据分页查询方法（查询全部、分页查询）。

模糊查询	@Query("SELECT g FROM Goods AS g WHERE g.name LIKE :#{'%' + #keyWord + '%'} AND g.dflag=:dflag") **public** List<Goods> findSplit(@Param(value = "keyWord") String keyWord, @Param(value = "dflag") Integer dflag, Pageable page); @Query("SELECT COUNT(g) FROM Goods AS g WHERE g.name LIKE :#{'%' + #keyWord + '%'} AND g.dflag=:dflag") **public** Long getSplitCount(@Param(value = "keyWord") String keyWord, @Param(value = "dflag") Integer dflag);
分页查询	@Query("SELECT g FROM Goods AS g WHERE g.dflag=:dflag") **public** List<Goods> findAllByDflag(@Param(value = "dflag") Integer dflag, Pageable page); @Query("SELECT COUNT(g) FROM Goods AS g WHERE g.dflag=:dflag") **public** Long getAllCountByDflag(@Param(value = "dflag") Integer dflag);

7.【mldnspring-mvc 项目】修改业务层实现子类中的 list 方法。

```
@Override
public Map<String, Object> list(String keyWord, int currentPage, int lineSize) {
    Map<String, Object> map = new HashMap<String, Object>();
    Sort sort = new Sort(Sort.Direction.DESC, "gid");                // 设置gid字段降序排列
    // 将分页与排序操作保存到Pageable接口对象中，通过DAO层调用方法，页数从0开始
    Pageable pageable = PageRequest.of(currentPage - 1, lineSize, sort);
    if (keyWord == null || "".equals(keyWord)) {                         // 查询全部操作
        map.put("allRecorders", this.goodsDAO.getAllCountByDflag(0));    // 数据统计
        map.put("allGoods", this.goodsDAO.findAllByDflag(0, pageable));  // 数据信息
    } else {                                                             // 模糊查询
        map.put("allRecorders", this.goodsDAO.getSplitCount(keyWord,0)); // 数据统计
        map.put("allGoods", this.goodsDAO.findSplit(keyWord,0,pageable)); // 数据信息
    }
```

```
        Map<Long, String> itemMap = new HashMap<Long, String>();              // 保存分类信息
        this.itemDAO.findAll().forEach((item)->{
            itemMap.put(item.getIid(), item.getTitle()) ;                    // 保存item信息
        });
        map.put("allItems", itemMap) ;                                        // 商品分类
        return map;
}
```

此时在业务层中，当进行全部商品信息查询时，将根据传入的 dflag 进行数据是否删除的判断，这样就可以实现数据逻辑删除处理了。

14.12.8　配置 Druid 数据源

Druid 是阿里巴巴提供的一款数据库连接池组件，该组件结合了常用数据库连接池的优点，并且追加了日志监控功能，可以直接利用 Web 控制台获取数据库的操作状态。

1.【mldnspring 项目】修改父 pom.xml 配置文件，追加 druid 相关依赖库。

属性配置	`<druid.version>1.1.8</druid.version>`

依赖库配置	`<dependency>`
	`<groupId>com.alibaba</groupId>`
	`<artifactId>druid</artifactId>`
	`<version>${druid.version}</version>`
	`</dependency>`

2.【mldnspring-mvc 项目】修改子 pom.xml 配置文件，引入 druid 依赖库。

```
<dependency>
    <groupId>com.alibaba</groupId>
    <artifactId>druid</artifactId>
</dependency>
```

3.【mldnspring-mvc 项目】修改 config/database.properties 配置文件。

database.druid.driverClassName=org.gjt.mm.mysql.Driver	# 定义数据库驱动程序名称
database.druid.url=jdbc:mysql://localhost:3306/mldn	# 数据库连接地址
database.druid.username=root	# 数据库连接用户名
database.druid.password=mysqladmin	# 数据库连接密码
database.druid.maxActive=1	# 数据库最大连接数
database.druid.minIdle=1	# 数据库最小维持连接数
database.druid.initialSize=1	# 数据库初始化连接
database.druid.maxWait=30000	# 数据库连接池最大等待时间
database.druid.timeBetweenEvictionRunsMillis=60000	# 配置间隔多久进行一次检测，检测需要关闭空闲连接，单位是ms
database.druid.minEvictableIdleTimeMillis=300000	# 配置连接在池中的最小生存时间，单位是ms

database.druid.validationQuery=SELECT 'x'	# 数据库状态检测
database.druid.testWhileIdle=true	# 申请连接时检测，如果空闲时间大于timeBetweenEvictionRunsMillis，执行validationQuery，检测连接是否有效
database.druid.testOnBorrow=false	# 申请连接时，执行validationQuery检测连接是否有效，该配置会降低性能
database.druid.testOnReturn=false	# 归还连接时，执行validationQuery检测连接是否有效，该配置会降低性能
database.druid.poolPreparedStatements=false	# 是否缓存preparedStatement，也就是PSCache。PSCache能提升支持游标的数据库性能，如oracle。MySQL下建议关闭
database.druid.maxPoolPreparedStatementPerConnectionSize=20	配置poolPreparedStatements缓存
database.druid.filters=stat	配置监控，统计拦截的filters。去掉后，监控界面SQL将无法统计

4.【mldnspring-jdbc 项目】修改 spring-database.xml 配置文件，使用 Druid 数据源（更换 DataSource 配置）。

```
<bean id="dataSource" class="com.alibaba.druid.pool.DruidDataSource" init-method="init">
    <property name="driverClassName" value="${db.druid.driverClassName}"/>    <!--驱动 -->
    <property name="url" value="${db.druid.url}"/>                            <!-- 地址 -->
    <property name="username" value="${db.druid.username}"/>                  <!-- 用户名 -->
    <property name="password" value="${db.druid.password}"/>                  <!-- 密码 -->
    <property name="maxActive" value="${db.druid.maxActive}"/>                <!-- 最大连接数 -->
    <property name="minIdle" value="${db.druid.minIdle}"/>                    <!-- 最小连接池 -->
    <property name="initialSize" value="${db.druid.initialSize}"/>        <!-- 初始化连接大小 -->
    <property name="maxWait" value="${db.druid.maxWait}"/>                <!-- 最大等待时间 -->
    <property name="timeBetweenEvictionRunsMillis"
        value="${db.druid.timeBetweenEvictionRunsMillis}" />              <!-- 检测空闲连接间隔 -->
    <property name="minEvictableIdleTimeMillis"
        value="${db.druid.minEvictableIdleTimeMillis}" />                 <!-- 连接最小生存时间-->
    <property name="validationQuery" value="${db.druid.validationQuery}" /> <!-- 验证 -->
    <property name="testWhileIdle" value="${db.druid.testWhileIdle}" />      <!-- 申请检测 -->
    <property name="testOnBorrow" value="${db.druid.testOnBorrow}" />        <!-- 有效检测 -->
    <property name="testOnReturn" value="${db.druid.testOnReturn}" />        <!-- 归还检测 -->
    <!-- 是否缓存preparedStatement（PSCache）。PSCache可以提升游标的性能，如Oracle -->
    <property name="poolPreparedStatements" value="${db.druid.poolPreparedStatements}" />
    <!-- 启用PSCache，必须配置大于0，当大于0时-->
    <property name="maxPoolPreparedStatementPerConnectionSize"
        value="${db.druid.maxPoolPreparedStatementPerConnectionSize}" />
    <property name="filters" value="${db.druid.filters}" />          <!-- 监控统计拦截的filters -->
</bean>
```

5．【mldnspring-mvc 项目】修改 web.xml 配置文件，追加 Druid 过滤器，以实现监控配置。

```
<filter>
    <filter-name>DruidWebStatFilter</filter-name>
    <filter-class>com.alibaba.druid.support.http.WebStatFilter</filter-class>
    <init-param>
        <param-name>exclusions</param-name>
        <param-value>*.js,*.gif,*.jpg,*.png,*.css,*.ico,/druid/*</param-value>
    </init-param>
</filter>
<filter-mapping>
    <filter-name>DruidWebStatFilter</filter-name>
    <url-pattern>/*</url-pattern>
</filter-mapping>
```

本程序进行 Druid 监控时，排除了不监控的操作后缀。

6．【mldnspring-mvc 项目】建立 Druid 控制台配置，修改 web.xml 文件，追加 Servlet 配置。

```
<servlet>
    <servlet-name>DruidStatView</servlet-name>
    <servlet-class>com.alibaba.druid.support.http.StatViewServlet</servlet-class>
</servlet>
<servlet-mapping>
    <servlet-name>DruidStatView</servlet-name>
    <url-pattern>/druid/*</url-pattern>
</servlet-mapping>
```

配置完成后，该项目将可以使用 Druid 进行数据库连接池管理。同时，可通过配置的 druid/* 利用浏览器进行访问。进行几次数据库操作后，通过 SQL 监控界面可观察到如图 14-19 所示的监控结果。

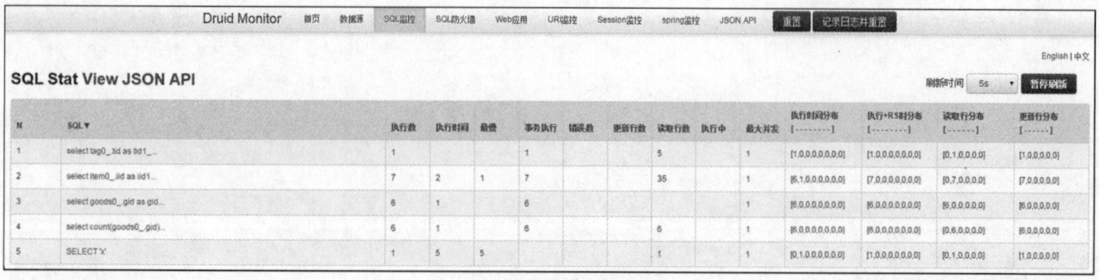

图 14-19　Druid 数据监听

14.13　本章小结

1．SpringMVC 是基于 Spring 开发框架的，可快速实现控制层处理操作，同时支持各种数据类型的转换。

2．SpringMVC 控制层中的方法可通过 ModelAndView 返回跳转路径信息与 request 属性，也可以利用字符串保存跳转路径。

3．在 SpringMVC 中，可利用 jackson 依赖库与@ResopnseBody 注解，将控制层中的数据以 Restful 形式返回。此操作是 SpringCloud 开发技术的实现基础。

4．SpringMVC 开发中，需要将 Web 资源保存到 WEB-INF 目录下，以实现安全访问。此时，可以通过控制器实现页面跳转；静态资源保存路径需要在配置文件中进行映射处理。

5．SpringMVC 支持拦截器，利用拦截器可以实现 Action 处理请求前的数据验证操作。

第 15 章

SpringSecurity

通过本章学习，可以达到以下目标：

1. 理解 SpringSecurity 的主要作用与技术实现。
2. 掌握 SpringSecurity 登录认证与相关登录处理操作。
3. 掌握 SpringSecurity 与数据库整合应用。
4. 掌握 UserDetailsService 业务接口使用。
5. 掌握 SessionManager、RememberMe、过滤器的使用。
6. 理解投票管理器与投票器的使用。

SpringSecurity 是 Spring 提供的一套认证与授权检测处理架构，利用它可以方便地实现登录认证与授权控制管理。本章将基于 SpringMVC 程序代码，为读者讲解 SpringSecurity 的应用。

提示：关于 SpringSecurity 框架。

SpringSecurity 是最早提供登录认证与授权检测的开发框架。虽然当下最流行的框架是 Shiro 开发框架（即 SSM 开发架构，由 Spring + Shiro + MyBatis 组成），SpringSecurity 已经很少出现在 Web 项目中，但对于流行的微架构（SpringCloud）而言，依然在大量使用 SpringSecurity。

15.1 SpringSecurity 简介

SpringSecurity 的前身是 Acegi Security 开发框架，它是一套完整的 Web 安全性应用解决方案，基于 SpringAOP 与过滤器实现安全访问控制。在 SpringSecurity 中，有两大核心主题。

☑ 用户认证（**Authentication**）：判断某个用户是否是系统的合法操作体，是否具有系统的操作权力。在进行用户认证处理中，核心信息为用户名与密码。

☑ 用户授权（**Authorization**）：一个系统中不同的用户拥有不同的权限（或称为角色），利用权限可以实现不同级别用户的划分，从而保证系统操作的安全。

为了实现安全管理，SpringSecurity 提供了一系列访问过滤器。所有访问过滤器都围绕着认证管理和决策管理展开，如图 15-1 所示。认证管理由 SpringSecurity 负责，开发者只需要掌握 UserDetailsService（用户认证服务）、AccessDecisionVoter（决策管理器）两个核心接口即可。

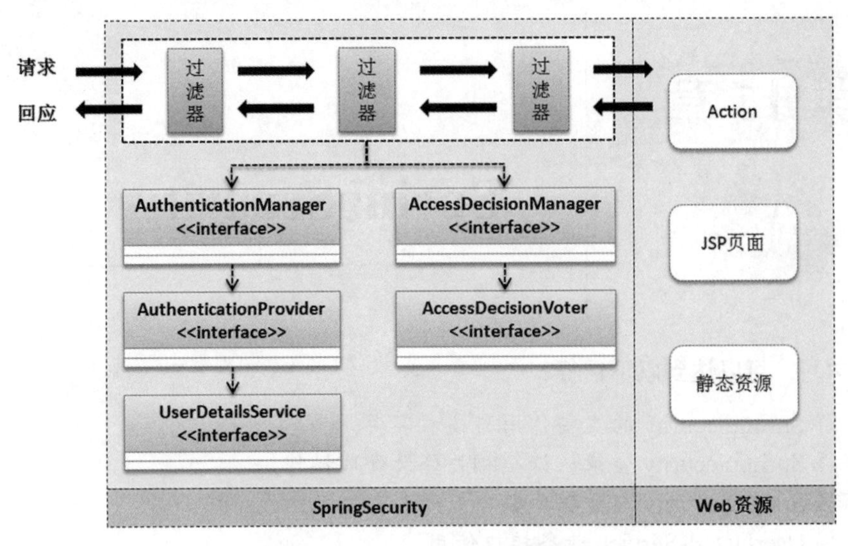

图 15-1 SpringSecurity 处理架构

SpringSecurity 中除了提供认证与授权外，还提供了 Session 管理、RememberMe（记住我）等常见功能。

15.2 SpringSecurity 编程起步

SpringSecurity 是一套独立组件，可在已经提供好的 Web 程序上整合执行，即开发者只需要编写核心的 Action 程序类，所有的认证与授权检测操作都可以通过后期配置实现。为了方便读者学习，本程序中定义的 Action 类将直接采用 Restful 风格进行内容展示。

1.【mldnspring-security 项目】创建一个 MessageAction 程序类，进行信息显示。

```
package cn.mldn.mldnspring.action;
import org.springframework.stereotype.Controller;
import org.springframework.web.bind.annotation.GetMapping;
import org.springframework.web.bind.annotation.RequestMapping;
import org.springframework.web.bind.annotation.ResponseBody;
@Controller                                  // 定义控制器
@RequestMapping("/pages/info/*")             // 定义访问父路径，与方法中的路径组合为完整路径
public class MessageAction {                 // 自定义Action程序类
    @GetMapping("/url")                      // 访问的路径为url.action
    @ResponseBody
    public Object url() {
        return "www.mldn.cn" ;
    }
}
```

本程序只实现了一个完整的 Action 程序类定义，未涉及任何的认证与授权检测操作。如果要为 /pages/info/url.action 访问路径配置安全访问，可以引入 SpringSecurity 组件。该组件利用

Web 过滤器对指定的请求拦截路径访问进行检测，检测通过后，可正常访问相应服务；如果检测失败，会显示相应的错误信息。其操作流程如图 15-2 所示。

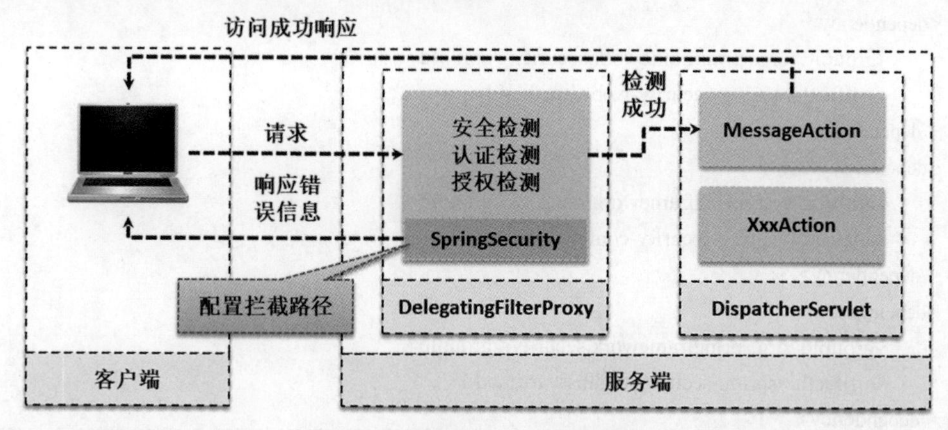

图 15-2　SpringSecurity 操作流程

由于 SpringSecurity 涉及的概念与配置较多，为了帮助读者更好地理解 SpringSecurity 组件的使用，本程序中将采用固定用户信息的模式进行操作。

2.【mldnspring-security 项目】修改父 pom.xml 配置文件，追加 SpringSecurity 相关依赖库管理。

```
<dependency>
    <groupId>org.springframework.security</groupId>
    <artifactId>spring-security-core</artifactId>
    <version>${spring.version}</version>
</dependency>
<dependency>
    <groupId>org.springframework.security</groupId>
    <artifactId>spring-security-web</artifactId>
    <version>${spring.version}</version>
</dependency>
<dependency>
    <groupId>org.springframework.security</groupId>
    <artifactId>spring-security-config</artifactId>
    <version>${spring.version}</version>
</dependency>
<dependency>
    <groupId>org.springframework.security</groupId>
    <artifactId>spring-security-taglibs</artifactId>
    <version>${spring.version}</version>
</dependency>
```

3.【mldnspring-security 项目】在子 pom.xml 文件中引入依赖库配置。

```
<dependency>
    <groupId>org.springframework.security</groupId>
```

```
        <artifactId>spring-security-core</artifactId>
    </dependency>
    <dependency>
        <groupId>org.springframework.security</groupId>
        <artifactId>spring-security-web</artifactId>
    </dependency>
    <dependency>
        <groupId>org.springframework.security</groupId>
        <artifactId>spring-security-config</artifactId>
    </dependency>
    <dependency>
        <groupId>org.springframework.security</groupId>
        <artifactId>spring-security-taglibs</artifactId>
    </dependency>
```

4.【mldnspring-security 项目】修改 web.xml 配置文件，追加过滤器配置。

```
    <filter>
        <filter-name>springSecurityFilterChain</filter-name>
        <filter-class>
            org.springframework.web.filter.DelegatingFilterProxy
        </filter-class>
    </filter>
    <filter-mapping>
        <filter-name>springSecurityFilterChain</filter-name>
        <url-pattern>/*</url-pattern>
    </filter-mapping>
```

本配置中，过滤器的访问路径为“/*”，表示所有的访问路径都需要进行 SpringSecurity 检测。这样当用户请求发送后，就会触发 SpringSecurity 检测机制，进行安全访问控制。

5.【mldnspring-security 项目】创建 spring/spring-security.xml 配置文件，并引入 security 命名空间，如图 15-3 所示。

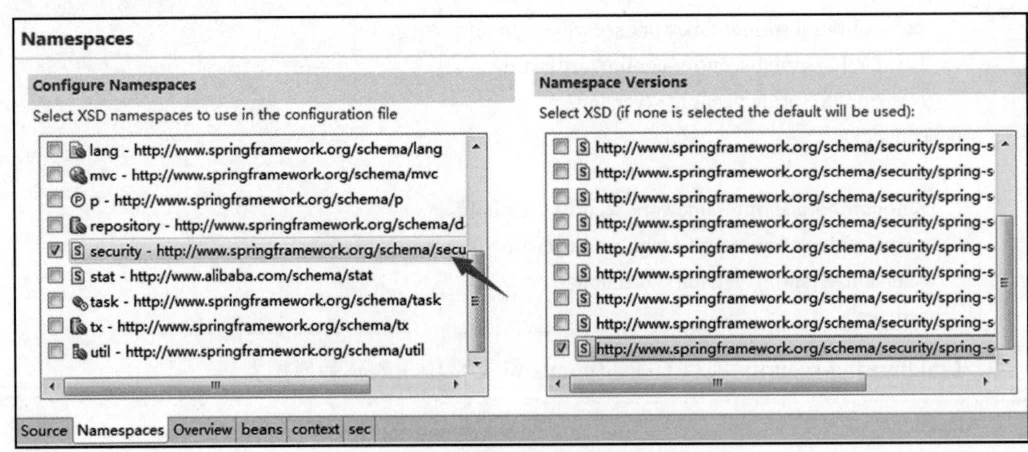

图 15-3　引入 security 命名空间

6.【mldnspring-security 项目】编辑 spring-security.xml 配置文件，在该配置文件中定义认证路径及用户信息。

```
<?xml version="1.0" encoding="UTF-8"?>
<beans xmlns="http://www.springframework.org/schema/beans"
    xmlns:xsi="http://www.w3.org/2001/XMLSchema-instance"
    xmlns:context="http://www.springframework.org/schema/context"
    xmlns:security="http://www.springframework.org/schema/security"
    xsi:schemaLocation="
        http://www.springframework.org/schema/security
        http://www.springframework.org/schema/security/spring-security-4.2.xsd
        http://www.springframework.org/schema/beans
        http://www.springframework.org/schema/beans/spring-beans-4.3.xsd
        http://www.springframework.org/schema/context
        http://www.springframework.org/schema/context/spring-context-4.3.xsd">
    <security:http auto-config="true">            <!-- 启用HTTP安全认证，并采用自动配置模式 -->
        <!-- 定义授权检测失败时的显示页面，一旦拒绝访问，将自动进行跳转 -->
        <security:access-denied-handler error-page="/WEB-INF/pages/error_page_403.jsp"/>
        <!-- 定义要拦截的路径，可以是具体路径，也可以使用路径匹配符设置要拦截的父路径 -->
        <!-- 表达式“hasRole(角色名称)”表示拥有此角色的用户才可以访问 -->
        <security:intercept-url pattern="/pages/info/**" access="hasRole('ADMIN')"/>
    </security:http>
    <security:authentication-manager>                    <!-- 定义认证管理器 -->
        <security:authentication-provider>               <!-- 配置认证管理配置类 -->
            <security:user-service>                      <!-- 创建用户信息 -->
                <!-- 定义用户名、密码（使用BCryptPasswordEncoder加密器进行加密）、角色信息（必须追加ROLE_前缀，否则无法识别） -->
                <!-- 用户名：admin，密码：hello，角色：ADMIN、USER -->
                <security:user name="admin" authorities="ROLE_ADMIN,ROLE_USER"
        password="{bcrypt}$2a$10$2y7higVhnHCn2L8//r/EVed2zi/LrQ.Y.svV.oeLqUM8xfUx5JWQC"/>
                <!-- 用户名：mldn，密码：java，角色：USER -->
                <security:user name="mldn" authorities="ROLE_USER"

    password="{bcrypt}$2a$10$vjXs780rO3rF8ZAXuBL4..c9icL4JDvr3sweCIU9y/QWiYlHgbKGa"/>
            </security:user-service>
        </security:authentication-provider>
    </security:authentication-manager>
</beans>
```

spring-security.xml 文件主要配置了要使用的用户认证与授权信息，且采用了自动配置模式（<security:http **auto-config="true"**>），所以并没有具体的登录页面，而是登录时自动为用户提供内置登录页。

 **提问：关于密码定义的一些问题。**

进行用户信息配置时，密码使用“{bcrypt}$2a$10$2…”形式进行了定义。这是什么含义？是如何生成的？

回答：bcrypt 是 Spring 提供的一种加密算法。

为了用户认证信息安全，密码访问时，SpringSecurity 使用标准接口 org.springframework.security.crypto.password.PasswordEncoder 定义加密处理标准。此接口及其常用子类如图 15-4 所示。

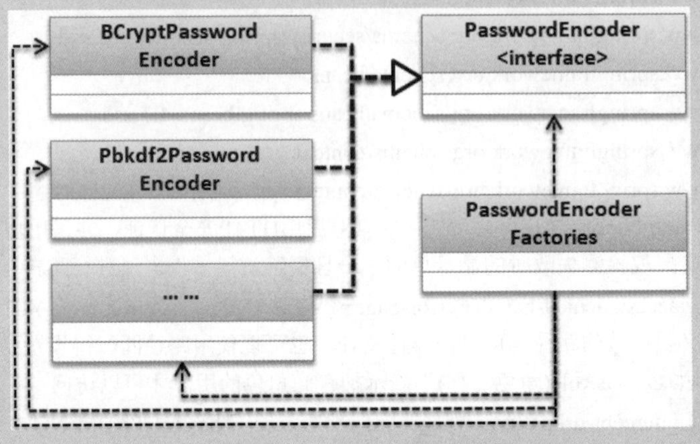

图 15-4　PasswordEncoder 继承结构

新版 SpringSecurity 推荐的加密算法为 bcrypt，所以这里定义密码时使用了{bcrypt}标记。为了方便管理不同的加密器，SpringSecurity 还提供了 PasswordEncoderFactories 工厂类，该类中注册了所有的加密器（有一些加密器已经不建议使用，如{noop}、{ldap}、{MD5}等）。

如果用户想定义自己的密码，可利用如下程序完成。

范例：使用 bcrypt 加密生成密码。

```
package cn.mldn.test;
import org.junit.Test;
import org.springframework.security.crypto.
        factory.PasswordEncoderFactories;
import org.springframework.security.crypto.
        password.PasswordEncoder;
public class TestPasswordEncoder {
    @Test
    public void testPassword() {
        String password = "hello" ;                              // 定义明文密码
        // 默认采用BCryptPasswordEncoder加密处理器
        PasswordEncoder passwordEncoder = PasswordEncoderFactories
            .createDelegatingPasswordEncoder();                  // 获取加密器实例
        String encode = passwordEncoder.encode(password);        // 加密
```

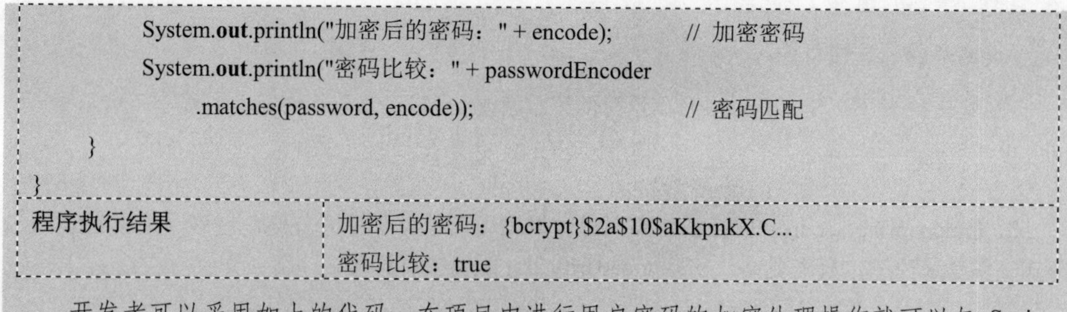

```
        System.out.println("加密后的密码：" + encode);          // 加密密码
        System.out.println("密码比较：" + passwordEncoder
                .matches(password, encode));                    // 密码匹配
    }
}
```

程序执行结果	加密后的密码：{bcrypt}$2a$10$aKkpnkX.C... 密码比较：true

开发者可以采用如上的代码，在项目中进行用户密码的加密处理操作就可以与 Spring Security 所提供的认证处理机制整合了。

在 SpringSecurity 配置文件里，最重要的是拦截路径与授权信息检测表达式（access 属性，该属性为 boolean 类型）。对于授权检测，常见配置如表 15-1 所示。

表 15-1 access 属性常用配置

编号	表达式	描述	范例
1	denyAll	拒绝全部访问（access=false）	access="denyAll"
2	permitAll	允许全部访问（access=true）	access="permitAll"
3	hasAnyRole(角色 1,角色 2,..)	拥有角色列表中任意一角色 access 内容为 true，否则为 false	access="hasAnyRole('ADMIN','USER')"
4	hasRole(角色)	拥有指定角色 access 内容为 true，否则为 false	access="hasRole('ADMIN')" 或者使用 and 或 or 进行逻辑计算
5	hasIpAddress(IP 地址)	将指定 IP 设置 true，否则为 false	access="hasIpAddress('192.168.31.0/24')" 192.168.31 网段的主机可以直接访问
6	isAnonymous()	是否为匿名访问	access="isAnonymous()"
7	isAuthenticated()	是否为认证访问	access="isAuthenticated()"
8	isRemberMe()	是否为 RememberMe 访问	access="isRemberMe()"
9	isFullyAuthenticated()	用户不是匿名，也不是 RememberMe 用户	access="isFullyAuthenticated()"
10	principal	允许直接访问的用户对象	access="principal"

7．【mldnspring-security 项目】程序配置完成后可以直接启动 Web 容器，当访问到 /pages/info/url.action 程序路径时会自动跳转到 /login 路径进行登录，如图 15-5 所示。由于此时该路径访问需要具有 ADMIN 角色的用户才可以访问，所以输入 admin/hello 账户信息，随后将显示如图 15-6 所示的界面。

图 15-5 默认登录页面

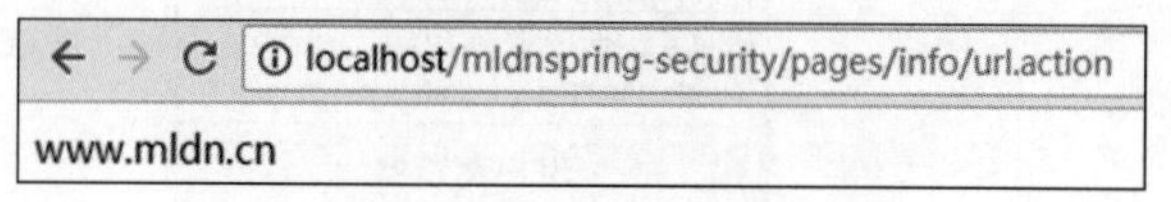

图 15-6 登录成功后跳转页面

 **提示：登录注销路径。**

在默认情况下，SpringSecurity 会提供登录表单，并且会自动将表单的提交路径设置为/login，如果登录后需要注销，可以使用/login?logout 路径进行注销。

8.【mldnspring-security 项目】如果用户觉得利用表单登录不太方便，也可以采用 http-basic 模式实现登录控制，只需要修改 spring-security.xml 配置文件即可。

```
<security:http auto-config="true"><!-- 启用HTTP安全认证，并且采用自动配置模式 -->
    <security:access-denied-handler error-page="/WEB-INF/pages/error_page_403.jsp"/>
    <security:intercept-url pattern="/pages/info/**" access="hasRole('ADMIN')"/>
    <security:http-basic/>                    <!-- 采用http-basic模式登录 -->
</security:http>
```

由于采用了 http-basic 模式进行登录控制，当用户需要进行认证处理时将不会跳转到登录表单，会出现如图 15-7 所示的弹出界面，在相应位置上输入用户名与密码即可正常访问。

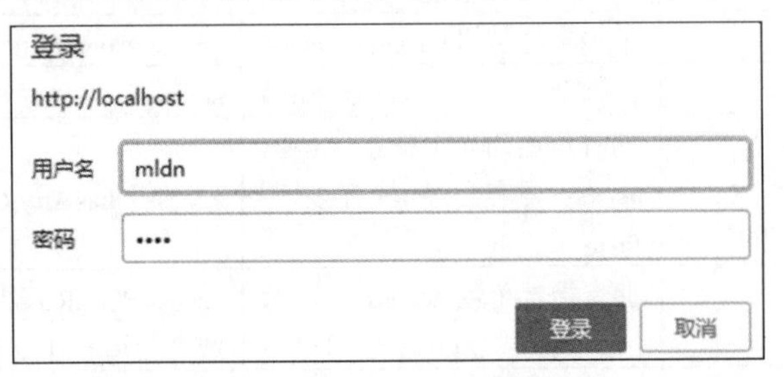

图 15-7　http-basic 模式登录

15.3　CSRF 访问控制

CSRF（Cross-Site Request Forgery，跨站请求伪造）是一种常见的网络攻击模式，攻击者可以在受害者完全不知情的情况下，以受害者身份发送各种请求（如邮件处理、账号操作等），且服务器认为这些操作属于合法访问。CSRF 攻击的基本操作流程如图 15-8 所示。

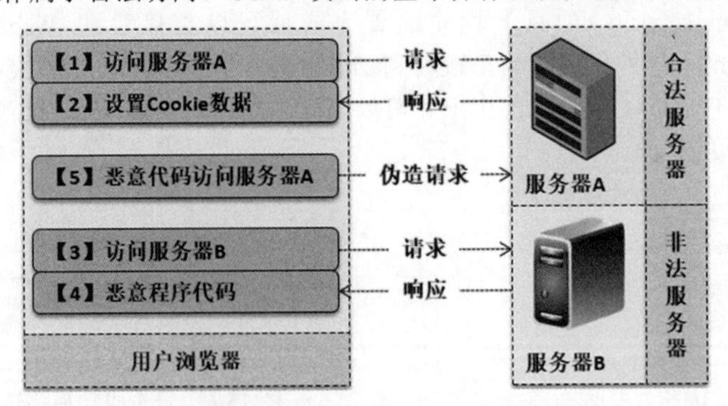

图 15-8　CSRF 攻击的基本操作流程

用户访问服务器 A 时，会在客户端浏览器中记录相应的 Cookie 信息，利用此 Cookie 进行用户身份的标注。在访问服务器 B 时，被植入了恶意程序代码，因此这些代码会在用户不知情的情况下以服务器 A 认证的身份访问其中的数据。

提示：CSRF 漏洞。

很多网站都存在 CSRF 访问漏洞。最早的 CSRF 漏洞是 2000 年提出的，但直到 2006 年才开始有人关注此漏洞。2008 年，国内外许多站点都爆发了 CSRF 漏洞安全问题，但直到今天还有许多网站存在 CSRF 漏洞。业界称 CSRF 为"沉睡的巨人"，其威胁程度可见一斑。

在实际开发中，有 3 种形式可以解决 CSRF 漏洞：验证 HTTP 请求头信息中的 Referer 信息，在访问路径中追加 token 标记，以及在 HTTP 信息头中定义验证属性。在 SpringSecurity 中可以采用 token 的形式进行验证，下面通过程序演示 CSRF 攻击的防范操作，本程序所采用的访问流程如图 15-9 所示。

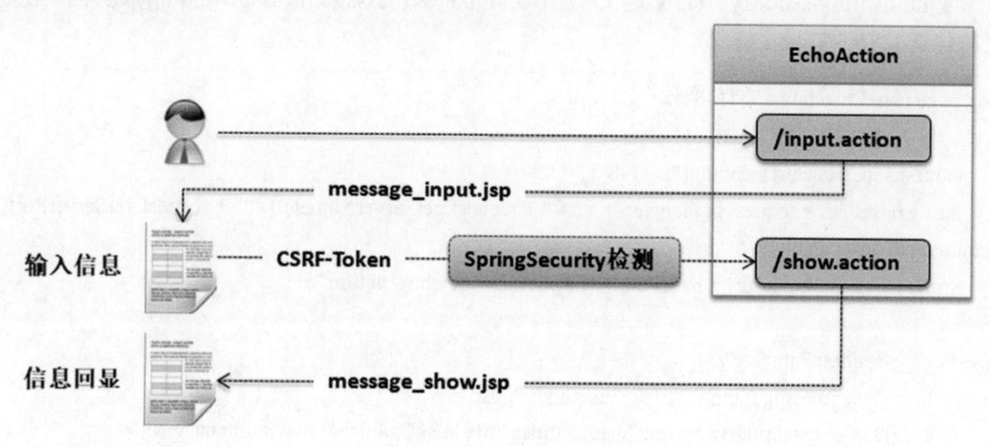

图 15-9　CSRF 访问处理

1.【mldnspring-security 项目】创建 EchoAction 程序类，主要负责信息输入页面的跳转与内容回显处理。

```
package cn.mldn.mldnspring.action;
import org.springframework.stereotype.Controller;
import org.springframework.web.bind.annotation.GetMapping;
import org.springframework.web.bind.annotation.RequestMapping;
import org.springframework.web.servlet.ModelAndView;
@Controller                                          // 定义控制器
@RequestMapping("/pages/message/*")                  // 访问父路径，与方法中的路径组合为完整路径
public class EchoAction {                            // 自定义Action程序类
    @RequestMapping("/show")                         // 访问路径
    public ModelAndView echo(String msg) {           // 接收请求参数
        ModelAndView mav = new ModelAndView("message/message_show") ; // 设置跳转路径
        mav.addObject("echoMessage", "【ECHO】msg = " + msg) ;          // 设置request数据
        return mav ;
```

```
    }
    @GetMapping("/input")                              // 访问路径
    public String input() {
        return "message/message_input";                // 设置跳转路径
    }
}
```

2.【mldnspring-security 项目】修改 spring-security.xml 配置文件，追加请求拦截路径。

```
<security:http auto-config="true"><!-- 启用HTTP安全认证，采用自动配置模式 -->
    <security:access-denied-handler error-page="/WEB-INF/pages/error_page_403.jsp"/>
    <security:intercept-url pattern="/pages/info/**" access="hasRole('ADMIN')"/>
    <security:intercept-url pattern="/pages/message/**" access="hasRole('USER')"/>
</security:http>
```

3.【mldnspring-security 项目】定义/WEB-INF/pages/message/message_input.jsp 页面，在表单定义时传送 CSRF-Token 信息。

```
<%@ page pageEncoding="UTF-8"%>
<%
    request.setCharacterEncoding("UTF-8") ;
    String basePath = request.getScheme() + "://" + request.getServerName() + ":" + request.getServerPort()
+ request.getContextPath() + "/";
    String message_input_url = basePath + "pages/message/show.action" ;
%>
<base href="<%=basePath%>" />
<form action="<%=message_input_url%>" method="post">
    消息内容：      <input type="text" name="msg" id="msg" value="www.mldn.cn"><br>
    <!-- 传递CSRF-Token信息，参数名称为_csrf，参数内容为随机生成的token数据 -->
    <input type="hidden" name="${_csrf.parameterName}" value="${_csrf.token}">
    <input type="submit" value="发送">   <input type="reset" value="重置">
</form>
```

在进行表单定义时，利用隐藏域实现了 CSRF-Token 信息的定义。图 15-10 演示了生成后的 token 内容。

```
<base href="http://localhost:80/mldnspring-security/" />
<form action="http://localhost:80/mldnspring-security/pages/message/show.action" method="post">
    消息内容： <input type="text" name="msg" id="msg" value="www.mldn.cn"><br>
    <!-- 传递csrf-token信息，参数名称为"_csrf"，参数内容为随机生成的token数据 -->
    <input type="hidden" name="_csrf" value="41294d86-814d-4822-bf19-6c290b7f2e57">
    <input type="submit" value="发送"> <input type="reset" value="重置">
</form>
```

图 15-10　CSRF-Token 内容（网页源代码）

4.【mldnspring-security 项目】定义/WEB-INF/pages/message/message_show.jsp 页面，回显输入内容。

```
<base href="<%=basePath%>" />
<h1>ECHO消息显示：${echoMessage}</h1>
```

至此，程序的基本流程开发完毕。用户进行信息输入时，如果发现没有 CSRF-Token 信息，将会跳转到<security:access-denied-handler>元素定义的错误信息显示页面。

5.【mldnspring-security 项目】虽然 CSRF-Token 可以解决 CSRF 攻击问题，但如果有些项目不需要处理 CSRF 漏洞，也可以通过配置的方式关闭 CSRF 校验。只要直接修改 spring-security.xml 配置文件中的<security:http>配置项即可。

```
<security:http auto-config="true"><!-- 启用HTTP安全认证，采用自动配置模式 -->
    <security:access-denied-handler error-page="/WEB-INF/pages/error_page_403.jsp"/>
    <security:intercept-url pattern="/pages/info/**" access="hasRole('ADMIN')"/>
    <security:intercept-url pattern="/pages/message/**" access="hasRole('USER')"/>
    <security:csrf disabled="true"/>    <!-- 关闭CSRF校验 -->
</security:http>
```

此时项目中，即便表单提交时没有 CSRF-Token 也可以正常访问。

15.4 扩展登录和注销功能

SpringSecurity 是一套完善的安全框架，为用户提供了基础的登录页面与注销路径，在实际的项目开发过程中登录与注销页面需要进行更加完善的页面设计，为此在 SpringSecurity 中用户也可以修改默认的登录与注销操作，自定义相关页面进行显示。

1.【mldnspring-security 项目】创建 /WEB-INF/pages/login.jsp 用户登录页面，该页面主要提供登录表单。

```
<%@ page pageEncoding="UTF-8"%>
<html>
<head>
<%
    request.setCharacterEncoding("UTF-8") ;
    String basePath = request.getScheme() + "://" + request.getServerName() + ":" +
                request.getServerPort() + request.getContextPath() + "/";
    String login_url = basePath + "mldn-login" ; // 登录表单提交路径，该路径需要配置
%>
<title>SpringSecurity安全框架</title>
<base href="<%=basePath%>" />
</head>
<body>
<h2>用户登录</h2>
<form action="<%=login_url%>" method="post">
    用户名：<input type="text" name="mid"><br>
```

```
    密   码：<input type="password" name="pwd"><br>
    <input type="submit" value="登录"><input type="reset" value="重置">
</form>
</body>
</html>
```

本页面只提供了用户登录信息的表单，由于 SpringSecurity 内部可以直接实现登录控制，所以此时定义的登录处理路径为/mldn-login（随后需要在 spring-security.xml 中配置）。

2.【mldnspring-security 项目】创建/WEB-INF/pages/welcome.jsp 页面，作为用户登录成功后的首页。

```
<%@ page pageEncoding="UTF-8"%>
<html>
<head>
<%
    request.setCharacterEncoding("UTF-8") ;
    String basePath = request.getScheme() + "://" + request.getServerName() + ":" + request.getServerPort()
+ request.getContextPath() + "/";
    String logout_url = basePath + "mldn-logout" ;                    // 注销路径
%>
<title>SpringSecurity安全框架</title>
<base href="<%=basePath%>" />
</head>
<body>
<h2>登录成功，欢迎您回来，也可以选择<a href="<%=logout_url%>">注销</a>！</h2>
<h3>更多内容请访问<a href="http://www.mldn.cn">MLDN-魔乐科技</a></h3>
<img src="mvcimages/mldn.png" style="width:500px;">
<img src="mvcimages/jixianit.png" style="width:500px;">
</body>
</html>
```

为了方便演示本程序，在登录成功页面提供注销路径/mldn-logout（该路径需要在 spring-security.xml 中配置）。

3.【mldnspring-security 项目】创建/WEB-INF/pages/logout.jsp 页面，作为注销成功后的显示页面。

```
<h2>注销成功，欢迎您再来！</h2>
<h3>更多内容请访问<a href="http://www.mldn.cn">MLDN-魔乐科技</a></h3>
```

4.【mldnspring-security 项目】由于所有 JSP 页面都保存在 WEB-INF 目录下，所以为了更方便访问页面，可以定义一个 GlobalAction 程序类，利用此程序类实现跳转。

```
package cn.mldn.mldnspring.action;
import org.springframework.stereotype.Controller;
import org.springframework.web.bind.annotation.RequestMapping;
```

```
@Controller                                             // 定义控制器
public class GlobalAction {                             // 定义全局Action类
    @RequestMapping("/loginPage")                       // 访问路径
    public String login() {                             // 登录表单路径
        return "login";                                 // 设置跳转路径
    }
    @RequestMapping("/welcomePage")                     // 访问路径
    public String welcome() {                           // 登录成功路径
        return "welcome";                               // 设置跳转路径
    }
    @RequestMapping("/logoutPage")                      // 访问路径
    public String logout() {                            // 登录成功路径
        return "logout";                                // 设置跳转路径
    }
}
```

为了方便访问路径，本程序使用@RequestMapping 注解进行定义，这样就可以使用 HTTP 各种请求方式进行访问，同时为了与访问路径有所区分，所有跳转的路径都使用了 xxxPage 的形式定义。

5.【mldnspring-security 项目】修改 spring-security.xml 配置文件，配置登录与注销。

```
<security:http auto-config="true"><!-- 启用HTTP安全认证，并且采用自动配置模式 -->
    <!-- 定义授权检测失败时的显示页面，一旦出现拒绝访问时将自动跳转 -->
    <security:access-denied-handler error-page="/WEB-INF/pages/error_page_403.jsp"/>
    <security:intercept-url pattern="/pages/info/**" access="hasRole('ADMIN')"/>
    <security:intercept-url pattern="/pages/message/**" access="hasRole('USER')"/>
    <!-- 登录成功后的首页，需要在用户已经认证后才可以显示 -->
    <security:intercept-url pattern="/welcomePage.action" access="isAuthenticated()"/>
    <security:csrf disabled="true"/>                        <!-- 关闭CSRF校验 -->
    <security:form-login                                    <!-- 配置表单登录 -->
        username-parameter="mid"                            <!-- 用户名请求参数 -->
        password-parameter="pwd"                            <!-- 用户名请求参数 -->
        authentication-success-forward-url="/welcomePage.action"    <!-- 登录成功页面 -->
        login-page="/loginPage.action"                      <!-- 登录表单路径 -->
        login-processing-url="/mldn-login"                  <!-- 表单提交路径 -->
        authentication-failure-forward-url="/loginPage.action?error=true"/> <!--失败页-->
    <security:logout                                        <!-- 注销路径配置 -->
        logout-url="/mldn-logout"                           <!-- 注销路径 -->
        logout-success-url="/logoutPage.action"             <!-- 注销成功页路径 -->
        delete-cookies="JSESSIONID"/>                       <!-- 注销后删除对应的Cookie数据 -->
</security:http>
```

创建完成后，SpringSecurity 认证检测失败后会自动跳转到/loginPage.action 路径，登录成功后会自动跳转到/welcomePage.action 路径，注销时会自动清除对应的 Cookie 数据，从而实现了自定义登录与注销操作。

15.5 获取认证与授权信息

用户认证成功后，SpringSecurity 会将用户的认证信息与授权信息保存在 Session 中，而保存的信息类型为 org.springframework.security.core.userdetails.User 类对象，此类的继承结构如图 15-11 所示。

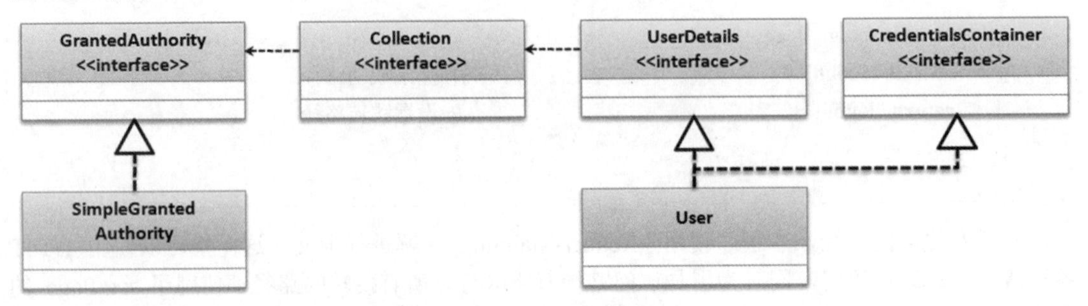

图 15-11 User 类继承结构

通过图 15-11 可以发现，SpringSecurity 中有两个核心接口保存用户信息：

☑ GrantedAuthority：保存授权信息。

☑ UserDetails：描述用户的详情与用户授权信息。

在 SpringSecurity 默认配置下，Spring 容器会自动帮助用户创建 User 类对象，并且将用户对应的认证信息与授权信息保存在 User 类对象中，要想获取这些信息可以采用表 15-2 所示的方法完成。

表 15-2 User 类常用方法

编号	方法	类型	描述
1	public User(String username, String password, boolean enabled, boolean accountNonExpired, boolean credentialsNonExpired, boolean accountNonLocked, Collection<? extends GrantedAuthority> authorities)	构造	实例化 User 类对象
2	public User(String username, String password, Collection<? extends GrantedAuthority> authorities)	构造	实例化 User 类对象
3	public String getUsername()	普通	获取用户名
4	public String getPassword()	普通	获取密码
5	public boolean isEnabled()	普通	账户是否可用
6	public boolean isAccountNonExpired()	普通	账户是否没有过期
7	public boolean isAccountNonLocked()	普通	账户是否没有被锁定

续表

编号	方法	类型	描述
8	public boolean isCredentialsNonExpired()	普通	认证信息是否没有失效
9	public java.util.Collection<GrantedAuthority> getAuthorities()	普通	获取所有的授权信息
10	public boolean isEnabled()	普通	账户是否可用

1.【mldnspring-security 项目】Action 中获取认证与授权信息时，所有认证数据都保存在 Authentication 接口实例中，所以要先获取 Authentication 接口对象，然后才可以通过 Authentication 得到 UserDetails 接口对象。

```
@RequestMapping("/welcomePage")                                          // 访问路径
public String welcome() {                                                // 登录成功路径
    Authentication authentication = SecurityContextHolder
        .getContext().getAuthentication() ;                              // 认证对象
    UserDetails userDetails = (UserDetails) authentication.getPrincipal() ; // 用户详情
    String username = userDetails.getUsername() ;                        // 获得用户名
    this.log.info("用户名：" + username);
    // 通过UserDetails对象可以获取当前用户的所有授权信息
    Collection<? extends GrantedAuthority> authorities = userDetails.getAuthorities() ;
    this.log.info("授权信息：" + authorities);
    return "welcome";                                                    // 设置跳转路径
}
```

2.【mldnspring-security 项目】在实际项目中，通常需要在 JSP 页面中获取相应的认证与授权信息，此时可以直接引入 SpringSecurity 的标签来获得。

```
<%@ taglib prefix="security" uri="http://www.springframework.org/security/tags" %>
<security:authorize access="isAuthenticated()">                <!-- 是否为认证过的用户 -->
    用户已经成功登录了！
</security:authorize>
<security:authorize access="hasRole('USER')">                  <!-- 是否拥有USER角色 -->
    拥有USER角色
</security:authorize>
<security:authorize access="hasRole('ADMIN')">                 <!-- 是否拥有ADMIN角色 -->
    拥有ADMIN角色
</security:authorize>
<h2>登录成功，欢迎“<security:authentication property="principal.username"/>”回来，
也可以选择<a href="<%=logout_url%>">注销</a>！ </h2>
```

本程序主要使用了两个标签，核心作用如下。

☑ **<security:authentication>**：获取认证信息，通过 Authentication 获取 UserDetails，得到用户名。

☑ **<security:authorize>**：授权信息，采用 Spring 表达式进行判断（与拦截路径判断一致）。

15.6 基于数据库实现用户登录

为了灵活管理用户的登录信息，在实际项目中需要将用户信息保存在数据库中，登录时利用数据库对认证信息进行检测。本节将为读者讲解 SpringSecurity 与数据库的认证整合处理。

15.6.1 基于 SpringSecurity 标准认证

SpringSecurity 本身可以直接利用配置文件实现用户认证与授权信息的查询处理，但是需要开发者在项目中配置好相应的数据库连接，同时由于 SpringSecurity 自身的查询约定，在定义查询语句时也需要对返回查询列的名称统一。

提示：关于本次基于数据库查询的操作。

为了尽可能帮助读者理解 SpringSecurity 中的自动查询处理支持，在本程序中将使用自定义表结构的形式完成（在查询的时候将为列定义别名以符合 SpringSecurity 查询要求），同时对于数据库的配置也将使用之前讲解过的 Druid 连接池。

为了方便管理用户认证与授权信息，在本程序中将采用如图 15-12 所示的数据表结构。

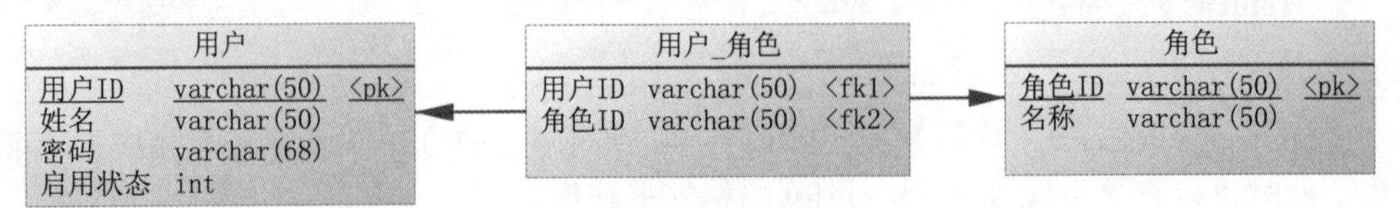

图 15-12 数据表结构

1.【mldnspring-security 项目】定义数据库创建脚本，该脚本信息的组成与之前固定认证信息的结构相同。

```
-- 删除数据库
DROP DATABASE mldn ;
-- 创建数据库
CREATE DATABASE mldn DEFAULT CHARACTER SET utf8 ;
-- 使用数据库
USE mldn ;
-- 创建用户表（mid：登录ID；name：真实姓名；password：登录密码；enabled：启用状态）
-- enabled取值有两种：启用（enabled=1），锁定（enabled=0）
CREATE TABLE member(
    mid         VARCHAR(50) ,
    name        VARCHAR(50) ,
    password    VARCHAR(68) ,
    enabled     INT(1) ,
    CONSTRAINT pk_mid PRIMARY KEY(mid)
```

```
) engine=innodb ;
-- 创建角色表（rid：角色ID，也是授权检测的名称；title：角色名称）
CREATE TABLE role (
    rid          VARCHAR(50) ,
    title        VARCHAR(50) ,
    CONSTRAINT pk_rid PRIMARY KEY(rid)
) engine=innodb ;
-- 创建用户角色关联表（mid：用户ID；rid：角色ID）
CREATE TABLE member_role (
    mid          VARCHAR(50) ,
    rid          VARCHAR(50) ,
    CONSTRAINT fk_mid FOREIGN KEY(mid) REFERENCES member(mid) ON DELETE
CASCADE ,
    CONSTRAINT fk_rid FOREIGN KEY(rid) REFERENCES role(rid) ON DELETE CASCADE
) engine=innodb ;
-- 增加用户数据（admin/hello、mldn/java）
INSERT INTO member(mid,name,password,enabled) VALUES ('admin','李兴华
','{bcrypt}$2a$10$2y7higVhnHCn2L8//r/EVed2zi/LrQ.Y.svV.oeLqUM8xfUx5JWQC',1) ;
INSERT INTO member(mid,name,password,enabled) VALUES ('mldn','魔乐科技
','{bcrypt}$2a$10$vjXs780rO3rF8ZAXuBL4..c9icL4JDvr3sweCIU9y/QWiYlHgbKGa',0) ;
-- 增加角色数据
INSERT INTO role(rid,title) VALUES ('ROLE_ADMIN','管理员') ;
INSERT INTO role(rid,title) VALUES ('ROLE_USER','用户') ;
-- 增加用户与角色信息
INSERT INTO member_role(mid,rid) VALUES ('admin','ROLE_ADMIN') ;
INSERT INTO member_role(mid,rid) VALUES ('admin','ROLE_USER') ;
INSERT INTO member_role(mid,rid) VALUES ('mldn','ROLE_USER') ;
-- 提交事务
COMMIT ;
```

2.【mldnspring-security 项目】修改 spring-security.xml 配置文件中的<security:authentication-provider>元素定义，将固定信息验证修改为 JDBC 的形式。

```
<security:authentication-manager>                    <!-- 定义认证管理器 -->
    <security:authentication-provider>               <!-- 配置认证管理配置类 -->
        <security:jdbc-user-service
            data-source-ref="dataSource"
            users-by-username-query="SELECT mid AS username,password,enabled
                                          FROM member WHERE mid=?"
            authorities-by-username-query="SELECT mid AS username, rid AS authorities
                                          FROM member_role WHERE mid=?"/>
    </security:authentication-provider>
</security:authentication-manager>
```

本程序主要使用<security:jdbc-user-service>元素配置数据库认证，属性作用如下。

☑ data-source-ref：定义要使用的数据源对象。

☑ users-by-username-query：用户认证查询，要求返回用户名（username）、密码（password）、启用状态（enabled），由于数据表中列的名称与查询要求不符，所以在查询时需要为查询列定义别名。

☑ authorities-by-username-query：用户角色查询，根据认证的用户名返回相应的角色信息，返回结构要求拥有用户名（username）、角色（authorities）两个信息。

15.6.2 UserDetailsService

在 SpringSecurity 中除了可以使用配置文件进行查询配置外，还可以由用户通过 UserDetailsService 接口标准实现自定义认证与授权信息查询处理，而使用 UserDetailsService 实现的查询会比配置文件定义查询更加灵活。UserDetailsService 接口如图 15-13 所示。

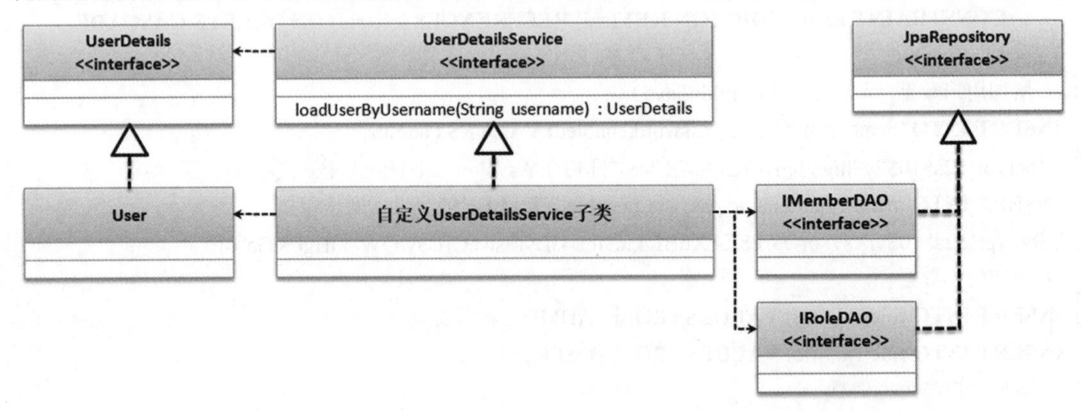

图 15-13　UserDetailsService 接口

在 UserDetailsService 接口里只有一个 loadUserByUsername 方法，此方法会根据用户名查询对应的用户信息与授权信息，而用户和授权信息由于都保存在数据库中，所以可利用 SpringDataJPA 实现查询操作。下面演示 UserDetailsService 接口的使用。

1. 【mldnspring-security 项目】定义用户信息数据层操作接口。

【PO】用户信息类（Member）

```java
package cn.mldn.mldnspring.po;
import java.io.Serializable;
import javax.persistence.Entity;
import javax.persistence.Id;
@Entity
@SuppressWarnings("serial")
public class Member
	implements Serializable {
	@Id
	private String mid ;
	private String name ;
	private String password ;
```

【DAO】IMemberDAO接口

```java
package cn.mldn.mldnspring.dao;
import org.springframework.data.jpa.repository.JpaRepository;
import cn.mldn.mldnspring.po.Member;
public interface IMemberDAO
	extends JpaRepository<Member, String> {
}
```

```
    private Integer enabled ;
    // setter、getter略
}
```

2. 【mldnspring-security 项目】定义授权信息数据层操作接口。

【PO】角色信息类（Role）

```
package cn.mldn.mldnspring.po;
import javax.persistence.Entity;
import javax.persistence.Id;
@Entity
@SuppressWarnings("serial")
public class Role
    implements Serializable {
    @Id
    private String rid ;
    private String title ;
    // setter、getter略
}
```

【DAO】IRoleDAO接口

```
package cn.mldn.mldnspring.dao;
import java.util.Set;
import org.springframework.data.jpa.repository.JpaRepository;
import org.springframework.data.jpa.repository.Query;
import org.springframework.data.repository.query.Param;
import cn.mldn.mldnspring.po.Role;
public interface IRoleDAO
    extends JpaRepository<Role, String> {
    /**
     * 根据用户ID查询对应的角色ID
     * @param mid 用户ID
     * @return 用户拥有的全部角色ID
     */
    @Query(nativeQuery=true,
        value="SELECT rid FROM member_role WHERE
mid=:mid")
    public Set<String> findAllByMember(
        @Param("mid") String mid) ;
}
```

3. 【mldnspring-security 项目】定义 UserDetailsService 接口子类，注入相应的数据层接口对象，实现数据查询。在对查询结果进行判断时，可以用 AuthenticationException 异常类抛出异常。AuthenticationException 异常类的常用子类如图 15-14 所示。

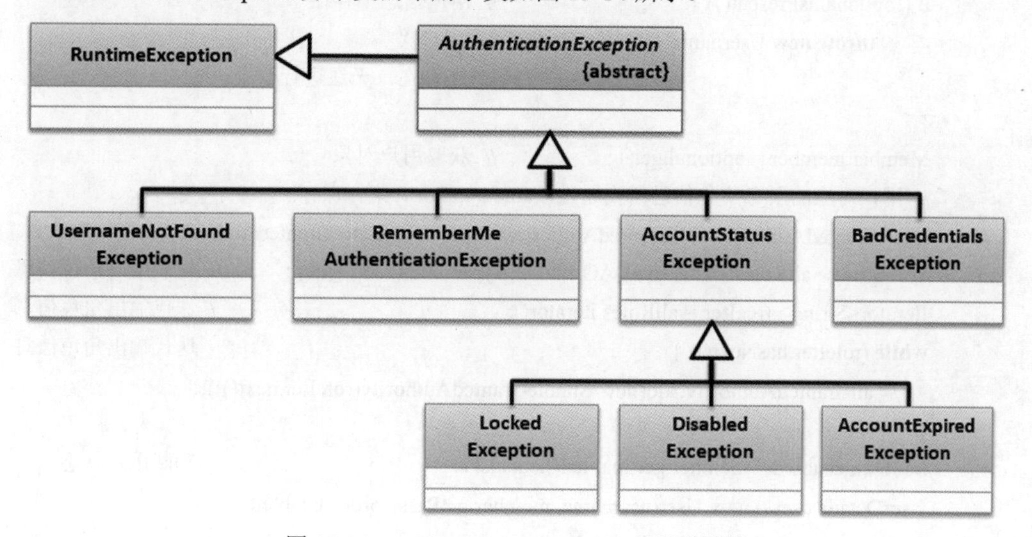

图 15-14 AuthenticationException 异常类结构

```
package cn.mldn.mldnspring.service;
import java.util.ArrayList;
import java.util.Iterator;
import java.util.List;
import java.util.Optional;
import java.util.Set;
import org.springframework.beans.factory.annotation.Autowired;
import org.springframework.security.core.GrantedAuthority;
import org.springframework.security.core.authority.SimpleGrantedAuthority;
import org.springframework.security.core.userdetails.User;
import org.springframework.security.core.userdetails.UserDetails;
import org.springframework.security.core.userdetails.UserDetailsService;
import org.springframework.security.core.userdetails.UsernameNotFoundException;
import org.springframework.stereotype.Service;
import cn.mldn.mldnspring.dao.IMemberDAO;
import cn.mldn.mldnspring.dao.IRoleDAO;
import cn.mldn.mldnspring.po.Member;
@Service(value="userDetailsService")                  // 注解配置，此名称要在spring-security.xml中使用
public class UserDetailsServiceImpl implements UserDetailsService {
    @Autowired
    private IMemberDAO memberDAO ;                   // 注入用户数据操作接口
    @Autowired
    private IRoleDAO roleDAO ;                       // 注入角色操作接口
    @Override
    public UserDetails loadUserByUsername(String username) throws UsernameNotFoundException {
        Optional<Member> optional = this.memberDAO.findById(username);   // 根据用户ID进行查询
        if (!optional.isPresent()) {                 // 用户信息不存在
            throw new UsernameNotFoundException("用户“" +
                            username + "”信息不存在，无法进行登录。");
        }
        Member member = optional.get() ;             // 获取用户对象
        // 用户对应的所有角色信息需要通过GrantedAuthority集合保存
        List<GrantedAuthority> allGrantedAuthority = new ArrayList<GrantedAuthority>();
        Set<String> allRoles = this.roleDAO.findAllByMember(username) ;        // 获取用户角色信息
        Iterator<String> roleIter = allRoles.iterator();                        // 迭代输出角色信息
        while (roleIter.hasNext()) {                                            // 迭代输出角色信息
            allGrantedAuthority.add(new SimpleGrantedAuthority(roleIter.next()));
        }
        boolean enabled = member.getEnabled().equals(1) ;                       // 判断用户状态
        UserDetails user = new User(username, member.getPassword(), enabled,
            true, true, true, allGrantedAuthority);
```

```
        return user ;                                          // 返回UserDetails对象
    }
}
```

4.【mldnspring-security 项目】修改 spring-security.xml 配置文件，使用 UserDetailsService 处理登录。

```
<security:authentication-manager>                    <!-- 定义认证管理器 -->
    <security:authentication-provider user-service-ref="userDetailsService"/>
</security:authentication-manager>
```

配置完成后，当前程序会使用 UserDetailsServiceImpl 实现子类进行用户认证信息与授权信息的获取，并且按照 SpringSecurity 的要求所有的信息都会包装在 UserDetails 接口对象中。

15.7　Session 管理

在系统管理中，为了用户信息的安全往往会对同一个账户的并发登录状态进行控制，所以在这种情况下往往需要对用户登录状态进行监听，即需要在内存中保存相应用户的 Session 列表，当出现账户重复登录的时候就可以进行指定 Session 的剔除操作。

1.【mldnspring-security 项目】修改 web.xml 配置文件，追加 Session 管理监听器。

```
<listener>        <!-- 定义Session管理监听器 -->
    <listener-class>
        org.springframework.security.web.session.HttpSessionEventPublisher
    </listener-class>
</listener>
```

2.【mldnspring-security 项目】修改 spring-security.xml 配置文件，取消自动配置，同时追加 Session 并发管理。

```
<security:http auto-config="false">                  <!-- 取消自动配置，才能启动Session管理 -->
    <security:session-management invalid-session-url="/loginPage.action">
        <security:concurrency-control                 <!-- 并发Session管理 -->
            max-sessions="1"                          <!-- 每个账户并发访问量 -->
            error-if-maximum-exceeded="false"         <!-- Session剔除模式 -->
            expired-url="/logoffPage.action"/>        <!-- Session剔除后的错误显示路径 -->
    </security:session-management>
    ... 相同配置项省略 ...
</security:http>
```

对于并发 Session 管理，需要考虑的是要剔除之前已登录的 Session 还是剔除之后登录的 Session 用户，这点可通过 error-if-maximum-exceeded 属性进行配置。该属性配置为 true，表示剔除新登录 Session 用户；为 false，表示剔除之前登录过的 Session 用户。

15.8 RememberMe

为了防止用户重复登录表单的填写，在实际项目中往往采用 RememberMe 功能，将用户登录信息暂时保存在 Cookie 中，这样每次访问时就可以通过请求 Cookie 信息获取用户登录状态。SpringSecurity 对这一功能提供了配置实现。

1. 【mldnspring-security】修改 login.jsp 页面，追加免登录组件（复选框）。

```
<form action="<%=login_url%>" method="post">
    用户名：<input type="text" name="mid"><br>
    密   码：<input type="password" name="pwd"><br>
    <input type="checkbox" id="remember" name="remember" value="true"/>下次免登录<br>
    <input type="submit" value="登录">
    <input type="reset" value="重置">
</form>
```

2. 【mldnspring-security 项目】修改 spring-security.xml 配置文件，追加 RememberMe 配置项。

```
<security:http auto-config="false">                    <!-- 取消自动配置 -->
    <security:remember-me                               <!-- 启用RememberMe功能 -->
        remember-me-parameter="remember"                <!-- 登录表单参数 -->
        key="mldn-lixinghua"                            <!-- Cookie加密密钥 -->
        token-validity-seconds="2592000"                <!-- 免登录失效（单位为s） -->
        remember-me-cookie="mldn-rememberme-cookie"/>   <!-- Cookie名称 -->
    相同配置项省略
</security:http>
```

本配置文件中定义了表单要使用的登录参数 remember，这样当用户登录后会自动在 Cookie 中提供 mldn-rememberme-cookie 内容（见图 15-15）。该 Cookie 的失效时间为 30 天，因此用户在 30 天年内重新打开浏览器不必再重复编写登录表单。

Name	Value	Domain	Path
JSESSIONID	19986DE111EF4C92DFADBEB2C05A68C3	localhost	/mldnspring-security
mldn-rememberme-cookie	YWRtaW46MTUzNzA1ODlwMjg0Njo0ZmJiZjg4...	localhost	/mldnspring-security

图 15-15　Cookie 中记录的免登录信息

3. 【mldnspring-security 项目】现在已经实现了 RememberMe 功能，此时的用户信息是在服务器内存中保存的。如果有需要，也可以将所有 RememberMe 信息在数据库中记录。此时需要使用如下的数据库创建脚本。

```
-- 使用数据库
USE mldn ;
-- 创建数据表保存免登录信息（数据表名称默认为persistent_logins）
CREATE TABLE persistent_logins (
```

```
    series              VARCHAR(64) ,
    username                    VARCHAR(100) ,
    token                       VARCHAR(64) ,
    last_used           TIMESTAMP ,
    CONSTRAINT pk_series PRIMARY KEY(series)
);
```

4.【mldnspring-security 项目】修改 spring-security.xml 配置文件，在 RememberMe 的配置中追加数据源设置。

```
<security:http auto-config="false">                        <!-- 取消自动配置 -->
    <security:remember-me                                  <!-- 启用RememberMe功能 -->
        remember-me-parameter="remember"                   <!-- 登录表单参数 -->
        key="mldn-lixinghua"                               <!-- Cookie加密密钥 -->
        token-validity-seconds="2592000"              <!-- 免登录失效（单位为s） -->
        remember-me-cookie="mldn-rememberme-cookie"        <!-- Cookie名称 -->
        data-source-ref="dataSource"/>                     <!-- 持久化保存数据源 -->
    相同配置项省略
</security:http>
```

在进行持久化保存时只需要按照数据表结构建立数据表，随后配置上要使用的数据源，这样在用户进行免登录选择的时候就可以自动将相应的信息保存在数据库中，即便服务器重新启动也不会丢失用户的免登录配置。

15.9　过滤器

SpringSecurity 的核心操作是依据过滤器实现的认证与授权检测，但是在 DelegatingFilterProxy 过滤器中为了方便进行认证与授权管理还提供了一套自定义的过滤链（见表 15-3）进行配置。同时这些过滤链拥有严格的执行顺序，才可以实现最终安全检测。

表 15-3　SpringSecurity-FilerChain

编号	配置器	过滤器	描述
1	OpenIDLoginConfigurer	OpenIDAuthenticationFilter	处理 OpenID（数字身份识别）授权请求
2	HeadersConfigurer	HeaderWriterFilter	在返回数据中添加 Security 头信息
3	CorsConfigurer	CorsFilter	跨域访问配置
4	SessionManagement Configurer	SessionManagementFilter	会话管理配置
5	PortMapperConfigurer	-	HTTP 与 HTTPS 访问端口重定向
6	JeeConfigurer	J2eePreAuthenticated ProcessingFilter	添加 Java EE 预授权处理机制支持
7	X509Configurer	X509AuthenticationFilter	添加 X509 预授权处理机制支持

续表

编号	配置器	过滤器	描述
8	RememberMeConfigurer	RememberMeAuthentication Filter	用户免登录配置
9	ExpressionUrl AuthorizationConfigurer	FilterSecurityInterceptor	访问权限判断，也是主要的 Filter
10	RequestCacheConfigurer	RequestCacheAwareFilter	请求信息缓存
11	ExceptionHandling Configurer	ExceptionTranslationFilter	处理 AccessDeniedException 及 AuthenticationException 异常
12	SecurityContext Configurer	SecurityContextPersistence Filter	SecurityContext 对象持久化过滤器，在请求最初阶段初始化并持久化该对象，在后续的过滤器中可以使用该对象来获取信息
13	ServletApiConfigurer	SecurityContextHolder AwareRequestFilter	在原始请求基础上包装后续使用的一些操作方法
14	CsrfConfigurer	CsrfFilter	跨域访问保护过滤器
15	LogoutConfigurer	LogoutFilter	登录注销过滤器
16	AnonymousConfigurer	AnonymousAuthentication Filter	匿名请求过滤器
17	FormLoginConfigurer	UsernamePassword AuthenticationFilter	表单登录处理过滤器
18	OAuth2LoginConfigurer	OAuth2AuthorizationRequest RedirectFilter	OAuth 认证访问过滤器
19	ChannelSecurity Configurer	ChannelProcessingFilter	通道访问安全过滤器，判断通道是否为 HTTP 或 HTTPS
20	HttpBasicConfigurer	BasicAuthenticationFilter	基础授权过滤器，其结果会保存在 SecurityContextHolder 中

下面通过自定义过滤器实现一个登录验证码的检测处理操作，由于验证码的检测需要结合用户登录处理，所以本次将直接继承 UsernamePasswordAuthenticationFilter 父类实现过滤器定义。

提示：使用 kaptcha 实现验证码。

本书为了方便将直接使用 Google 开源的验证码组件 kaptcha，该组件的核心配置如下。

1．修改 pom.xml 配置文件，追加 kaptcha 组件依赖。

```
<dependency>
    <groupId>com.github.axet</groupId>
    <artifactId>kaptcha</artifactId>
    <version>0.0.9</version>
</dependency>
```

2．为方便配置建立一个 KaptchaConfig 配置类，进行 DefaultKaptcha 类对象的创建。

```
@Configuration
public class KaptchaConfig {
    @Bean
    public DefaultKaptcha captchaProducer() {
        DefaultKaptcha captchaProducer = new DefaultKaptcha();
        Properties properties = new Properties();
        properties.setProperty("kaptcha.border", "yes");                          // 边框
        properties.setProperty("kaptcha.border.color", "105,179,90");
        properties.setProperty("kaptcha.textproducer.font.color", "red");
        properties.setProperty("kaptcha.image.width", "125");                     // 宽度
        properties.setProperty("kaptcha.image.height", "45");                     // 高度
        properties.setProperty("kaptcha.textproducer.font.size", "35");           // 大小
        properties.setProperty("kaptcha.session.key", "captcha");                 // 属性名称
        properties.setProperty("kaptcha.textproducer.char.length", "4");          // 长度
        properties.setProperty("kaptcha.textproducer.font.names",
            "宋体,楷体,微软雅黑");                                                 // 字体
        Config config = new Config(properties);                                   // 配置类
        captchaProducer.setConfig(config);                                        // 保存配置
        return captchaProducer;
    }
}
```

本程序中，通过 properties.setProperty("kaptcha.session.key", "captcha"); 语句设置了验证码的名称。

3．在 GlobalAction 类中追加一个验证码的显示路径。

```
    @RequestMapping(value = "/RandomCode")
    public ModelAndView kaptcha() {
        HttpServletRequest request = ((ServletRequestAttributes)
RequestContextHolder.getRequestAttributes()).getRequest();
        HttpServletResponse response = ((ServletRequestAttributes)
RequestContextHolder.getRequestAttributes()).getResponse();
        HttpSession session = request.getSession();
        response.setHeader("Pragma", "No-cache");              // 不缓存数据
        response.setHeader("Cache-Control", "no-cache");       // 不缓存数据
        response.setDateHeader("Expires", 0);                  // 不失效
        response.setContentType("image/jpeg");                 // MIME类型
        String capText = captchaProducer.createText();         // 获取验证码上的文字
        // 将验证码上的文字保存在Session中
        session.setAttribute(Constants.KAPTCHA_SESSION_KEY, capText);
        String code = (String) session.getAttribute(Constants.KAPTCHA_SESSION_KEY);
        this.log.info("验证码为:" + code);
```

```
            BufferedImage image = this.captchaProducer.createImage(capText);// 图像
            try {
                    OutputStream output = response.getOutputStream() ;
                    ByteArrayOutputStream bos = new ByteArrayOutputStream();
                    ImageIO.write(image, "JPEG", bos);             // 图像输出
                    byte[] buf = bos.toByteArray();
                    response.setContentLength(buf.length);
                    output.write(buf);
                    bos.close();
                    output.close();
            } catch (Exception e) {}
            return null ;
    }
```

上述程序中，session.setAttribute(Constants.**KAPTCHA_SESSION_KEY**, capText); 语句非常关键，作用是保存 Session 中的属性名称，随后的代码将通过此属性名称获取生成的验证码数据以实现与输入验证码的匹配。

4. 验证码设置在根路径上显示，所以还需要在 spring-security.xml 配置文件中追加拦截路径。

```
    <security:intercept-url pattern="/**" access="permitAll"/>
```

为了方便，本处只列出了核心代码，完整代码可以参考对应项目中的程序文件。

1. 【mldnspring-security 项目】修改 login.jsp 页面，追加验证码输入框。

```
<form action="<%=login_url%>" method="post">
    用户名：<input type="text" name="mid"><br>
    密   码：<input type="password" name="pwd"><br>
    验证码：<input type="text" maxlength="4" size="4" name="code">
            <img src="RandomCode.action"><br>
    <input type="checkbox" id="remember" name="remember" value="true"/>下次免登录<br>
    <input type="submit" value="登录"><input type="reset" value="重置">
</form>
```

2. 【mldnspring-security 项目】创建 UsernamePasswordAuthenticationFilter 子类，以实现验证码检测。

```
package cn.mldn.mldnspring.filter;
import javax.servlet.http.HttpServletRequest;
import javax.servlet.http.HttpServletResponse;
import org.springframework.security.authentication.AuthenticationServiceException;
import org.springframework.security.authentication.UsernamePasswordAuthenticationToken;
import org.springframework.security.core.Authentication;
```

```
import org.springframework.security.core.AuthenticationException;
import org.springframework.security.web.authentication.UsernamePasswordAuthenticationFilter;
import com.google.code.kaptcha.Constants;
public class ValidatorCodeUsernamePasswordAuthenticationFilter
        extends UsernamePasswordAuthenticationFilter   {
    private String codeParameter = "code" ;                                  // 验证码参数名称
    @Override
    public Authentication attemptAuthentication(HttpServletRequest request, HttpServletResponse response)
throws AuthenticationException {
        String captcha = (String) ((HttpServletRequest) request).getSession()
                .getAttribute(Constants.KAPTCHA_SESSION_KEY);           // 获取生成的验证码
        String code = request.getParameter(this.codeParameter) ;            // 获取输入验证码
        String username = super.obtainUsername(request).trim() ;            // 取得用户名
        String password = super.obtainPassword(request).trim() ;            // 取得密码
        UsernamePasswordAuthenticationToken authRequest = new
            UsernamePasswordAuthenticationToken(username, password);        // 生成认证标记
        super.setDetails(request, authRequest);                             // 设置认证详情
        // 当没有输入验证码，没有生成验证码时或者验证码不匹配时会提示错误
        if (captcha == null || "".equals(captcha) || code == null ||
            "".equals(code) || !captcha.equalsIgnoreCase(code)) {
            request.getSession().setAttribute("SPRING_SECURITY_LAST_USERNAME", username);
            throw new AuthenticationServiceException("验证码不正确！");
        }
        return super.getAuthenticationManager().authenticate(authRequest) ;  // 父类做认证处理
    }
    public void setCodeParameter(String codeParameter) {
        this.codeParameter = codeParameter;
    }
}
```

3. 【mldnspring-security 项目】此时要采用的是自定义的登录认证过滤器，所以最好的做法是单独配置一个登录控制操作，替换原始的**<security:form-login>**配置项，具体配置项如下。

☑ 定义一个新的登录处理终端，实现认证控制。

```
<bean id="authenticationEntryPoint"
    class="org.springframework.security.web.authentication
                .LoginUrlAuthenticationEntryPoint">
    <constructor-arg index="0" value="/loginPage.action"/>
</bean>
```

☑ 定义登录成功处理 Bean。

```
<bean id="loginLogAuthenticationSuccessHandler"
```

```
    class="org.springframework.security.web.authentication
        .SavedRequestAwareAuthenticationSuccessHandler">
    <property name="defaultTargetUrl" value="/welcomePage.action" />
</bean>
```

☑ 定义登录失败处理 Bean。

```
<bean id="simpleUrlAuthenticationFailureHandler"
    class="org.springframework.security.web.authentication
        .SimpleUrlAuthenticationFailureHandler">
    <property name="defaultFailureUrl" value="/loginPage.action?error=true" />
</bean>
```

☑ 由于此时需要单独配置登录操作，所以还需要为认证管理器设置一个 ID，方便引用。

```
<security:authentication-manager id="authenticationManager"><!-- 定义Security认证管理器 -->
    <security:authentication-provider user-service-ref="userDetailsService"/>
</security:authentication-manager>
```

☑ 配置新定义的过滤器 ValidatorCodeUsernamePasswordAuthenticationFilter，配置好相应的路径处理类对象，与认证管理器引用。

```
<bean id="validatorCode"
class="cn.mldn.mldnspring.filter.ValidatorCodeUsernamePasswordAuthenticationFilter">
    <property name="authenticationSuccessHandler" ref="loginAuthenticationSuccessHandler"/>
    <property name="authenticationFailureHandler" ref="simpleUrlAuthenticationFailureHandler"/>
    <property name="authenticationManager" ref="authenticationManager"/>
    <property name="filterProcessesUrl" value="/mldn-login"/>
    <property name="usernameParameter" value="mid"/>
    <property name="passwordParameter" value="pwd"/>
</bean>
```

☑ 修改<security:http>元素配置，配置新的认证终端处理，同时需要在登录前使用验证码过滤器。

```
<security:http auto-config="false" entry-point-ref="authenticationEntryPoint">
    <security:custom-filter ref="validatorCode" before="FORM_LOGIN_FILTER"/>
    相同配置项省略
</security:http>
```

至此可以实现验证码检测，当验证码检测不通过时将不会进行用户信息认证操作。

15.10 SpringSecurity 注解

对于访问路径的安全检测除了拦截路径的配置外，还可以通过注解的形式进行控制层或业

务层中指定方法的访问验证，在 SpringSecurity 中提供的注解有如下两组。

☑　@Secured：该注解为早期注解，可以直接进行角色验证，但是不支持 SpEL 表达式。

☑　@PreAuthorize / @PostAuthorize：该注解支持 SpEL 表达式，其中@PreAuthorize 在方法执行前验证，而@PostAuthorize 在方法执行后验证，一般使用较少。

1. 【mldnspring-security 项目】由于所有注解都要写在 Action 或业务层中，所以修改 spring-mvc.xml 配置文件，添加 SpringSecurity 注解的启用配置，同时删除 spring-security.xml 配置文件在<security:http>元素中定义的请求拦截路径。

```
<security:global-method-security          <!-- 启用SpringSecurity注解功能 -->
      pre-post-annotations="enabled"      <!-- 启用@PreAuthorize / @PostAuthorize功能 -->
      secured-annotations="enabled"/>     <!-- 启用@Secured功能 -->
```

2. 【mldnspring-security 项目】修改 GlobalAction 类中的 welcomePage 路径。该路径要求认证过的用户才可以访问。

```
@PreAuthorize("isAuthenticated()")
@RequestMapping("/welcomePage")               // 访问路径
public String welcome() {                     // 登录成功路径
      return "welcome";                       // 设置跳转路径
}
```

由于需要使用 SpEL 表达式，所以在进行认证检测时直接使用了@PreAuthorize 注解。这样，当进行访问时如果当前用户没有登录过，则会跳转到登录表单页面；如果用户登录过，就可以直接访问。

3. 【mldnspring-security 项目】修改 EchoAction 程序类，追加角色判断。

```
@Controller                                                    // 定义控制器
@RequestMapping("/pages/message/*")                            // 访问父路径，与方法中的路径组合为完整路径
public class EchoAction {                                      // 自定义Action程序类
    @PreAuthorize("hasRole('ADMIN')")
    @RequestMapping("/show")                                   // 访问的路径
    public ModelAndView echo(String msg) {                     // 接收请求参数
          ModelAndView mav = new ModelAndView("message/message_show") ; // 设置跳转路径
          mav.addObject("echoMessage", "【ECHO】msg = " + msg) ;        // 设置request数据
          return mav ;
    }
    @PreAuthorize("hasRole('ADMIN')")
    @GetMapping("/input")                                      // 访问路径
    public String input() {
          return "message/message_input";                      // 设置跳转路径
    }
}
```

本程序利用注解为具体的 Action 方法进行了授权检测，当不具有指定角色用户访问时会自动跳转到错误页显示。

15.11 投票器

对于安全访问，除了使用拦截路径形式进行访问控制外，还可以利用决策管理器根据实际业务实现安全访问。利用决策管理器中的投票机制，可以决定是否需要进行授权控制。

在之前的访问控制中使用过两个投票器：角色投票器（RoleVoter）与认证投票器（AuthenticatedVoter）。所有的投票器都会被访问决策管理器所管理，这些类之间的结构如图 15-16 所示。

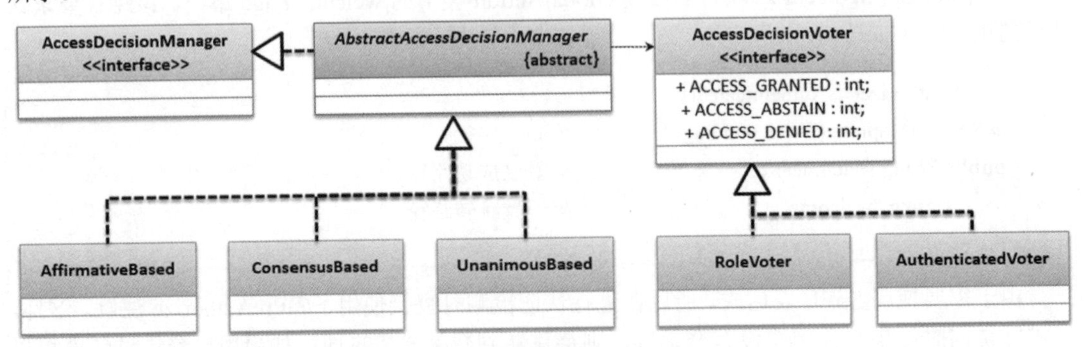

图 15-16 投票器组成结构

通过图 15-16 可以发现，决策管理器的父接口为 AccessDecisionManager，同时在此接口中定义了投票的 3 种状态：赞成（ACCESS_GRANTED）、弃权（ACCESS_ABSTAIN）、反对（ACCESS_DENIED）。在 SpringSecurity 中定义了 3 个常用的投票管理器，具体作用如下。

- ☑ org.springframework.security.access.vote.AffirmativeBased：一票通过，只要有一个支持就允许访问。
- ☑ org.springframework.security.access.vote.ConsensusBased：半数以上支持票数就可以访问。
- ☑ org.springframework.security.access.vote.UnanimousBased：全票通过后才可以访问。

如果要实现一个自定义的投票器并且可以被投票管理器所管理，那么该投票类需要实现 AccessDecisionVoter 父接口，在 AccessDecisionVoter 接口中定义的方法如表 15-4 所示。

表 15-4 AccessDecisionVoter 接口方法

编号	方法	类型	描述
1	**public boolean** supports(ConfigAttribute attribute)	普通	通过配置的属性，判断是否需要提供投票器支持。返回 true，表示使用该投票器
2	**public boolean** supports(Class<?> clazz)	普通	根据类型，判断是否需要提供投票器支持
3	**public int** vote(Authentication authentication, Object object, Collection<ConfigAttribute> attributes)	普通	投票处理，可以返回 ACCESS_ GRANTED、ACCESS_ABSTAIN、ACCESS_DENIED 三类结果

15.11.1 AccessDecisionVoter

下面采用自定义投票器的方式实现本地访问控制。本例中，通过 localhost（127.0.0.1）访问的用户不需要登录就可以直接进行操作。

1. 【mldnspring-security 项目】建立一个 IP 地址的投票器。

```
package cn.mldn.mldnspring.config;
import java.util.Collection;
import java.util.Iterator;
import org.springframework.security.access.AccessDecisionVoter;
import org.springframework.security.access.ConfigAttribute;
import org.springframework.security.core.Authentication;
import org.springframework.security.web.authentication.WebAuthenticationDetails;
public class IPAddressVoter implements AccessDecisionVoter<Object> {
    private static final String LOCAL_FLAG = "LOCAL_IP";          // 需要判断的访问标记
    @Override
    public boolean supports(ConfigAttribute attribute) {      // 如果有指定配置属性，则执行投票
        return attribute != null
                && attribute.toString().contains(LOCAL_FLAG);
    }
    @Override
    public boolean supports(Class<?> clazz) {                  // 对所有访问类均支持投票
        return true;
    }
    @Override
    public int vote(Authentication authentication, Object object,
            Collection<ConfigAttribute> attributes) {
        if (!(authentication.getDetails()
                instanceof WebAuthenticationDetails)) {        // 如果不是来自Web访问
            return AccessDecisionVoter.ACCESS_DENIED;          // 拒绝该用户访问
        }
        // 通过认证信息获取用户的详情内容，该内容类型为WebAuthenticationDetails
        WebAuthenticationDetails details = (WebAuthenticationDetails)
                authentication.getDetails();
        String ip = details.getRemoteAddress() ;               // 取得当前操作的IP地址
        Iterator<ConfigAttribute> iter = attributes.iterator() ; // 获取每一个配置属性
        while (iter.hasNext()) {                               // 循环每一个配置属性
            ConfigAttribute ca = iter.next() ;                 // 获取属性
            if (ca.toString().contains(LOCAL_FLAG)) {          // 如果在本地执行
                if ("0:0:0:0:0:0:0:1".equals(ip) || "127.0.0.1".equals(ip)) {    // 本机访问
                    return AccessDecisionVoter.ACCESS_GRANTED ;  // 访问通过
                }
```

```
            }
        }
        return AccessDecisionVoter.ACCESS_ABSTAIN ;                    // 弃权不参与投票
    }
}
```

本程序中设置了一个新的访问控制标记 LOCAL_IP，如果拦截路径上使用了此访问类型，同时又属于本机直接访问的情况，将不会进行认证处理，可以直接使用。

2. 【mldnspring-security 项目】由于需要引入新的投票器，所以此时需要修改 spring-security.xml 配置文件，定义新的投票管理器，且配置相应的投票器。

```
<bean id="accessDecisionManager"
    class="org.springframework.security.access.vote.AffirmativeBased">        <!-- 管理器 -->
    <constructor-arg name="decisionVoters">                                   <!-- 投票器 -->
        <list>
            <!-- 定义角色投票器，进行角色验证 -->
            <bean class="org.springframework.security.access.vote.RoleVoter" />
            <!-- 定义认证投票器，用于判断用户是否已经认证 -->
            <bean class="org.springframework.security.access.vote.AuthenticatedVoter" />
            <!-- 自定义投票器，如果是本机IP则不进行过滤 -->
            <bean class="cn.mldn.mldnspring.config.IPAddressVoter" />
            <!-- 开启表达式支持，这样就可以在进行拦截时使用SpEL -->
            <bean class="org.springframework.security.web.
                            access.expression.WebExpressionVoter"/>
        </list>
    </constructor-arg>
</bean>
```

在本次定义的决策管理器中一共设置了 3 个投票器，由于使用的是 AffirmativeBased 管理器，所以只要有一个投票器投出赞成票，就可以进行访问。

3. 【mldnspring-security 项目】修改 spring-security.xml 配置文件中的<security:http>配置项，引入自定义的访问决策管理器，同时在需要验证的路径上定义 LOCAL_IP 标记。

```
<security:http
    auto-config="false"
    entry-point-ref="authenticationEntryPoint"
    access-decision-manager-ref="accessDecisionManager">
    <security:intercept-url pattern="/pages/message/**"
        access="hasRole('USER') or hasRole('LOCAL_IP')"/>
    相同配置项省略
</security:http>
```

配置成功后，当通过 localhost 或者 127.0.0.1 访问/pages/message/** 路径中的信息时，将不再受到角色限制。如果是远程访问，则依然需要按照传统方式登录认证后才可以访问。

15.11.2 RoleHierarchy

为了进一步完善授权访问的级别层次，SpringSecurity 提供了角色继承概念，即使用者可定义继承的层次关系，这样，拥有更高级别角色的用户可以直接访问低级别角色的信息。

 **提示：关于角色继承的描述。**

假设在 spring-security.xml 中有如下两个拦截路径。

```
<security:intercept-url pattern="/pages/message/input.action"
        access="hasRole('USER')"/>
<security:intercept-url pattern="/pages/message/show.action"
        access="hasRole('ADMIN')"/>
```

此时两个访问路径分别配置了两种角色，按照之前的定义，假设 ROLE_ADMIN 是最高管理员权限，在这样的配置下如果一个用户只拥有 ROLE_ADMIN 角色，依然无法访问 ROLE_USER 对应的信息。配置了角色层次关系后，即便拥有 ROLE_ADMIN 的用户没有 ROLE_USER 角色，也可以访问 ROLE_USER 角色对应的信息。

为了实现这一操作，需要先删除 member_role 表中 admin 用户对应的 ROLE_USER 角色信息。

在 SpringSecurity 中，对于角色继承提供了 org.springframework.security.access.hierarchicalroles.RoleHierarchy 接口，此接口定义如下。

```
public interface RoleHierarchy {
    /**
     * 获取全部授权信息
     * @param authorities  定义的授权访问列表
     * @return 所有可用的授权访问列表
     */
    public Collection<? extends GrantedAuthority> getReachableGrantedAuthorities(
            Collection<? extends GrantedAuthority> authorities);
}
```

此接口定义了获取全部可用控制权限的方法，SpringSecurity 框架还提供了一个基础的实现子类 RoleHierarchyImpl，下面将利用此类结合配置文件，实现角色继承。配置文件中，角色继承的配置格式为：角色 1 > 角色 2 > ... > 角色 n。

1. 【mldnspring-security 项目】修改 spring-security.xml 配置文件，配置角色继承关系。

```
<!-- 定义角色继承，由于是固定信息，本次将使用RoleHierarchyImpl -->
<bean id="roleHierarchy"
    class="org.springframework.security.access.hierarchicalroles.RoleHierarchyImpl">
    <property name="hierarchy">                          <!-- 定义继承层次关系 -->
        <value>                                          <!-- 定义角色层次 -->
```

```
            ROLE_ADMIN > ROLE_USER
        </value>
    </property>
</bean>
```

2. 【mldnspring-security 项目】由于所有的角色检测操作都通过表达式进行配置，所以需要在 WebExpressionVoter 投票器中进行角色继承配置，修改 spring-security.xml 配置文件。

```
<bean id="accessDecisionManager"
    class="org.springframework.security.access.vote.AffirmativeBased">    <!-- 管理器 -->
    <constructor-arg name="decisionVoters">                                <!-- 投票器 -->
        <list>
            <!-- 定义角色投票器，进行角色验证 -->
            <bean class="org.springframework.security.access.vote.RoleVoter" />
            <!-- 定义认证投票器，用于判断用户是否已经认证 -->
            <bean class="org.springframework.security.access.vote.AuthenticatedVoter" />
            <!-- 定义角色继承投票器，此投票器可以在注解中使用 -->
            <bean class="org.springframework.security.access.vote.RoleHierarchyVoter">
                <constructor-arg ref="roleHierarchy"/>     <!-- 引用角色继承配置 -->
            </bean>
            <!-- 由于需要通过SpEL定义访问继承关系，所以要在Web表达式投票器中配置角
色继承定义 -->
            <bean class="org.springframework.security.web.access.
                    expression.WebExpressionVoter">
                <property name="expressionHandler">     <!-- 定义表达式处理器 -->
                    <bean class="org.springframework.security.web.access.
                            expression.DefaultWebSecurityExpressionHandler">
                        <!-- 引用角色继承配置 -->
                        <property name="roleHierarchy" ref="roleHierarchy" />
                    </bean>
                </property>
            </bean>
        </list>
    </constructor-arg>
</bean>
```

配置完成后，当用户拥有了 ROLE_ADMIN 角色就可以访问 ROLE_USER 对应的资源。

15.12 基于 Bean 配置

SpringSecurity 的所有核心配置，除了可以利用配置文件实现外，也可以采用 Bean 配置完成。SpringSecurity 中提供了 WebSecurityConfigurer 接口，开发者只需要实现此接口或继承

WebSecurityConfigurerAdapter 抽象类，即可实现配置。SpringSecurity 的基本定义结构如图 15-17 所示。

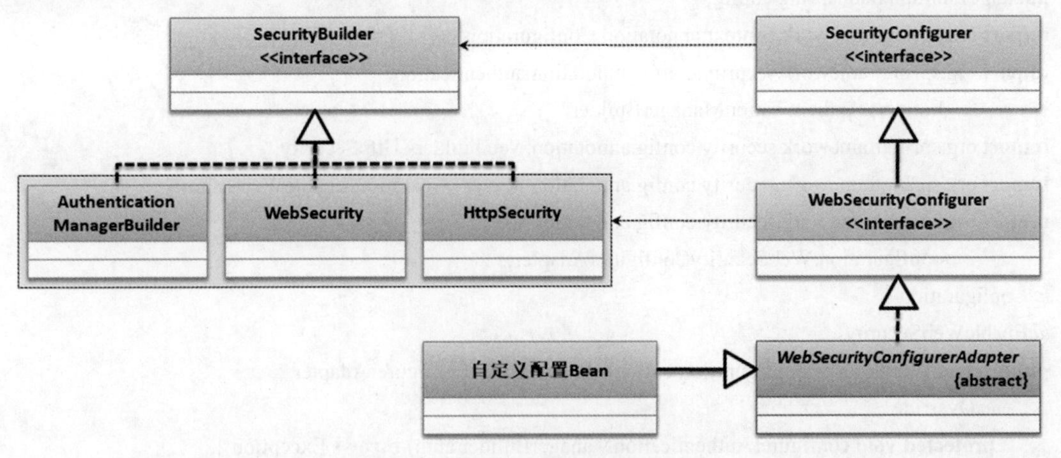

图 15-17　Bean 配置 SpringSecurity

自定义 SpringSecurity 配置类时，可以根据需要覆写 WebSecurityConfigurerAdapter 抽象类中的 configure 方法，如表 15-5 所示。

表 15-5　configure 方法

编号	方法	类型	描述
1	protected void configure(AuthenticationManagerBuilder auth) throws Exception	普通	认证管理器配置
2	protected void configure(HttpSecurity http) throws Exception	普通	HTTP 安全访问配置
3	public void configure(WebSecurity web) throws Exception	普通	Web 安全访问配置

下面将通过 Bean 配置，实现与之前相同的配置效果。

提示：关于 Bean 配置 SpringSecurity 的意义。

虽然 SpringSecurity 中的所有配置项都可以基于 XML 配置实现，但是考虑到读者对于微服务架构的学习，本书专门安排了基于 Bean 的管理配置，这样就可以有效地与 SpringBoot、SpringCloud 框架整合开发。关于微架构的相关内容读者可以参考笔者出版的《Java 微服务架构实战》。

15.12.1　基础配置

下面将通过 Bean 实现一个固定认证信息的登录、注销、认证与授权控制的配置操作。需要注意的是，如果通过配置类定义 SpringSecurity 配置，需要满足如下两项。

- ☑ 配置 Bean 需要设置在扫描路径中，并且需要使用@Configuration 注解声明。
- ☑ 由于该 Bean 主要负责 SpringSecurity 配置，所以需要在类定义中使用@EnableWeb Security 注解。

范例：定义 WebSecurityConfiguration 配置类，进行 SpringSecurity 基础配置。

```
package cn.mldn.mldnspring.config;
import org.springframework.context.annotation.Configuration;
import org.springframework.security.config.annotation.authentication.
        builders.AuthenticationManagerBuilder;
import org.springframework.security.config.annotation.web.builders.HttpSecurity;
import org.springframework.security.config.annotation.web.configuration.EnableWebSecurity;
import org.springframework.security.config.annotation.web.
        configuration.WebSecurityConfigurerAdapter;
@Configuration
@EnableWebSecurity
public class WebSecurityConfiguration extends WebSecurityConfigurerAdapter {
    @Override
    protected void configure(AuthenticationManagerBuilder auth) throws Exception {
        // 进行用户角色配置时不需要追加ROLE_前缀，系统会自动添加
        auth.inMemoryAuthentication()                                  // 固定认证信息
            .withUser("admin")                                         // 用户名
            .password("{bcrypt}$2a$10$2...")                           // 登录密码
            .roles("USER","ADMIN") ;                                   // 角色
        auth.inMemoryAuthentication()                                  // 固定认证信息
            .withUser("mldn")                                          // 用户名
            .password("{bcrypt}$2a$10$v...")                           // 登录密码
            .roles("USER") ;                                           // 角色
    }
    @Override
    protected void configure(HttpSecurity http) throws Exception {
        http.csrf().disable() ;                                        // 禁用CSRF验证
        // 配置拦截路径的匹配地址与限定授权访问
        http.authorizeRequests()                                       // 配置认证请求
            .antMatchers("/pages/message/**").access("hasRole('ADMIN')")// 授权访问
            .antMatchers("/welcomePage.action").authenticated()        // 认证访问
            .antMatchers("/**").permitAll() ;                          // 任意访问
        // 配置HTTP登录、注销与错误路径
        http.formLogin()                                               // 登录配置
            .usernameParameter("mid")                                  // 用户名参数配置
            .passwordParameter("pwd")                                  // 用户名参数配置
            .successForwardUrl("/welcomePage.action")                  // 登录成功路径
            .loginPage("/loginPage.action")                            // 登录表单页面
            .loginProcessingUrl("/mldn-login")                         // 登录路径
            .failureForwardUrl("/loginPage.action?error=true")         // 登录失败路径
            .and()                                                     // 配置连接
            .logout()                                                  // 注销配置
```

```
            .logoutUrl("/mldn-logout")                                // 注销路径
            .logoutSuccessUrl("/logoutPage.action")                   // 注销成功路径
            .deleteCookies("JSESSIONID")                              // 删除Cookie
            .and()                                                    // 配置连接
            .exceptionHandling()                                      // 认证错误配置
            .accessDeniedPage("/WEB-INF/pages/error_page_403.jsp") ;  // 授权错误页
    }
}
```

本程序实现了一个基本的 SpringSecurity 使用配置，主要使用了两个 configure 方法。

☑ configure(AuthenticationManagerBuilder auth)：配置认证用户。这里使用 auth.inMemory Authentication 方法声明了两个用户，并为其分配了相应角色。

☑ configure(HttpSecurity http)：配置拦截路径、登录、注销与授权错误。

15.12.2　深入配置

SpringSecurity 中除了基础的登录认证外，还有 UserDetails、Session 管理、过滤器、访问注解等相关配置。这些配置也可以直接通过 Bean 实现管理。

范例：自定义认证处理操作。

```
@Configuration
@EnableWebSecurity
@EnableGlobalMethodSecurity(prePostEnabled = true, securedEnabled = true)    // 启用注解配置
public class WebSecurityConfiguration extends WebSecurityConfigurerAdapter {
    @Autowired
    private DataSource dataSource ;                                             // 数据源
    @Autowired
    private UserDetailsService userDetailsService ;                             // 用户服务
    @Autowired
    private SavedRequestAwareAuthenticationSuccessHandler successHandler ;      // 成功页
    @Autowired
    private SimpleUrlAuthenticationFailureHandler fialureHandler ;              // 失败页
    @Autowired
    private SessionInformationExpiredStrategy sessionExpiredStrategy ;          // Session失效策略
    @Autowired
    private UsernamePasswordAuthenticationFilter authenticationFilter ;         // 认证过滤器
    @Autowired
    private JdbcTokenRepositoryImpl tokenRepository ;                           // token存储
    @Autowired
    private LoginUrlAuthenticationEntryPoint authenticationEntryPoint ;
    @Override
    protected void configure(AuthenticationManagerBuilder auth) throws Exception {
        auth.userDetailsService(this.userDetailsService) ;                      // 基于数据库认证
    }
```

```
    @Override
    protected void configure(HttpSecurity http) throws Exception {
        http.csrf().disable() ;                                              // 禁用CSRF验证
        http.httpBasic().authenticationEntryPoint(this.authenticationEntryPoint) ;
        // 进行拦截路径的匹配地址配置与授权访问限定
        http.authorizeRequests()                                             // 配置认证请求
            .antMatchers("/pages/message/**").access("hasRole('ADMIN')")     // 授权访问
            .antMatchers("/welcomePage.action").authenticated() ;            // 认证访问
        // 进行http注销与错误路径配置，登录操作将由过滤器负责完成
        http.logout()                                                        // 注销配置
            .logoutUrl("/mldn-logout")                                       // 注销路径
            .logoutSuccessUrl("/logoutPage.action")                          // 注销成功路径
            .deleteCookies("JSESSIONID","mldn-rememberme-cookie")            // 删除Cookie
            .and()                                                           // 配置连接
            .exceptionHandling()                                             // 认证错误配置
            .accessDeniedPage("/WEB-INF/pages/error_page_403.jsp") ;         // 授权错误页
        http.rememberMe()                                                    // 开启RememberMe
            .rememberMeParameter("remember")                                 // 表单参数
            .key("mldn-lixinghua")                                           // 加密key
            .tokenValiditySeconds(2592000)                                   // 失效时间
            .rememberMeCookieName("mldn-rememberme-cookie")                  // Cookie名称
            .tokenRepository(this.tokenRepository) ;                         // 持久化
        http.sessionManagement()                                             // Session管理
            .invalidSessionUrl("/loginPage.action")                          // 失效路径
            .maximumSessions(1)                                              // 并发Session
            .expiredSessionStrategy(this.sessionExpiredStrategy) ;           // 失效策略
        http.addFilterBefore(this.authenticationFilter,                      // 追加过滤器
                UsernamePasswordAuthenticationFilter.class);
    }
    @Override
    public void configure(WebSecurity web) throws Exception {
        web.ignoring().antMatchers("/index.jsp") ;                           // 忽略的验证路径
    }
    @Bean
    public UsernamePasswordAuthenticationFilter getAuthenticationFilter() throws Exception {
        ValidatorCodeUsernamePasswordAuthenticationFilter filter = new
ValidatorCodeUsernamePasswordAuthenticationFilter() ;
        filter.setAuthenticationManager(super.authenticationManager());      // 认证管理器
        filter.setAuthenticationSuccessHandler(this.successHandler);         // 登录成功页面
        filter.setAuthenticationFailureHandler(this.fialureHandler);         // 登录失败页面
        filter.setFilterProcessesUrl("/mldn-login");                         // 登录路径
        filter.setUsernameParameter("mid");                                  // 参数名称
        filter.setPasswordParameter("pwd");                                  // 参数名称
```

```
        return filter ;
    }
    @Bean
    public LoginUrlAuthenticationEntryPoint getEntryPoint() {
        return new LoginUrlAuthenticationEntryPoint("/loginPage.action") ;
    }
    @Bean
    public SavedRequestAwareAuthenticationSuccessHandler getSuccessHandler() {// 认证成功处理
        SavedRequestAwareAuthenticationSuccessHandler handler = new
                SavedRequestAwareAuthenticationSuccessHandler() ;
        handler.setDefaultTargetUrl("/welcomePage.action");                  // 成功页面
        return handler ;
    }
    @Bean
    public SimpleUrlAuthenticationFailureHandler getFialureHandler() {       // 认证失败处理
        SimpleUrlAuthenticationFailureHandler handler = new
                SimpleUrlAuthenticationFailureHandler() ;
        handler.setDefaultFailureUrl("/loginPage.action?error=true");        // 失败页面
        return handler ;
    }
    @Bean
    public JdbcTokenRepositoryImpl getTokenRepository() {                     // 持久化Cookie
        JdbcTokenRepositoryImpl tokenRepository = new JdbcTokenRepositoryImpl();
        tokenRepository.setDataSource(this.dataSource);
        return tokenRepository;
    }
    @Bean
    public SessionInformationExpiredStrategy getSessionInformationExpiredStrategy() {
        // Session失效策略，同时配置失效后的跳转路径
        return new SimpleRedirectSessionInformationExpiredStrategy("/logoffPage.action");
    }
}
```

本程序针对之前的配置，追加了 RememberMe（包括数据库持久化）、验证码检测和并发 Session 访问控制。考虑到 Spring 开发的标准型，将所有可能使用到的对象以 Bean 的形式进行配置，随后根据需要进行注入。

15.12.3　配置投票管理器

SpringSecurity 配置类提供了投票管理器，按照下面的顺序实现投票器的配置即可。本程序依然使用 AffirmativeBased 投票管理器，该投票管理器需要通过构造方法配置所有的投票器对象。

范例：配置投票管理器。

```
@Configuration
@EnableWebSecurity
@EnableGlobalMethodSecurity(prePostEnabled = true, securedEnabled = true)          // 启用注解配置
public class WebSecurityConfiguration extends WebSecurityConfigurerAdapter {
    // 省略重复配置代码
    @Autowired
    private RoleHierarchy roleHierarchy ;                                            // 角色继承
    @Autowired
    private SecurityExpressionHandler<FilterInvocation> expressionHandler ;          // 表达式处理器
    @Autowired
    private AccessDecisionManager accessDecisionManager ;                            // 投票管理器
    @Override
    protected void configure(HttpSecurity http) throws Exception {
        http.authorizeRequests().accessDecisionManager(this.accessDecisionManager) ;
        // 省略重复配置代码
    }
    @Bean
    public RoleHierarchy getRoleHierarchy() {                                        // 角色继承配置
        RoleHierarchyImpl role = new RoleHierarchyImpl() ;                           // 角色继承
        role.setHierarchy("ROLE_ADMIN > ROLE_USER");                                 // 继承关系
        return role ;
    }
    @Bean
    public AccessDecisionManager getAccessDecisionManager() {
        // 将所有用到的投票器设置到List集合中
        List<AccessDecisionVoter<? extends Object>> decisionVoters = new ArrayList<>() ;
        decisionVoters.add(new RoleVoter()) ;                                        // 角色投票器
        decisionVoters.add(new AuthenticatedVoter()) ;                               // 认证投票器
        decisionVoters.add(new RoleHierarchyVoter(this.roleHierarchy)) ;             // 角色继承投票器
        WebExpressionVoter webVoter = new WebExpressionVoter() ;
        webVoter.setExpressionHandler(this.expressionHandler);
        decisionVoters.add(webVoter) ;                                               // 表达式解析
        AffirmativeBased access = new AffirmativeBased(decisionVoters) ;             // 定义投票管理器
        return access ;
    }
    @Bean
    public SecurityExpressionHandler<FilterInvocation>
        getSecurityExpressionHandler() {                                             // 配置表达式
        DefaultWebSecurityExpressionHandler expressionHandler = new
                DefaultWebSecurityExpressionHandler() ;
        expressionHandler.setRoleHierarchy(this.roleHierarchy);                      // 设置角色继承
```

```
        return expressionHandler ;
    }
}
```

由于投票管理器中需要考虑到表达式的支持，所以在本程序创建投票管理器对象时，依然在表达式配置类中注入了角色继承关系，最终的投票管理器需要通过 HttpSecurity 类对象完成配置。

15.13　本章小结

1．SpringSecurity 是基于 Spring 开发框架实现的认证与授权检测框架，可以帮助开发者方便地进行登录认证与授权管理。

2．SpringSecurity 主要是基于配置实现的安全访问，开发者需要明确地定义好认证用户、拦截地址，同时也需要配置好相应的登录与注销路径。

3．CSRF（Cross-Site Request Forgery）是 Web 开发中较为常见的安全隐患，在 SpringSecurity 中良好地解决了此安全漏洞，可以利用 token 标记的形式解决。

4．UserDetailsService 是 SpringSecurity 实现的用户处理业务，利用此接口可实现数据库中的用户认证与授权信息读取，但是需要谨记的是，在 SpringSecurity 框架中只要认证成功就会自动获取授权信息。

5．Session 管理依靠 HttpSessionEventPublisher 监听器实现，可以通过配置方便地实现重复用户的剔除操作。

6．使用 RememberMe 功能可以减少登录表单的使用，同时也可以结合数据库进行 RememberMe-Token 存储。

7．SpringSecurity 中允许用户自定义访问控制器，这样可以实现更多的业务处理能力。在定义访问控制器时，如果要使用 SpEL 进行授权检测，则一定要配置 WebExpressionVoter 投票器。

8．SpringSecurity 除了基于 XML 配置文件管理外，也支持 Bean 配置管理，此配置形式在 SpringBoot 或 SpringCloud 开发框架中较为常见。